Life on Ice

Life on Ice

A History of New Uses for Cold Blood

JOANNA RADIN

The University of Chicago Press   Chicago and London

The University of Chicago Press, Chicago 60637
The University of Chicago Press, Ltd., London

Published 2017
Printed in the United States of America

30 29 28 27 26 25 24 23 22 21   2 3 4 5

ISBN-13: 978-0-226-41731-8 (cloth)
ISBN-13: 978-0-226-44824-4 (e-book)
DOI: 10.7208/chicago/9780226448244.001.0001

Library of Congress Cataloging-in-Publication Data

Names: Radin, Joanna, author.
Title: Life on ice: a history of new uses for cold blood / Joanna Radin.
Description: Chicago: The University of Chicago Press, 2017. | Includes bibliographical references and index.
Identifiers: LCCN 2016033177 | ISBN 9780226417318 (cloth: alk. paper) | ISBN 9780226448244 (e-book)
Subjects: LCSH: Frozen blood. | Blood—Cryopreservation. | Medicine—Research—History—20th century. | Cryopreservation of organs, tissues, etc.—Moral and ethical aspects. | Medical anthropology.
Classification: LCC QH324.9.C7 R33 2017 | DDC 362.17/84—dc23 LC record available at https://lccn.loc.gov/2016033177

⊗ This paper meets the requirements of ANSI/NISO Z39.48-1992 (Permanence of Paper).

To MHG,

who warms my heart

Blood is not the simple fluid it was once thought to be.

DOUGLAS M. SURGENOR, "BLOOD"

The utopian vision can and must do without men of flesh and blood. After all, *there is no such place.*

ROBERTO FERNANDEZ RETAMAR, "CALIBAN"

Contents

Preface: Frozen Spirits

On a cold, gray morning in February of 2010, I drove through the snow to the State University of New York at Binghamton, a research institution in the south central part of the state. In the car, I listened as a journalist described her new book, which told the story of cells salvaged by a scientist at Johns Hopkins University Medical School from a patient in the 1950s. This patient, an African American tobacco-farming woman, had an especially virulent form of cervical cancer. The cells from her cancer were transformed, without her knowledge, into what would become one of the most important biomedical technologies of the century: HeLa.

That year science writer Rebecca Skloot's *The Immortal Life of Henrietta Lacks* would become a mass-market best seller. The story of HeLa, the cell line, was already well known to historians of life science.[1] But no one was prepared for the extent to which the story of Henrietta Lacks, the person, would become an international sensation.[2] The ghost of Henrietta Lacks pushed into the spotlight the otherwise esoteric subject of the role that preserved human body parts have played in contemporary science and medicine. Op-eds and academic articles were written. Rumors, which would later become reality, had already begun to circulate that Oprah Winfrey would play Lacks in the movie. Lacks's specter was also present during the month I lived in Binghamton.

I was there to work at the Serum Archive maintained by the university's Program of Biomedical Anthropology and to learn how long-frozen samples of human blood were being given new life in a genomic age. Serum is the liquid component of blood, which was an especially important research material for biologists and epidemiologists before scientists had the ability to analyze DNA. In Binghamton, tens of thousands of serum samples—extracted during the Cold War from members of communities described at the time as "primitive" and destined to disappear—were, in the early years of the twenty-first century, being prospected for fragments of DNA that could be used to answer questions about health, identity, and kinship.

Though none of the samples in Binghamton were cell lines, each of the thousands of blood samples preserved in the freezers at the Serum Archive were extracted from individuals who had and still have what sociologist Avery Gordon calls "complex personhood."[3] Ghosts, Gordon has argued, are social actors too; they have their own issues and concerns. The spirits of both the collected and their collectors were also present in the Serum Archive. They, like Lacks's ghost, unsettle ideas about scientific practice and even about what it can mean to be alive.

Today, members of certain indigenous communities want their ancestors' blood removed from these low-temperature crypts. At the same time, the scientists who serve as stewards of this cold blood are committed to maintaining the vital legacies of their deceased mentors. Listening to these ghosts can quickly become a cacophony of demands and seemingly irreconcilable desires.

This book is a response to the ghosts who haunt the archives of human biology. It tells the history of efforts to freeze blood drawn from members of human communities seen, during the Cold War, as destined to disappear. It is a story of salvage and salvation that casts a bright light on the often invisible forms of labor and value that have contributed to the creation of a vast global biomedical infrastructure and its transformations through time. This infrastructure is made up of both technologies for maintaining low temperatures and human-derived tissues of all kinds, including materials as varied as blood samples, tumors, umbilical cords, and even embryos.

These frozen populations, these time-traveling resources, are at once both more and less than human, maintained in order to be reconstituted, deployed and sometimes even destroyed, in ways that continue to surprise those who have contributed to their creation and maintenance. The pages that follow explain the unique cultural and technical circumstances that gave momentum to, and are beginning to thaw, forms of life frozen during biomedicine's ice age.

Introduction: Within Cold Blood

When HIV/AIDS emerged as a pandemic of global proportions, epidemiologists asked where and when it began. They believed that understanding the biological history of the disease, particularly its origins, could help them better characterize the virus, which might help to contain its spread or even to identify a cure. In 1985 they found an answer. In a freezer. Maintained in the United States, this freezer held 672 samples of blood collected in the Belgian Congo in 1959. One tested positive for antibodies to HIV-1 and has for over thirty years remained the oldest such biological trace of the disease ever documented.[1]

That vial of blood had been frozen for posterity after initial use in a study of the blood group genetics of African populations living near Léopoldville, now Kinshasa, Congo. The scientists who preserved it imagined that blood collected for one set of uses might one day reveal new forms of knowledge. Only a few milliliters of blood, that sample was created at a time when antibiotics were still largely regarded as magic bullets, making it possible to imagine that new infections could be quickly curtailed if only they could be identified early enough. The ravages of HIV/AIDS, Ebola, and even multi-drug-resistant tuberculosis were virtually unknown. In other words, a blood sample collected in 1959 was thawed a quarter century after it had been collected, to study a virus that no scientist knew existed when it was first frozen. More recently, this same sample has been thawed yet again, this time to sequence the fragments of viral DNA also preserved within.[2]

This book examines how and why frozen blood samples—in particular those collected from colonial or newly postcolonial regions in the decades after World War II—became a resource for biomedical science.[3] It is a history of a phenomenon known as biobanking, in which bits of tissues from humans and nonhumans are stored at very low temperatures for future research.[4] Most of the millions of tissues preserved in biomedicine's freezers today have been obtained from people during their visits to physicians, but the roots of the contemporary biobank grew from the work of researchers who sought to connect the field with the laboratory and the clinic at a moment when America was emerging as a global power during the Cold War. At the same time that nuclear weapons were being stockpiled, evidence of biological variation was being accumulated and stored in earnest. Human biologists—a network of experts in biological anthropology, population genetics, and epidemiology—began to express concerns about the destructive effects of atomic energy, chemical pollutants, and urbanization on our species. They believed that molecular approaches would facilitate a global stocktaking of the biosphere and establish baselines against which Euro-Americans could measure the extent to which the negative by-products of modernity had mutated *their own* bodies. The urge to create a total archive "before it was too late," a common refrain in scientific and popular literature, would be compromised if important sources of data were not secured.[5] Access to technologies of preservation—refrigerators, freezers, dry ice, and liquid nitrogen—enabled some human biologists to imagine blood as a potent resource for securing the future of a particular kind of universal human, which Donna Haraway has critiqued as "man the hunter."[6] Freezing would make it possible for blood to be redirected in time: no longer circulating only in human bodies, but as part of a global infrastructure to support the rise of biomedical science.[7]

In this Cold War context, cold temperature was not only a material technology that took on new relevance in biomedicine. It was also fashioned into a thermodynamic metaphor used to justify the sampling of such communities. In the late 1950s, French anthropologist Claude Lévi-Strauss repurposed Victorian-era concerns about entropy and degeneration to articulate a distinction between "hot" and "cold" societies.[8] He argued that so-called hot societies like his own were dynamic sources of novelty and innovation, whereas those he characterized as "cold" absorbed and neutralized change.[9] In this formulation, which echoed sentiments of his memoir *Tristes Tropiques*, translated as *World on the Wane*, the heat of modernity was thawing the cultural practices that characterized anthropologists' cherished objects of study: the "primitive."[10]

Despite their own vocal critiques of scientific racism, an influential group of human biologists adopted this sensibility, and in the process renewed and repurposed the Enlightenment ideal of the "noble savage" to serve the atomic age.[11] They operated under the assumption that the machinery of cold storage could serve as a temporal and thermal prosthetic for maintaining the biology of so-called cold societies. They adopted the freezer as a time capsule, a means of making a biological freeze-frame for the future, where it might assume great value even if no one could say for sure what that value might be, let alone who should decide. Within cold blood, then, these specific mid-twentieth-century concerns about contamination, disequilibrium, and loss were also preserved and persist in the contemporary low-temperature infrastructure of biomedicine and life science.

The Technoscience of Life at Low Temperature

In biomedicine, cold storage—using dry ice, mechanical refrigerators and freezers, and liquid nitrogen—is often regarded as inert technology that supports the more dynamic work of analyzing molecules and finding cures.[12] Infrastructure—large-scale, layered, complex relational systems of embedded standards—as sociologist Susan Leigh Star has recognized, is similarly taken to be the framework that is "forgotten, the background, frozen in place."[13] An infrastructure made of refrigerators, freezers, and their biological contents might seem doubly destined to be ignored. However, it is the ability to hold still biological substances at various degrees of low temperature that has enabled such materials to become incredibly mutable and mobile, able to be manipulated, relocated, and recombined to answer questions other than the ones for which they were initially extracted from the body. Examining the history of the biobank in terms of low-temperature tissue-based infrastructure makes it possible to know how ideas about what life is and how it has been valued have changed and continue to change over time.

The concepts and practices involved in making life physiologically amenable to the human-engineered low-temperature environment emerged from the thermodynamic interactions of biology and industry, medicine and the military, from the nineteenth century through the Cold War.[14] Technologies of cold storage, developed initially for the preservation of the flesh of dead livestock and later for the maintenance of their living gametes, found their way into the biomedical laboratory through multiple channels. This deeper history of scientific refrigeration and freezing is the focus of chapter 1. It provides a historical milieu for thawing ideas about infrastructure such that it can be better understood as a dynamic process of recursive reconfigurations of the relationship between materials, politics, epistemologies, and values through space and time.

In the early twentieth century, Hannah Landecker has argued, "freezing came to serve as a central mechanism [used] both within individual laboratories and companies and within the biological research community more generally to standardize and stabilize living research objects that were by their nature in constant flux."[15] During the Cold War, human biologists began to adopt technologies of cold storage, which they regarded as tools for suspending animation. They idealized and put these technologies to work as antientropic machines, temporal prostheses, and environments of artificial stability. The value of freezing, especially for those who created collections of salvaged human blood, was located in the potential for deferring use and for reuse for purposes known and as yet unknown. Like Lord Kelvin's thermodynamic theorizing, which in the nineteenth century had been preoccupied with minimizing waste, freezing purportedly rare specimens of blood would prevent the untapped knowledge contained within from being squandered.[16] The lower the temperature, the slower a material was thought to decay, enabling it to be preserved over previously unfathomable timescales. As one scientist boasted in the early 1960s, "If the coefficient of decay in storage is of the same order as that for several enzymatic reactions and if the principle holds at very low temperatures, the decay which takes place in 3 weeks in material stored at +2°C would take 15 years at –78°C (in a bath cooled by dry ice), and some 53,000 years at –196°C (in liquid nitrogen)."[17]

The scientist who made this grand assertion, a Catholic priest and biophysicist named Basile Luyet, was the "father" of cryobiology, defined as the science of frosty life. Beginning in the 1930s, he cultivated a cosmology of cold that revolved around efforts to understand what he called "latent life."[18] For Luyet, latency—a form of suspended animation—was a liminal space between active life and certain death that could be used to probe the ambiguous boundary between the two states. *Latency* would also come to refer to untapped or concealed potential of life or life forms that had been redirected in time through the use of low temperature. The ambiguities implied by the word *latency* have been exploited to describe a huge range of concealed forces. Among them, Freud's theorization of sexual development; the lag between infection and symptoms of infectiousness in diseases like tuberculosis; Marx's figuration of "latent capital" as the time between the sale of a commodity and its purchase; and Edward Said's "latent orientalism," in which he exposed the racism in the assumption that to be other is to be backward.[19]

In the realm of life science, Luyet's efforts to draw lines around what life was and was not were quickly complicated by the very ways practices of freezing and thawing unsettled Euro-American assumptions about what it meant to be alive. The technical effort to maintain blood acquired from human bodies, in particular, was often accompanied by a great deal of labor to manage social and moral concerns about its appropriate use. For this reason, the history of freezing of biological entities (and pieces and immunological traces of them) also requires attention to the emergence of new social and cultural formations within and beyond the realm of biomedicine.[20] Defining "life," and for that matter, "death" has always been as much a historical and anthropological problem as a scientific one.

That a Catholic priest is considered the progenitor of cryobiology is a reminder that science and religion, too, are historically grounded binaries that do not capture the messy and indistinct realities that accompany efforts to locate life within a body or to preserve life outside of that body.[21] The history of technical efforts to freeze blood, to transform flesh itself into a resource for the future, are literally and figuratively intermingled with Christian salvation histories, which persist—along with other concealed or cryptic forms of life—as ghosts in the infrastructure of biomedicine.

Temporalities of Salvage

By the 1960s, refinements in the ability to freeze and thaw tissues at will had allowed techniques of cryopreservation to outrun the epistemological foundations of cryobiology propounded by Luyet and his disciples. The practical applications that grew out of cryobiology's experimental research agenda were quickly adopted with dramatic consequences by experts in disciplines from cattle breeding to genetics, physical anthropology, biomedicine, and public health. Cryopreservation contributed to reorganizing the temporal imaginaries of these fields in ways that tacked between the deep past and the indefinite future.[22] Much as the refrigerated railroad car supported the restructuring of agriculture and practices of capitalist trading at the turn of the twentieth century, the perfection of even lower-temperature technologies was crucial to the larger rearrangement of the temporal politics and practices of human biology and biomedicine in the years that stretched from the ragged ends of colonialism to the present.[23]

Toward the end of the Second World War, for instance, innovations in cold storage had allowed blood to be shipped to soldiers in the Pacific, and it soon became apparent this infrastructure, also known as a "cold chain," could be deployed in reverse. During the Cold War, frozen blood came to function as a literal and figurative connective tissue for reweaving the relationship between various disciplines within the sciences as well as between the United States and various colonial and newly postcolonial regions.[24] For human biologists who set their sights on the study of human communities whom they believed represented an idealized version of a universal human past, the ability to salvage the blood of so-called primitive peoples would contribute to constructing knowledge to manage a new world order.

Amid the intensified thermal politics created by the detonation of the atomic bomb, scientists began to confront the long-term problems posed by the iatrogenic by-products of innovation, including radiation but also chemical pollution and a growing global population. Perhaps the greatest problem was figuring out how to measure the impact of these potentially toxic forces, a process that would unfold over years, decades, and even centuries. In this sense, the Cold War was not only a time of nuclear standoffs between the United States and the Soviet Union. It was an epoch characterized by anxiety about new temporal horizons of risk.[25] This took the form of a rush to establish baselines, to salvage bodily evidence that could be used to distinguish the less polluted human past from its ambiguously contaminated future.

This book, then, is also a history of what I call "salvage biology," a twentieth-century corollary to a longer project of salvage anthropology.[26] The blood of members of so-called primitive groups was thought to contain a quarry of potentially invaluable information that would reveal itself as new molecular techniques emerged. A widespread assumption that these relics of the past were in danger of disappearing imbued the endeavor with a sense of urgency. In chapter 2, I examine this discourse of salvage as a form of anticipation about the creation of a new blood-based infrastructure for managing future risks to population health.[27] In the circumpolar North, ideas about the ability to salvage and preserve blood to serve the biomedical "as yet unknown" emerged out of the United States' military's mineral and medical prospecting efforts. Alaska Native peoples' bodies—along with their lands—were mined to support the ascendancy of America as a superpower not only in the realm of politics but also in life science. I describe how the Yale epidemiologist John Rodman Paul used his experience collecting blood in the vicinity of a United States naval base in the American territories of the far north to justify a broader program of serological surveillance that was ultimately adopted by the World Health Organization.

To Paul and other scientists interested in epidemiology and ecology among remote human communities—those who might be reservoirs of dangerous emerging infections *as well as* of helpful concealed adaptations to existing infections—harnessing artificial cold appeared to be an ingenious way of both crystallizing a problem, the loss of time, and providing its solution. It was expected that, if properly preserved, blood samples could and would be mined repeatedly, each time identifying novel elements, including ones not even anticipated by those who created the collection. The ability to reconstitute the collection, to make different constituents of individual samples work together, made it a dynamic and generative epistemic system as productive as the "experimental" ones that were taking hold in the molecular lab.[28] In this sense, the freezer filled with blood came to serve as what information theorist Geoff Bowker has referred to as an "artificial memory system" where what was archived were "not facts, but disaggregated classifications that can at will be reassembled to take the form of facts about the world."[29] Despite epidemiologists' claims that serological techniques were expanding the possibilities for medicine and public health, they often dismissed other forms of local knowledge not produced in the laboratory, including subjects' own memories of epidemics. Only certain accounts of life and death were seen as legitimate and relevant to the effort to create a global infrastructure for epidemiological surveillance.

Chapter 3 considers a different temporal dimension of this anticipatory orientation by examining how, for human geneticists who also sought to make population-scale collections of frozen blood, the enterprise was admixed with powerful retrospective emotions of nostalgia, guilt, and regret. University of Michigan human geneticist James Neel extended the reach of Cold War American science when he enrolled the Atomic Energy Commission to fund literal stocktaking of "primitive" peoples around the world.[30] Their purported isolation and untimeliness—the violence of which is expressed through the label "primitive"—situated their bodies as unique and precarious sources of value. Salvaging their blood would be used to save evidence of lives from the past. These were communities worth studying, in Neel's opinion, because, unlike the survivors of the atomic bombing of Hiroshima and Nagasaki he had worked with previously, they were biologically naïve relics whose bodies could provide controls or baselines for calculating radiation risk.

Human biologists used the concept of the baseline to construct the "primitive" as the uncontaminated normal standard by which the citizen of modernity could measure his own pollution by technoscientific society. It was a temporal marker that served to distinguish the concealed risks of the postnuclear era from the supposedly more easily detectable harms of earlier industrial and preindustrial ages. Moreover, population geneticists believed that adaptations present in the blood of their "primitive" subjects would, in time, be revealed as the products of natural selection to particular environments. As ionizing radiation from nuclear tests and potential nuclear warfare threatened to scramble these signals from the past, many scientists invested in the importance of creating an archive of evidence of what it had been like to be a human "relying on his biological endowment."[31] These scientists were members of organizations that had been created after the Second World War, including the World Health Organization and the decade-long International Biological Program (IBP). A worldwide survey of biological variation, such as that supported by the IBP, would be a means for enacting scientific internationalism in the service of salvaging a fleshy record of universal humanity.[32]

Collecting, Maintaining, Reusing, and Returning

The importance of diplomacy in this project of scientific internationalism became clear when Neel and other human biologists made use of a National Science Foundation–sponsored research vessel, the *Alpha Helix*, to enable them to travel to reach communities they understood to be situated differently in time from themselves. The *Alpha Helix* was designed as a floating laboratory to connect the molecular sciences with the field. The three human biology expeditions undertaken with the help of the *Alpha Helix*, each of which focused on the collection of blood in the Amazon and Melanesia, are the subject of chapter 4. These expeditions were emblematic of the technical, diplomatic, and interpersonal challenges scientists faced as they attempted to navigate the cold chain, including what happened when elements of the frozen infrastructure they were attempting to create broke down. Following the *Alpha Helix* across space and across time reveals new kinds of "ships"—kinship, ownership, stewardship—that sail to the core of efforts to cope with the promise and peril of biomedical innovation predicated on access to human body parts.

Those who collected and froze blood often did not know precisely what it would reveal, despite their insistence that it was imperative to collect and freeze this precious substance before members of "cold" societies became "hot." In ways that anticipate current enthusiasm for "big data," a domain in which questions are often articulated only after an answer has been discerned, these collectors saw themselves as participating in a new mode of doing science in which their vast accumulations of research materials would generate the hypotheses. Population-level assemblages of frozen blood, which were often subdivided and sent to labs specializing in different analytic techniques, created a spatially and temporally distributed network for producing biomedical and epidemiological knowledge.

This was how Baruch Blumberg and Carleton Gajdusek—both of whom ran labs at the National Institutes of Health—conducted the respective research that won them each the Nobel Prize in 1976. Blumberg, using blood samples that he and others had collected from members of geographically isolated groups in the Pacific, came to understand that hepatitis B had a viral etiology. Gajdusek, a veteran collector who sailed on the *Alpha Helix* to collect blood, had earlier used brains collected from members of the Fore in the Papua New Guinea highlands to solve the mystery of the degenerative disease kuru—later deemed a new life form, a prion.

These and other researchers maintained a faith in the progressive nature of their scientific endeavor. It seemed inevitable that, over time, new ways of analyzing blood would enable stored samples to reveal their secrets, either through the identification of new constituents or through the comparison of new configurations of samples. It was a form of reductive holism, motivated by a dream of being able to piece together molecular-level insights into something approximating a total archive of human variation.

Scientists' claims for the importance of maintaining access to research materials have been and continue to be predicated on ideas about the concealed potential of preserved blood, a different form of latency than that examined by Luyet.[33] What has never been clear was how anyone might detect when a future in which the true value of the sample could be revealed had arrived. The Greeks had a term for this kind of indeterminate revelation or time lapse—*kairos*, which complemented and complicated another form of temporality—*chronos*. *Kairos* is a way of understanding biomedical faith in the latent—present, yet presently absent—potential of salvaged blood samples that justified their removal from the regularized flow of *chrono*logical time. It is particularly helpful for demonstrating how narratives of technological progress come to be animated as much by desires for salvation and redemption as by forms of scientific rationality. It is also helpful for demonstrating how those narratives can be disrupted or reconceived.

Writing in the decades that practices of cryopreservation were being refined through efforts to preserve blood, historian Reinhart Koselleck argued that modernity brought with it progress but also a new experience of accelerated time. Koselleck's insight was that this accelerating progress "betokened a present compounded of many layers of time, a simultaneity of the nonsimultaneous."[34] His effort to understand what he saw as a rupture created by modernity yielded a mode of historical thinking that embraced the complexities of temporality and the need for a plurality of points of view in any effort to make historical knowledge—be it by scholars working with paper archives or with frozen tissue. The dynamic relationship between what Koselleck referred to as the "space of experience" and the "horizon of expectation" help, as historian John Zammito has argued, to "give purchase on the paradoxes *latent* in the dimensions of time—present, past, and future—as well as to underwrite interpretive 'fusion of horizons' with the radically alien."[35]

The multiple and divergent fates met by frozen collections has revealed unexpected mutations induced by relocating life in space and time.[36] Practices of freezing and thawing a given sample of blood are also acts of ontologizing; choosing to change the phase of a substance from a solid to a liquid is also a way of deciding what it is, which enables it to accrete new meanings while also retaining old ones.[37] It has been a form of power to make the body multiple, which complicates efforts to make decisions about the definition and administration of life and death.[38]

Blood is charged with strong valences grounded in its protean capacities.[39] It has long been regarded as both a biological and a social fluid.[40] Even before the Cold War, blood was used as tool for regulating populations based on race.[41] This was especially true for Indigenous North Americans, where blood quantum became tied to questions of sovereignty and identity.[42] As DNA emerged as an especially valuable object of knowledge in the 1990s, this tendency has recurred.[43] Molecular practices contribute to characterizing elements that are concealed within blood as well as to the construction of ideas about citizenship and identity.[44] The experience of encountering latent life—be it suspended or concealed—was, and continues to be, characterized by the discovery of stowaways, carriers, and other entities that become legible and meaningful at different and unexpected moments.[45]

The meanings and forms of value generated from blood and other body parts are neither stable nor evenly distributed. Warwick Anderson's book *The Collector of Lost Souls* documented mutual and uneven transformations that occurred between American scientists, including Carleton Gajdusek, and the Fore Pacific highlanders from whom they collected brains to investigate kuru.[46] The distance between Gajudsek's lab and his field site in Papua New Guinea facilitated his efforts to bracket off intimate and subjective relationships with his subjects from his quest for scientific acclaim.[47]

Gajudsek died in 2008, but the blood he collected from the Fore and dozens of other populations around the world persists. Neel (who died in 2000) and Blumberg (in 2011) have also been outlived by their collections, which are both more valuable and more vulnerable than ever. Before he died, Blumberg appealed—unsuccessfully—to his colleagues through the pages of *Science* to find an heir to his vital legacy: freezers filled with tens of thousands of human and nonhuman blood serum samples.[48] When otherwise ephemeral tissues are able to persist over time—even beyond the lifespan of their collectors—who will become responsible for their maintenance? Who will get to decide what constitutes a worthy use of these precious and, ultimately, finite materials?[49] This may be the most important question connected to the reuse of frozen blood, not least of all because each act of thawing degrades the sample and involves some degree of destruction. The freezer, a seemingly mundane machine for stilling life, has increasingly become the focus of contestation over what it means to be dead.[50]

In the twenty-first century, the salvaged biological materials that comprise this frozen infrastructure—the networked constellation of freezers filled with blood and other kinds of tissue—continue to produce new forms of biomedical knowledge and new biological and social forms of life. Frozen blood has been important for studies of human variation and ancestry, making it also a resource for the booming market in what is sometimes referred to as "recreational" genomics.[51] It is being used to document changes in the presence of toxic chemicals in the environment and has been positioned as a resource for the emerging field of proteomics.[52] Scientists have recovered from within cold human blood samples forms of nonhuman life, including microbes, and have figured out how to extract and sequence their DNA. As a result, blood that was initially collected as a means of understanding human ecology has become a microcosm unto itself, harboring life forms that are of interest on their own terms.

At the same time, some members of some of the indigenous communities who have persisted in the world as their blood persists in biomedicine's freezers do not approve of its use in answering questions they have not themselves asked. They do not believe all of these new questions serve their futures; rather, the answers to such questions may put those futures in jeopardy. Certain descendants of those whose blood has been collected have come to see the freezer as an insecure environment from which the remains of their ancestors must be rescued. They want it back and have demonstrated that they too are capable of salvage. They too are capable of disrupting the chronological flow of time that maintains, as one recent scientific freezer advertisement claimed, "the future, inside." As their blood becomes more and more potentially productive of biomedical futures, certain Indigenous people have struggled to reconcile the social and moral needs of their communities with those of science. In the contemporary frozen archive, blood has continued to fulfill its scientific potential as a biomedical research object even as it provokes a reassessment of the ethical and political dimensions of knowledge making involving human subjects in the post–Cold War period, the focus of chapter 5.

Futures, Inside

I too am a collector. As a historian, especially one writing about the creation and politics of a peculiar kind of archive—one made of blood samples, rather than documents—I am acutely aware of the fragmentary and potentially violent nature of any effort to reconstruct the past. Interestingly enough, this dimension of my own practices of making knowledge is analogous to that of the scientists about whom I write. What we know is intimately linked to why we want to know it as well as how we go about doing so. This is one of the fundamental insights of the history and social study of science, technology, and medicine.

Recently, some historians have sought to reclaim their power to speak to the present by plunging deeper into the past or by embracing biological methods that purport to confirm, once and for all, how it really was.[53] This book takes a different approach. It argues that before it is possible to rescale the time span of history, it is necessary to engage with histories of temporality itself, the complex and culturally produced ways of imagining and existing in time that shape the questions that can be asked by any kind of historical enterprise, be it drawn from manuscript or blood-based archives. The recognition that the past is neither singular nor fixed has led latency to be adopted by other postwar historians as a method of being attentive to the unspoken events and concealed emotions that shape our assumptions about the present.[54]

It is my hope that practicing scientists will read this book to gain a better understanding of how the techniques, machines, assumptions, and research materials that may seem mundane today are the product of historically specific choices that have had unexpected consequences. I also hope that members of indigenous communities and others interested in understanding how and why science came to value their body parts as research materials will be better informed in their efforts either to contest, participate, or even to reimagine such enterprises.

Freezing is a dynamic process that requires a great deal of energy, including technical, emotional, and ethical labor. Similarly, each act of thawing changes relationships in ways that may dissolve and reconstitute boundaries between insides and outsides, humans and non-humans, and even between life and death. The recurrence of efforts to find new uses for old

1 "The Future, Inside." REVCO advertisement, circa 2010. Biomedicine's molecular future swirls out when the freezer door is opened. The ultralow −86°C REVCO freezer shown here has since been discontinued. Reprinted with permission of Thermo Fisher Scientific.

blood raises fundamental questions about animacy, including what gets to count as life and whether or not that is the same thing as being alive, what kind of life matters, where it matters, when, and to whom.[55] In their purported ability to suspend and restore animation, practices of freezing and thawing actually redistribute vitality, enabling those who know how and where to look for it to detect forms of life at registers that might not have otherwise been visible. In practice, this includes the detection of an array of molecules, microorganisms, and ancestral spirits, each of which makes claims on the futures generated by the freezer.

The history of practices of freezing bodily substance is important for the forms of movement and social mutation it makes visible. Freezing blood has enabled it to circulate across place and time, but it has also enabled multiple different groups to identify it as a source of liveliness in terms that exceed the language of molecular biology.[56] The persistence of bodily substance in time makes clear that life and ideas about the forces that give it vitality are—like infrastructures—constantly in flux, even when they appear to be frozen. In the twenty-first century, whether or not cryopreserved tissues can be considered life has become more, not less, contentious.[57] The status of a frozen biological material becomes particularly salient when the substance in question has the reproductive potential of an embryo or a human cell line. What about a strand of DNA? What about a blood serum sample that is thought to contain no cells or DNA at all? Molecular explanations may not be equally relevant to all those with stakes in defining life. Who gets to decide?

Will scientists be able to better create ways of reckoning with the unexpected challenges to their authority posed by forces that are described in terms of ethics, spirituality, or morality? Considering the frozen milieus created by cold storage technologies as literal "in-between" zones rather than spaces of stasis may lead to a dissolution or rearrangement of restrictive binaries like living and dead, human and nonhuman, matter and spirit, organism and machine. The freezer and the larger infrastructures it makes possible are sites for examining the potential of such machines themselves to do more than serve inert matter available to be used by humans to master nature.[58] Freezing, when understood as a technology for recalibrating human awareness of the complexity of scales of life and of time, rather than merely as a technology for deferring death and decay, can contribute to the making of biomedical knowledge that supports flourishing not only for unknowable futures but for a greater array of lives in the here and now.

PART I

The Technoscience of Life at Low Temperature

ONE

Latent Life in Biomedicine's Ice Age

Anything could be taking place inside the gray brick building behind a strip mall in Rockville, Maryland, a suburb of Washington, DC. A fourteen-foot-high tank of liquid nitrogen—permanently parked in one of a dozen spots that also serve an office machine supply depot and a carpentry center—provides the only clue that inside are nearly five million frozen human tissue samples. Since 1977 this building, known as the Biomedical Research Institute (which today goes by its acronym, BRI), has served as a low-temperature warehouse with, in its own words, "capabilities to store and distribute precious biomaterials [to] help speed research and hasten the discovery of cures and treatments."[1] This looks like 174 freezer chests set at –80°C and 45 large liquid nitrogen tanks at –196°C, each of which help to maintain human blood samples associated with private and government research projects, including clinical trials related to HIV/AIDS.

A quarter mile down the road is the official headquarters of this repository: a red brick building that contains a suite of laboratories that have come to be devoted to research on schistosomiasis, an infectious agent also known as bilharzia, or snail fever. (In addition to its human specimens, BRI also maintains hundreds of thousands of schistosomes at various stages of their life cycles.)[2] In the lobby, the realism of a life-sized mural of a family knee-deep in the murky waters of the global South is punctuated by a biology-textbook-style illustration of the transmission and complex life cycle of the schistosome, which radically changes form as it develops. A set of doors, painted to look like they lead into a thatched hut, open instead into a suite of offices and gleaming laboratories.

Together these two buildings comprise the American Foundation for Biomedical Research and Biomedical Research Institute, known only as BRI, a not-for-profit organization that currently describes itself as "committed to improving global health research."[3] Nothing about the activities of the BRI is especially unique. It is by no means the only site of research on infectious disease in Bethesda. Nor does it maintain the largest low-temperature biospecimen repository in the state. That honor belongs to the National Institutes of Health, located down the road.[4]

BRI is, however, the one of the first low-temperature biomedical research and storage facilities in the world, dating back to its founding in Wisconsin in the late 1940s as the American Foundation for Genetic Research. It is one node in a vast and distributed frozen infrastructure that serves a range of knowledge projects connected to understanding life and its limits. It is both singular and mundane, remarkable in the scope of its mission and typical in the materiality of its functioning. BRI is a place where bits of bodies are held in crystalline stillness. Sometimes they are removed from the freezer to continue the investigations for which they were initially preserved. In other instances, they are thawed to reveal previously concealed or cryptic life forms such as microorganisms. Both of these possibilities contribute to the biomedical value of maintaining frozen tissues, despite the energy required to use cold storage technologies in the service of suspending or slowing animation and decay.

Before there was an industry based around the preservation of frozen human tissues, there was cryobiology, the study of icy or frosty life. Cryobiology was a mid-twentieth-century instantiation of the much older question of whether or not life could be stopped and restarted at will. In 1958 a biologist named David Keilin, then at the Motelno Institute for Research in Parasitology at the University of Cambridge, England, addressed the Fellows of the Royal Society. In his talk, titled "The Problem of Anabiosis or Latent Life," Keilin explained how the emergence of new technologies—including both mechanical freezers and liquid gases—was advancing research on the "state of an organism when its metabolic activity is at its lowest ebb," but not too low to be subsequently restored.[5]

Keilin reviewed the long history of experimental inquiry into this biological state, which had appeared to some early modern Christian observers to involve a form of resurrection. The Catholic priest Lazzaro Spallanzani wrote in the eighteenth century of the phenomenon Keilin referred to as latency: "It confounds the most accepted ideas of animality; it creates new ideas, and becomes an object no less interesting to the researches of the naturalist than to the speculation of the profound metaphysician."[6] The question of "whether life under certain conditions may be a discontinuous process" was, Keilin argued, one of the oldest questions, "reflected in almost all religions, in some legends and even in fairy stories." Latency, in providing a way of figuring life beyond limits, oriented cryobiology towards secular answers to questions about immortality that had previously been dealt with in the realm of religion.[7]

In terms that were both more spectacular but also more commonplace than Keilin's, a lyrical essay published in 1922 in the *Scientific Monthly* had attempted to translate the significance of "latent life." Describing the phenomenon as a "halfway house" between active life and death, the writer elaborated with a string of metaphors that, when rattled off in succession, strove to evoke a sense of the limitless potential contained within life itself:

> In living matter, the molecular whirl is at its intensest [*sic*]; in latent life the molecular whirl is for a time arrested; in death the molecular whirl has been stopped forever. In life the dancers are in the mazes of an elaborate figure; in latent life each individual is standing stock still; in death every dancer has fallen over. In latent life the weights of the protoplasmic clock have been seized by a mysterious hand; in death they have descended to their full extent and cannot be wound up again, for the cord is broken. In latent life there is only a stoppage, in death the end has been reached. In life, "the sands of time" are running out rapidly; in latent life the stream has stopped; in death the sand is all in the lower globe.[8]

Keilin, in his 1958 talk, revisited and refined the many different kinds of experimental approaches around the mysterious nature of latent life by proposing his own term, "cryptobiosis."[9] When an organism was in cryptobiosis, an observer would be unable to judge whether or not restoring it to life would be possible—unless, of course, the observer had intimate knowledge as to whether the form or structures that supported metabolic activity had been maintained. In Keilin's view, which looked forward to the potential of the emerging science of cryobiology, understanding the biophysical properties of life—its internal form or structure—could demystify the seemingly spectral ability of its function to be restored when it emerged from the crypt of the freezer.

The ability to access and harness low temperature, in addition to contributing to the elucidation of the fundamental nature of life, had already begun to enable researchers to displace and disperse biological matter through space and time. Cryopreservation, as this practice came to be known, made it possible to experience vitality as containing more and different forms of potential than were immediately apparent. The realization that some life forms could conceal still other life forms—a kind of cryptozoology—added to freezing's potential to generate new forms of knowledge, value, and life itself.[10] Both of these properties of latency—icy stillness and the secret or concealed, which I examine in terms of the cryo and the crypto—gave form to biomedicine's ice age.

Cryobiology is an exemplary kind of technoscience—a mode of knowledge production that exceeds and subverts distinctions between science and technology, between nature and culture, and between matter and spirit.[11] In the twenty-first century, the freezer filled with biospecimens has emerged as a secular reliquary of latent life, an organic machine that produces biological and social innovations and, perhaps, even revelations through its abilities to preserve and reorient biological matter through time.[12] Life as an object of inquiry becomes most tangible when it subverts the boundaries of knowledge to become useful to science.[13]

In Search of Lost Time

The single most interesting detail about BRI might be that its founder was a biophysicist and Catholic priest. Basile Luyet, who was known within the worldwide order of Missionaries of St. Francis de Sales and to many of his students as Father Luyet, has often been remembered, also, as a father of cryobiology.[14] Luyet's career, which involved the creation of institutional formations including laboratories, journals, and international networks for the circulation of knowledge about latent life, culminated in the creation of the blood-based storage facility to exploit its potential, BRI. Tracing the arc of his career makes visible the historical coordinates of the frozen infrastructure of biomedicine and life science.

Luyet began his interests in cryobiology looking not to play God but to make sense of the mysteries of the universe created by God. For Luyet, cryobiology could provide a means of situating the mutability of life as intrinsic to its sanctity. Even before he became associated with BRI in 1956, when it was still known as the American Foundation for Genetic Research, and several years before the term *cryobiology* was coined, Luyet was an evangelist for a secular cosmology that would account for the history of life and its potential in terms of low temperature. There is no evidence that he ever commented explicitly on how his religious commitments informed his scientific ones, yet his repeated efforts to reinforce the boundary between the two belie their entanglement.

2 Basile Luyet. Reprinted with permission of Florimont Archives.

Luyet's ideas about cold were indebted to the insights of nineteenth-century thermodynamic theorists who had reached the conclusion, as he put it, "that what gives our senses the impression of cold or warm is merely the velocity of motion of the molecules" and that, therefore, there is a lower limit of temperature at which the molecules stop moving: absolute zero.[15] By the 1960s Luyet had broadened his appreciation of thermodynamic theory to acknowledge that it had also yielded innovations that transformed industrial production. He observed and accepted that cryobiology as a nascent discipline was characterized by similar patterns, including "the intensification and organization of the research and the accumulation of data on the multiple aspects of preservation at low temperature."[16]

Born in Saviese, Switzerland, in 1897, Luyet demonstrated early on the qualities that would characterize him as an inveterate experimenter and institution builder. In his youth he was an amateur ethnologist who established the journal *Cahiers de valaisans de folklore*, dedicated to salvaging the traditions and culture of his beloved alpine homeland. At the age of twenty-four he joined the order of St. Francis de Sales. Soon after, he enrolled at the university in Geneva, where he was its first Catholic priest to be granted a doctorate. Technically, he was granted two doctorates: one in "natural sciences" and the other in physics.[17] This work, which involved applying theories of material structure to living beings, earned him Yale's Seesel biology prize in 1928. The prize supported him for a year of postgraduate work at Yale in 1929, where he became interested in the biological definition of life.

His path into this morass was to start by asking what biological life was *not.* He reasoned that since death was the destruction of life, he would approach the matter from the purview of the end or near-end of life. At Yale, he worked in the laboratory of Ross Granville Harrison, where he conducted experiments on the ability of ultraviolet light to induce genetic mutations in fungi spores.[18] These interests soon brought him to the Rockefeller Institute in New York, where as a visiting fellow he became acquainted with the famous physician and experimentalist Alexis Carrel.

Carrel's first experiments with such precarious forms of life—which he referred to as "latent"—involved freezing, thawing, and transplanting pieces of dog artery.[19] In a 1910 article, Carrel explained that "a tissue is in latent life when its metabolism becomes so slight that it cannot be detected, and also when its metabolism is completely suspended. Latent life means, therefore, two different conditions, unmanifested actual life and potential life."[20] The former, "a normal stage in the evolution of all organisms as they progress towards death," was in Carrel's view a temporary and inevitable phenomenon, analogous to ideas then being articulated in studies of sexual development in the human sciences; its telos, or temporal goal, was determined but not yet actualized.[21] The latter condition consisted of a "suspension of all actual vital processes," leading Carrel to attest that tissues in this state, "could be preserved outside of the body for an indefinite period of time," and thus, through the use of technologies of cold storage, could reveal new properties and uses of life.[22] Citing Paul Ehrlich's success in freezing tumors and subsequently demonstrating that they would grow again when thawed, Carrel deemed it "reasonable . . . that protoplasm can be placed in such a condition that life does not exist actually, but merely potentially."[23]

Carrel was by no means the first to entertain the possibilities of latent life. Latency had long been considered of great metaphysical and biological significance.[24] As Keilin would explain to his peers at the Royal Society, natural philosophical interest in the subject was exemplified by Antoine Leeuwenhoek's seventeenth-century study of organisms including the tardigrade, which had the ability to become reanimated after long periods of being dehydrated without necessarily being exposed to low temperatures.[25] By the nineteenth century, the French physiologist Claude Bernard had elevated latency to a central role, defining it as the cessation of metabolism in an organism—a halting of the biochemical activity in the *milieu interieur.*[26]

The regulation of this internal environment was the organism's means of compensating for variations in the environment that existed beyond its body. In this view, organisms strove to achieve a form of homeostasis between the two, internal and external. Bernard understood latency to be "the state of an absolute chemical indifference" characterized by the "suppression of all interrelationships" between an organism and the environment in which it persisted.[27] The stability of the internal milieu of the organism itself was the fundamental condition of its independence or autonomy. For Bernard and, later, for Walter Cannon, the first professor of physiology at Harvard, blood was the mechanism by which this internal equilibrium was maintained.[28]

It was only at the turn of the twentieth century that this epistemology of latency inspired programmatic research with freezing blood and pieces of organisms, ranging from cells to organs and arteries, extracted from the *milieu interieur* but kept alive in a prosthetic external environment.[29] This kind of manipulation was a clear extension of the engineering ideal that was then taking hold in biology, ranging from Jacques Loeb's demonstration of artificial parthenogenesis to Carrel's many experiments with transplantation, transfusion and, later, tissue culture.[30] The engineering ideal, an incipient form of technoscience, collapsed distinctions between biology and technology. It did so by transforming life into a set of substrates to be cultured and manipulated--what the philosopher Martin Heidegger, writing about technology at this time, would have recognized as a "standing reserve."[31] Latent life became the raw material out of which experimentalists created knowledge, technologies, and still more forms of life.

In Russia, P. I. Bakhmet'ev was also experimenting with inducing in insects and bats a phenomenon akin to latency: anabiosis, a state that is neither life nor death.[32] So tantalizing were the commercial possibilities of the ability to freeze life that Bakhmet'ev's expertise attracted the attention of the Moscow Refrigeration Committee. The committee sought to support his research in the hopes of extending the distribution of live fish and caviar—a plan that ground to a halt with Bakhmet'ev's own untimely death in 1913.[33] Though the years following the Bolshevik Revolution were not good ones for anabiotic (or many other kinds of biological) research, utopian fiction writers in Leninist Russia, like those in the United States, repeatedly linked suspended animation to the conviction that through the unwavering progress of science—not religion—death could be conquered, manipulated, or at least inevitably postponed.[34]

Creating Cold

In his early experiments with the latent life of arteries, Alexis Carrel had determined that "the best method of preservation consists of placing the vessels . . . in an ice box, the temperature of which is slightly above the freezing point."[35] But an ice box, a technology as basic as it sounds, proved to be an unreliable instrument. Fluctuations in temperature, even of a few degrees, could prove disastrous when dealing with fragile tissues, the form of which needed to be maintained in order for life to be restored after suspension. Carrel was frustrated by the difficulty of inducing the cold-generating component of his experimental apparatus to maintain a fixed temperature. The inability to produce precise and reliable temperatures placed a technical constraint on studies of latent life, not least of all because it made it impossible to ensure that the fragile structures contained within a given tissue would be uniformly preserved.

Technologies for producing and sustaining a stable environment of artificial cold were still relatively rare in the early twentieth century. The most dramatic changes in the ice and refrigeration industry happened between the 1870s and 1920s, culminating in the precision control of artificial low temperature.[36] Between 1924 and 1934, physicists Leo Szilard and Albert Einstein applied for twenty-nine German patents, most of which, as historian Jonathan Rees has noted, dealt with home refrigeration.[37] It was demands from producers of food, not physicists, that would eventually lead to the first technologies for producing specific and dependable low temperatures. The refrigerator, and later the freezer, like so many other technologies before it, would find its way to the lab via the farm and the kitchen.[38]

Consumers, however, did not immediately take to artificial refrigeration. They needed to be persuaded of its value and the social transformations that would accompany new modes of acquiring and consuming food. In the late nineteenth century, some Americans even viewed man-made ice as blasphemous: akin to playing God.[39] They believed that to manufacture and sell the frozen state was to tamper with the order of nature. It was also not uncommon to regard ice as evil matter to be transcended.[40] Though Romantic thinkers sought to recast ice as a vehicle and revelation of vital energy, the new thermodynamic theory, in its alignment with Darwinian ideas about extinction and heat death of the universe, coupled ice with apocalypse.[41]

The arrow of time derived from the second law of thermodynamics: the entropy of the world strives for a maximum. It was in 1865 that Rudolf Clausius had first described this phenomena in technical terms, as well as what would become accepted as the first law of thermodynamics, that the energy of the world is constant. These laws were verified and cultivated by William Thomson, later known as Lord Kelvin, and by 1870 thermodynamics had its own ontology, laws, concepts, and research program.[42] Those thinkers and tinkerers who subscribed to this agenda were fixated on harnessing knowledge of energy in the service of reducing waste, which guided its applications in the realm of industry.[43]

A young professor at Munich's Polytechnic School named Carl von Linde was at the forefront of integrating the theoretical insights of thermodynamics with his engineering-oriented refrigeration design.[44] Largely through the efforts of Linde, science, technology, and industry were combined to produce an approach to refrigeration that was intended to bootstrap the beer brewing industry in late nineteenth-century Germany. Linde's contribution was essentially a technoscientific one, to erase the distinction between engineering and science by applying the laws of thermodynamics to his commercial innovation.[45]

The research and development program that Linde pioneered was an important aspect of "technical thermodynamics," predicated on engineering the removal of heat from a system. This enabled him to more efficiently standardize and export his ideas for commercial benefit. It also represented a form of practical rationality that, as historian Mikael Hård has argued, participated in the larger social project of the disenchantment of the world.[46] In this view, while the immediate goal of scientizing refrigeration was to control production, it represented a transformation in the relationship between humans and their machines, namely one oriented around their conception as servants of human intention.[47] By virtue of the labor they metabolized, nineteenth-century social theorists were moved to regard machines as "frozen spirit."[48] Linde, ever the enterprising industrialist, sought to exploit his refrigerating machines, famously claiming that he and his peers "could not master the weather, [but] they were able to control the temperature."[49]

Nowhere was this technical control more dramatic than in his development of the first continuous process for the liquefaction of air. Linde demonstrated the importance of recursive relations between experiment and apparatus as well as between science and technology.[50] Linde's contemporary, Paul Becquerel, who preserved plant seeds for several months at ultralow temperatures and subsequently induced them to germinate, refined Claude Bernard's definition of latency to mean life in state of suspended animation. To conduct these experiments, which were done at –235°C (the temperature of liquid helium), Becquerel had to gain access to technologies for producing such ultralow temperatures. In the first decade of the twentieth century, he traveled to Leiden, where Kamerlingh Onnes was searching for absolute zero. In this and many other instances, efforts to achieve ever-lower temperatures required the development of new technology, which subsequently found uses beyond the laboratory.[51] Becquerel, less concerned with industrial applications than Linde, allowed himself to speculate beyond the life span of humans on earth, observing of the gradual heat death of the universe implied by thermodynamics that keeping germs at very low temperatures would be the "most efficient way for the preservation of life on earth as long as possible."[52]

In Linde's case, the special apparatus he developed for liquefying air led him to expand the market for industrial gases overseas, making the science of producing extremely low temperatures also a technique for improving industrial productivity. The United States' huge steel industry created a market for liquid oxygen used in welding, which inspired Linde to set up a manufacturing outpost in 1907 in Tonawanda, near Buffalo, New York (chosen due to its proximity to the inexpensive hydroelectric energy source of Niagara Falls). In 1914, Linde established research laboratories there to study the properties of gases at low temperatures, including liquid nitrogen. (During World War I, Linde Air Products joined with four other American companies to become Union Carbide, preeminent in the development of the petrochemical industry.) In this way, Linde—along with Onnes's industrial lab in Leiden—was at the forefront of the establishment of refrigeration-oriented research centers, many of which were established after World War II in Europe as well as in South Africa and Australia.[53]

Other techniques and technologies for the production of artificial cold were created during the period between the two wars. Chief among them was solid carbon dioxide.[54] Also known as dry ice, the raw carbon dioxide required for its production came from the by-products of industry, particularly from fermentation and lime processing. Early experiments with cold, such as Carrel's, relied upon dry ice (usually introduced into a water bath) to produce temperatures in the range of –70°C to –80°C. It is for this reason that –80°C later became a standard set point for commercial deep freezers used in many scientific labs.

In the United States, the development of mechanical refrigeration actually stimulated the market for dry ice. The two technologies could be used together in the "cold chain" that connected farm to household and field to lab. The "cold chain" is a term that has long been used to describe the links needed to maintain low temperature for preserved materials as they circulate.[55] Clarence Birdseye's commercial fast freezing technique—which he proudly claimed was inspired by his observation of Inuit fish preservation practices while working as a naturalist in Labrador, Newfoundland—contributed greatly to stimulating the market for frozen fruits and vegetables.[56]

Birdseye recognized a phenomenon that would later be more systematically investigated by cryobiologists: that cell walls of perishable food items did not degrade significantly when frozen rapidly.[57] Food items, despite having already been harvested, began to be referred to in terms of freshness. Freshness indexed a frozen material's connection to place; it was a rhetorical strategy for persuading consumers that the food they were eating was as alive as it had been at the moment it had been picked. During the food shortages of the First World War, the freshness promised by refrigeration also became a strategy for encouraging thrift.[58]

By the 1930s, many American and European households had refrigerators, but demand for freezers that could maintain lower temperatures grew along with Birdseye's empire between 1935 and 1940.[59] In subsequent years, the global network dependent upon this suite of new technologies for producing and maintaining artificial cold expanded dramatically.[60] Cryoinfrastructure spread in tandem with the electricity grids that began to crisscross the rural regions of the United States and major metropoles around the world.[61] Nicola Twilley has evocatively described this artificial "coldscape" as the "unobtrusive architecture of man's unending struggle against time, distance, and entropy itself."[62]

Companies like Electrolux sought to colonize the tropics' developing markets for otherwise perishable commodities. Expeditions conducted by Admiral Richard Byrd in Antarctica and sponsored by the Kelvinator refrigeration company, stoked interest in the possibilities for freezing perishable substances at home and in the lab using mechanical refrigeration. One such experiment involved the use of Kelvinator machines aboard his ship, *The Bear*, to preserve food for the crew and to bring back samples of native flora and fauna encountered during his travels.[63] The kinds of technologies that were developed during the interwar period—which were not always created specifically to serve biological research—would stoke efforts to study life at low temperature, not least of all because any scientist who had a food refrigerator or freezer at home could begin to experiment with slowing decay.

3 This advertisement, one of many produced in the 1930s, situates the mechanical (in this case, gas) refrigerator as an agent of empire. It reads, in Portuguese, "In five parts of the world, and especially around the Portuguese Empire, all speak well of the refrigerator." Six different indigenous cultural "types" embrace the refrigerator as a welcome sign of modernity against a backdrop of a globe illuminating the many territories still under Portuguese influence, including regions of Asia, Africa, Melanesia, Europe, and India (not shown).

Cold Shock

Even once the technologies comprising the scientific cold chain had been created, there was at least one major roadblock on the path to perfect preservation: the danger of the thaw. If obtaining the requisite and reliable low temperatures to freeze posed one set of challenges, unfreezing them created still others. Defrosting frozen life was at once an experimental and a practical problem made more complex by the ways it displaced questions of spirit and the soul.

Carrel and others had begun experimenting with stockpiling skin and corneas at temperatures near freezing, but they were reluctant to store tissues at or below zero degrees Celsius. Upon removal from the frozen state, the cells were often reduced to slurry.[64] This phenomenon was described as "cold shock"—the recognition of a trauma to the organism upon thawing.[65] "Cold shock," also known as "thermal shock," remains poorly understood, even in the twenty-first century.[66] In the 1930s, Luyet and others speculated that the damage had to do with the formation of ice crystals tearing through cells as they formed, which did not become apparent until it was time to defrost.[67]

The desire to characterize and ultimately circumvent this damage drew Luyet more deeply into the study of life and temperature. He believed he had found his holy grail when, in the summer of 1937 in the library at the Woods Hole Marine Biological Laboratory in Massachusetts, he came across a reference to the work of a German physicist named Gustav Tammann. In 1898 Tammann had written that the molecules of carbon compounds could be solidified in an amorphous state of glass without being rearranged as sharp crystals.[68] Tammann referred to this amorphous solid-state phenomenon as "vitrification." His work suggested to Luyet that if one could freeze a life form quickly enough and coldly enough, perhaps using liquid gases, the formation of ice crystals could be avoided and the indefinite extension of life would become feasible. Luyet set immediately to work, using frog sperm (the favored experimental material of the eighteenth-century Catholic priest Lazzaro Spallanzani).[69] Pursuit of vitrification or "rapid freezing" became Luyet's new calling.[70]

Luyet's 1937 encounter with ideas about vitrification was timely. That year, the world's first blood bank was established in Chicago at the Cook County Hospital. Prior to this time, transfusions could only be conducted with fresh blood taken from donors immediately before it was needed, a practice also pioneered by Carrel.[71] This changed when physician Bernard Fantus, who had helped establish the Blood Preservation Laboratory at Cook County Hospital, used sodium citrate to help stabilize blood for short-term storage.[72] Fantus appropriated one of the household refrigerators that were finding their way into hospitals. While his innovation allowed for the stockpiling of blood, it would still have to be disposed of after only a few weeks.[73]

In the late 1930s, however, Luyet was less interested in the potential biomedical applications of vitrification than he was in mastering rapid freezing as an experimental technique. He was working at the Jesuit St. Louis University in Missouri and had just launched a new journal called *Biodynamica*, described as a foray into the "experimental study of working hypotheses on the nature of life" at the intersections of physical chemistry and biology.[74] His aim was to import the physical sciences' practice of theoretical speculation, backed up by experimental verification, into biology. His wager was that "the processes we call life" emanated from a relatively small fraction of the "formed" or "structural elements" of the cell such as the genes or chromosomes.[75] The logo he chose for the journal was of his own design, two slightly overlapping circles, each drawn around obscure iconography. This image served as a cosmogram, a way of understanding how Luyet saw the organization of the natural universe.[76] It was a representation that carefully sidestepped metaphysical or spiritual questions.

Luyet described his cosmogram in the first issue, which included articles he had written between 1934 and 1938. To explain the circle on the right, he addressed biochemists, stating that to those "who may not be conversant with the 'Knighthood of Biology,' we would mention that the drawing on the right represents what is called a 'mitotic figure,'" in process of cell division. In addressing "those biologists who have never been 'initiated' into the 'Order of Physical Chemistry,'" he explained that the drawing on the left "represents the modern conception of an atom of carbon, consisting, it is thought, of a positively charged nucleus indicated by the sign + (it is not a cross!), and of 6 negative electrons arranged in 2 orbits about this nucleus."[77] Luyet's emphatic clarification of the positively charged nucleus suggests that he was concerned his priestly status might dilute his authority to speak in the register of science.

Though there is no indication that he wished to break with the church, the first article in the journal, "Working Hypotheses on the Nature of Life," laid out two programmatic questions that suggested Luyet's own dissatisfaction with answers provided by that institution: "1) What is the essential intrinsic mechanism of the vital processes? 2) What, in the living beings, are the elements really endowed with life?"[78] The final article in the first issue of *Biodynamica*, "Lower Limit of Vital Temperatures: A Review," was coauthored with Marie Pierre Gehenio, a nun and senior researcher at St.

BIODYNAMICA

A Scientific Journal for the Elaboration and the Experimental Study of Working Hypotheses on the Nature of Life

Edited by
BASILE J. LUYET
Saint Louis

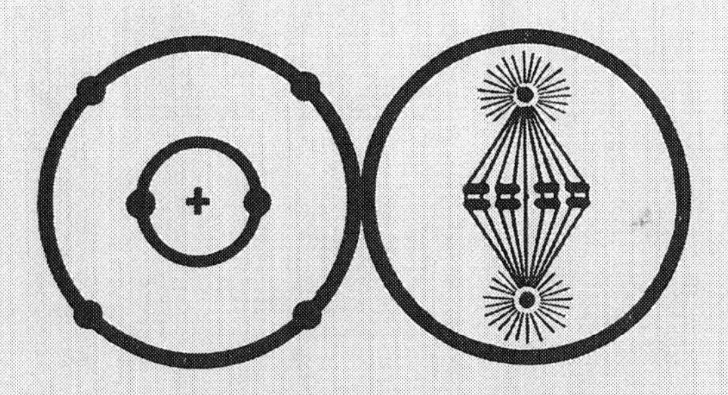

VOL. 1

Nos. 1-33

1934-1938

4 The "cosmogram" that accompanied the first issue of *Biodynamica*. Personal collection of author.

5 Sister Pierre Gehenio, Basile Luyet's collaborator, at work at the University of St. Louis in her clerical garb. Reprinted with permission of Florimont Archives.

Louis University, and intended as a prologue to this new science of life at low temperature. Luyet and Gehenio further formalized their study of life at low temperature by publishing a comprehensive review of all prior work on the subject in 1940. *Life and Death at Low Temperature* reached back to experiments undertaken by Leeuwenhoek and Spallanzani. The book documented and critiqued centuries' worth of efforts to revive a huge range of organisms and came to be regarded as a founding textbook of cryobiology.[79]

Though both Luyet and Gehenio worked and made public appearances in their clerical garb (Gehenio clothed in her habit and Luyet in his collar), Luyet never commented explicitly on the religious or even the moral implications of his research. Luyet's biographer suggests that while he often insisted on a connection between his scientific work and religious belief, in practice he strenuously avoided mixing explanatory schema.[80] Luyet sought to respect a boundary between science and religion that he had not created but did not try to undermine.[81] He would recognize Vatican II as having modernized Catholicism, though he believed it marginalized any efforts to coordinate knowledge and faith.[82] Perhaps the biggest historical mystery of the problem of "cold shock" was the ambivalent role of the priest who, in seeking to understand deep epistemological questions regarding "the elements really endowed with life," contributed to the opening of human life to processes of manipulation and industrialization.

Bloodshed

Despite the proliferation of techniques that emerged during the interwar period, the scientific uses of refrigeration in the decades leading up to the Second World War were relatively small and idiosyncratic in scale. The needs of wartime, and the innovations it spawned, would play an important role in bringing refrigeration and freezing fully into the domain of biomedicine, initially through the scaling up of efforts to preserve blood plasma and, later, penicillin and polio vaccine.[83]

Refrigeration and freezing companies—whose primary market had previously been the preservation of foodstuffs—were conscripted into the war effort. REVCO, for example, whose name would become synonymous with low-temperature scientific preservation after the war, had begun in 1938 as a manufacturer of ice cream vending machines (REVCO stands for "Refrigerated Vending Company").[84] The REVCO monomat dispensed cups of ice cream, usually vanilla (the duomat sold two flavors, vanilla and chocolate). During World War II, REVCO was commissioned by the US government to manufacture its first −40°C low-temperature freezer for the aircraft industry. After the war, REVCO continued to manufacture deep-freezers, built-in refrigerators, and heat pumps and soon was manufacturing exclusively scientific refrigeration equipment, becoming the supplier of choice for many life scientists.

As Cold War set in, US government agencies began to provide generous support to low-temperature research across the sciences.[85] The biggest funders were the Atomic Energy Commission, the Office of Naval Research, the Air Force Office of Scientific Research and Development, the Department of Commerce, and, somewhat later, the National Science Foundation. The increase in funding for low-temperature research was due in part to the fact that the use of supercold temperatures had relevance to the development of new kinds of weaponry, including the hydrogen bomb. It also stemmed from the realization that improvements in the understanding of the persistence of life at low temperature would have implications for the creation of repositories for blood in the event of a future nuclear war.

Statistics on the astronomical amount of blood that would be required for transfusions abounded, fueling efforts to find ways to stockpile the substance for long periods of time. One estimate held that 250,000 pints of blood would be needed within the first four days following the atomic bombing of an American city. These calculations provided what the trade journal *Refrigerating Engineering* declared to be an "ominous" need for refrigeration. The journal predicted that the country would require as many as two thousand blood donation centers, supported by thousands of refrigerators. The editor was moved to add a maudlin note: "An excellent market if you are fortunate enough to be around to get the business."[86]

Anxieties about a shortage of blood would grow over the next few decades. R. Keith Cannan, former executive director of the Atomic Bomb Casualty Commission and chairman of the Division of Medical Sciences at the National Academy of Sciences, remarked in the early 1960s that the then "relatively short storage period" of twenty-one days for blood intended for transfusion "imposes serious operational and economic problems," indicating that there had been an ongoing effort to develop methods that would enable the extension of this period.[87]

Cannan had also begun to be exposed to ethical and diplomatic challenges involved with preserving human tissues intended for purposes of research, not transplantation. Around the same time he had expressed concern about the need to extend the storage period for blood, he became responsible for managing the return of organs autopsied from Japanese victims of the bombing of Hiroshima.[88] The Japanese materials were preserved in formaldehyde and not frozen, but contestations over the endurance of such tissues in American repositories foreshadowed those that would emerge in full force in the twenty-first century, described in chapter 5.

Closely tracking scientific, technical, and political changes, though not necessarily moral ones, Luyet eagerly accepted opportunities to spread the secular gospel of cold as they presented themselves. One such opportunity arrived in 1948, when he received an invitation to participate in a Harvard conference on blood preservation. The meeting was organized for the US military, the National Research Council, the National Institutes of Health, and the Red Cross by physical chemist Edwin Cohn. The goal was to make improvements on the ability to provide whole blood to transfuse soldiers injured in far-flung theaters of war, and perhaps also civilians at home.[89]

During the war, Cohn's system of scaling up the fractionating of blood into its component parts had been responsible for saving thousands of lives.[90] When the needs of war waned, Cohn's methods enabled the transformation of civilian blood into a valuable substrate for the production of pharmaceutical commodities.[91] Luyet impressed the notoriously imperious Cohn and was brought on board to improve the long-term storage of blood. Specifically, he was meant to apply his ideas about the vitrification process of rapid freezing to the transformation of human blood into a biomedical resource that could be stockpiled indefinitely, available whenever and however it might be needed.

Luyet had not actually worked with blood of any sort prior to his 1948 encounter with Cohn. His research had been limited to materials like onionskins, tardigrades, and vinegar eels, which had long been the chosen materials for experiments in latent life. A young physician named P. J. Schmidt was recruited from New York to assist Luyet. Liquid nitrogen provided the requisite low temperatures and industrially bred cattle bodies served as the experimental material. Schmidt recalled of the yearlong effort, "Days would begin with early morning trips to a nearby slaughterhouse to catch blood spurting from the slit throat of a cow swinging from a huge hook."[92] This messy activity was followed by meticulous afternoon recording of evidence of freezing injury to blood cells that had been immersed in liquid nitrogen. The shift to blood as an experimental substrate allowed Luyet to make enormous strides in his study of latent life, most significantly because he could use red cell breakdown, also known as "hemolysis," as a way of quantifying cell life and death. Although he learned a great deal about freezing injury, he still could not prevent its occurrence in his experiments with vitrification.

The Biophysics of Blood

Across the Atlantic a team of British reproductive biologists at the Mill Hill Laboratory outside London, drawing on Luyet's work, had found a way to begin to overcome the danger of the thaw. In 1949, Chris Polge, Audrey Smith, and Alan Parkes published a paper indicating that the addition of glycerol to fowl sperm could prevent hemolysis after having been frozen at −79°C (the temperature of dry ice).[93] This transformative insight was actually the result of a fortuitous mistake. They had not intended to test the effects of glycerol—a simple, viscous sugar alcohol compound—but had mistakenly used a bottle mislabeled as "fructose."[94] As early as 1946, French biologist Jean Rostand, they later learned, had also explored the ability of glycerol to protect cells. Unlike Polge, Parkes, and Smith, he did not apply the finding.[95]

From the early 1940s, like Luyet, biologist Audrey Smith had also been conducting wide-ranging experiments in low-temperature biology, including work on cold shock, removal of water, velocity of temperature change, effects of extracellular crystallization, intracellular freezing, the use of cryoprotective substances, duration of survival at low temperatures, and the effect of cold on enzyme activities.[96] The discovery of glycerol as a cryoprotectant was a watershed: it stimulated interest in the study of the fundamental properties of life at low temperature even as money began to flow toward specific biomedical applications of cryopreservation. Few understood how glycerol worked to protect cells, but it would soon be revealed that during the winter, many insects have glycerol concentrations between 2 and 25 percent, suggesting that nature had scooped technoscience.[97]

It would be another British scientist, a biophysicist named James Lovelock, who, in 1953, would suggest that the concentration of salts during freezing were responsible for the destruction of blood cells.[98] The year before, he had collaborated with Polge on experiments in freezing bull sperm and became interested in how glycerol worked to prevent hemolysis.[99] He reasoned that the damage to cells during freezing and thawing was not, as Luyet and many others had surmised, due to ice crystals but to the increase in salt concentration that occurred when ice separated as pure water. Thawing cells from the frozen state subjected them to stress great enough to break open from their salt-damaged state. Glycerol, he concluded, acted as a salt buffer and prevented salt concentrations from rising above a critical level.[100] Lovelock delved still deeper into the study of thermal shock, which he interpreted as the destructive impact of changes in the physical environment.[101] Lovelock—who would later be described as "probably the first serious cryobiologist with real biophysical credentials"—saw the crux of the problem as one of maintaining equilibrium in a closed system.[102]

While Luyet was collaborating with Cohn, Dr. George Hyatt, an orthopedic surgeon at the Naval Medical Center in Bethesda, was helping to establish the US Navy Tissue Bank, which focused on efforts to preserve human bone and tissues for clinical use. It was the first of its kind in that it operated outside of the hospital setting. It also set standards for cryopreservation, as well as immunological principles of tissue transplantation.[103] The first European tissue bank was established around the same time in Czechoslovakia.[104]

In England, David Keilin was himself experimenting with a collection of blood samples preserved in glass containers at room temperature. He had inherited these materials from his mentor, serologist George Nuttall, who in 1904 had published an early study called *Blood Immunity and Blood Relationship.* Nearly fifty years later, the hemoglobins preserved within demonstrated remarkable hardiness. Though they were completely deoxygenated, Keilin was surprised to observe that many structural properties of the hemoglobins were "indistinguishable from freshly collected blood samples."[105] What was significant about the endurance of the hemoglobins in vitro was that they by far surpassed the normal lifespan of hemoglobin in vivo, which is only about 120 days.

This revelation inspired Keilin to investigate other realms in which serological studies had been performed on accidentally preserved or incidentally maintained specimens, including Egyptian mummies and mammoths frozen in the Siberian permafrost. He even spoke, in 1958, of researchers who had recovered a particular organic acid in Baltic amber. Keilin concluded that despite "the attempts by several generations of workers to submit an organism to . . . temperatures very near absolute zero, although of great biological interest," such low temperatures "were probably not essential for the demonstration that life may be a discontinuous process."[106] For Keilin, the key was the maintenance of structure: "Only when the structure is damaged or destroyed," he stated "does the organism pass from the state of anabiosis or latent life to that of death."[107]

Nevertheless, cryopreservation was quickly emerging as the dominant practice for maintaining life in the latent state. Harold Meryman, a young physician and biophysicist who would become Luyet's protégé, became a key figure in the activities of the navy's research on the cryopreservation of human biologics, especially blood. In 1951, he spent a year in Korea and Japan researching the effects of frostbite on troops. Afterward, he joined the department of biophysics at Yale as a research fellow sponsored by the American Cancer Society, ultimately returning to the Naval Medical Research Institute. There, Meryman was put in charge of the first US-made electron microscope, an instrument that was central to what Nicholas Rassmusen called the "biophysics bubble" and that played an important role in early studies of freezing injury on living tissues.[108]

It was at the Naval Medical Research Institute in Bethesda that Meryman first encountered Luyet. Using the electron microscope, Meryman had developed a technique for replicating the surfaces of frozen specimens, and his boss, another biophysicist named Kenneth Cole, made it a point to introduce Luyet when he visited. Luyet subsequently invited Meryman to an important London symposium on freezing and drying.[109] Several years later, while following in Luyet's footsteps at the Biophysics department at Yale, Meryman could state that the "existing and potential applications of freezing, both as a means of preservation and as a vehicle for the suspension and study of transient phenomena, are legion."[110]

Meryman augmented Luyet's cosmology of latent life, extending it to include not only the study of the nature of life at low temperature but the manifold uses of such a state. He predicted that with more knowledge about the mechanisms of freezing and thawing injury, "the usefulness of low temperatures in biology should experience considerable development . . . [offering] exciting potentialities for biological research, not only for the purpose of indefinite preservation, but to provide a true state of suspended animation for the study of transient phenomena which can in this way be interrupted and immobilized for biological eternity in the solid state."[111] Meryman helped extend Luyet's inquiries into the fundamental nature of life to efforts to make biological matter a resource for biomedicine.

So evocative was this new richly endowed science of cold that those who might not have otherwise thought to work with low temperatures were drawn to the field.[112] As scientists took up Meryman's call and made efforts to figure out the mechanics of blood at low temperature, they also experimented with various types of freezing systems including freeze-drying, mechanical refrigeration, and liquid nitrogen. Each of these techniques offered different registers or degrees of cold, in part because the degree at which blood was frozen was considered an "essential ingredient" in its preservation.[113]

In general, the lower the temperature, the slower a tissue could be expected to decay. Preserving cells intact over the long term required the ultralow temperatures provided by liquid nitrogen, but many other serological analyses could be successfully performed on plasma stored at –20°C degrees, a temperature supported by mechanical REVCO refrigerators, or even at room temperature, as Keilin had discovered for hemoglobins. Tissues for which the ultimate disposition was unclear often wound up in a middle zone, at –80°C, particularly if they had initially come to the laboratory on dry ice.

Lovelock, for one, continued to seek out new substances that could be used to preserve cells when frozen at low temperature. In 1959, he and colleague M. W. Bishop introduced a new cryoprotectant, called dimethyl sulfoxide (DMSO).[114] One of the advantages of DMSO over glycerol was that it penetrated red blood cells more rapidly, and less of the solute was required to afford protection against freezing damage. These benefits were mitigated by the fact that thawed blood containing DMSO was marked by a distinctly unpleasant rotten-egg odor that persisted for several days.[115]

In Lovelock's autobiography he remarked of his early contributions to cryobiology that, "This knowledge [of salinity] stayed with me and when later the idea of Gaia, a self-regulating Earth, first came into my mind, I began to wonder how the salinity of the sea had always kept below five per cent . . . it is one of the puzzles posed by the notion of Gaia."[116] Studies of the biophysics of blood contributed to the rise of research into the fundamentals of life at low temperature and its uses in the realm of biomedicine but also facilitated the export and scaling up of ideas about regulation of the *milieu interieur* of the body to the planet.[117]

A Brave and Frosty New World

Luyet continued to promote his cosmology of cold during the Cold War within and beyond academia. He achieved international renown for experiments that echoed Carrel's infamous "immortal chicken heart."[118] News of his success in freezing a chicken embryo heart using liquid nitrogen, which continued to beat when thawed, reached Australia with the headline "Embryo Is Frozen to Make Life."[119] The ability to control the cessation and animation of metabolic processes became conflated, at least in the mass media, with the ability to conjure life de novo.

In 1952 Luyet was profiled in *Time* magazine. The article, titled "Deep Freeze," speculated on the fantastic possibilities for freezing life. Extrapolating from Luyet's more modest studies, the author suggested that "men might be put into deep-freeze and revived thousands of years later. At the very least, spermatozoa from exceptional males could be saved to fertilize females of the future."[120] Geneticist Herman J. Muller, a powerfully controversial voice in mid-twentieth-century science as well as in public culture, weighed in on the idea of such an enterprise, arguing that "here we have nothing to lose, but we and the world have everything to gain."[121] It was concern with this kind of hubris that had led Aldous Huxley to write his novel *The Island*, a counterpart to the better-known *Brave New World*, both of which deal with the creation of new castes of society based on access to sperm from centralized banks of frozen genetic stock.[122]

Though Huxley meant his work to sound a warning about what could come of this revolutionary kind of control over life, the transformative practical potential of the technologies he described resonated with many.[123] Philosopher Hannah Arendt, in her 1958 book *The Human Condition*, expressed concern that scientists were using frozen germplasm to manipulate the possibilities for human life without adequately reckoning with their responsibilities to society.[124] Reproductive biologist Alan Parkes, one of those who contributed to the glycerol breakthrough, who is credited with coining the term "cryobiology," attributed his own interest in the field to a 1942 paper by two reproductive endocrinologists (including Gregory Pincus, who would help develop the first birth control pill). Parkes would later play an active role in promoting the selective and indefinite perpetuation of life through the Galton Foundation, which was set up in 1968 to support study of the "conservation, evolution and progress of mankind."[125]

This 1942 paper, which documented the failure to revive sperm after freezing and thawing, was a rare example of the publication of unsuccessful efforts.[126] In Parkes's recollection, however, there was a consolation prize. This account of null results led him to speculate that if it were possible to "reduce a cell to a state of suspended animation at very low temperature and thaw it again without damage, it would be potentially immortal at the low temperature."[127] This was a definition of immortality that captured the essence of latent life in biomedicine's ice age: alive only in potential, but not capable of being declared dead due to the circumstances of its low-temperature milieu.

News of Parkes, Smith, and Polge's success in freezing sperm had traveled far and wide, especially in the realm of agribusiness. It first caught the attention of an entrepreneur named Rockefeller Prentice, the owner of the American Breeders Service—a cattle production business Prentice had founded in Madison, Wisconsin, in 1941. In 1949 Prentice purchased eighty acres on the outskirts of the city to support the newly created American Foundation for the Study of Genetics, a not-for-profit corporation devoted to research on the science of breeding.

Dr. Elwin Willett, an animal physiologist, was its first director of research. Willett's work, conducted in tandem with a biochemist and a low-temperature physiologist, focused on testing the feasibility of egg transplants in dairy cows, which involved killing heifers in order to harvest their ova.[128] The innovations made by Polge and his colleagues led Prentice to redirect attention to sperm. "Since one of the advantages of artificial insemination is the capability of getting thousands rather than dozens of calves from each sire," Prentice adopted the view that "the importance of that sire being truly genetically superior is obvious."[129] The industrial farm became a site for the commingling of cryobiological and eugenic thought and practice.[130]

Early experiments with artificial insemination of cattle were constrained by the inability to preserve semen other than in a liquid form for a few days at just above the freezing point, around 35°F to 38°F.[131] "Thus," as was pointed out in a corporate history of the American Breeders Service, "semen had to be shipped usually on an every-other-day basis to the field technicians scattered across the country where the cows were located."[132]

In 1953, Prentice recruited Chris Polge as a consultant to work on the fundamentals that would enable more robust commercial application of these cryopreservation technologies. Later that year, in fact, the first calf—appropriately named "Frosty"—was born in North America from semen frozen by Prentice's staff.

Polge sold Prentice on his expertise by showing a short film he had made depicting the freezing process, "the passage of time being indicated by a madly flicking calendar," and the thawing and insemination practices leading to the first calf being born from frozen semen.[133] Prentice was reportedly so compelled by the film that he asked for a copy, which he had edited to include a sequence at the end, reflecting the birth of Frosty, with the slogan "Progress, Progress the American way!" This declaration of American innovation touched a nerve with the British scientists. Polge and his colleagues decided to apply for a patent of the glycerol process. It was refused in the United States on the grounds that the Frenchman Rostand's 1946 paper technically preempted the British team's findings. This was disappointing but not a total defeat. If the British could not patent the technology, then at least no one in the United States could either.[134]

Prentice, though, was after more than patents. He wanted to scale up process and already had the infrastructure to do so. He identified Luyet as someone experienced enough in low-temperature biology to help develop the technique of freezing bull sperm to a practicable level. Luyet accepted an invitation to work in Madison while on sabbatical from St. Louis University. Recruited by the promise of facilities, financial support, and the freedom to pursue his research without any other commitments, in 1956 Luyet decamped to Wisconsin. Luyet instructed that his new laboratory be built in dimensions all using the golden ratio, in keeping with his belief that the structure of matter was what produced vital phenomenon, which he likened to an architecture.[135]

The state-of-the art laboratory, built to his specifications, had multiple workspaces and forms of instrumentation, including cold rooms, freezer rooms, dark rooms, electron microscopy, and X-ray diffraction. Meryman, who also dabbled in research with bovine artificial insemination, marveled that Luyet's new laboratory "contained walk-in refrigerators at 4°C; housed within one of them was a large walk-in freezer at –20°C, and located within that freezer was yet another that operated at –40°C."[136] Luyet was soon directing the American Foundation for the Study of Genetics' new Departments of Biophysics and Low Temperature Biology. In Madison, the odd couple of Prentice, "the wealthy worldly lawyer-businessman, and Luyet, the unworldly, cleric-scientist, prospered."[137]

Soon after Luyet became established in Madison, Prentice invited George Beadle, an expert in gene regulation and, in 1958, a Nobel laureate, to the lab. Beadle recommended that Luyet should assume the directorship of the foundation. In January of that year the American Foundation for the Study of Genetics was changed to the American Foundation for Biological Research (AFBR) and, accordingly, shifted its focus from breeding to cryobiology, with Luyet as director of research and education. Under Luyet's leadership, the laboratory in Madison became one of the largest and most important centers for cryobiological research in the world.[138]

Prentice's team had also been working on the problem of keeping biologics cold as they traveled between headquarters and various farms. The need to improve mobile cold storage—a crucial link in the cold chain—had become imperative. Investigations were begun into the use of mechanical refrigeration and, while it would be useful for storage at a fixed location, the cumbersome nature of such machines was not easy to translate into field use. Traditional gas refrigerators—which were quickly becoming obsolete—were technically portable, but liquid nitrogen offered the possibility for ease in facilitating the connections between field and lab.[139] Storing nitrogen in the liquid state was more efficient than maintaining it as a gas (the volume reduction between gaseous nitrogen at atmospheric pressure and the liquid state is about seven-hundred-fold). Not only could it provide extremely low temperatures, it was inert and nonflammable, making it safe for the biologics it was protecting and the field technicians handling it.

Experts at the Linde cryobiology lab in Tonawanda, New York, were recruited to the American Breeders Service's team. Together, they succeeded in producing a mobile cold-storage canister that combined both a new insulation material and a vacuum to support liquid nitrogen. This prototype, which became available for use in the field in 1956, allowed for a two-week holding period that enabled it to be "transported from farm to farm as the local inseminator made his rounds."[140] Pleased with the success of linking farm and lab across the American Midwest, American Breeders Service converted to this system completely in 1958. Though the company had invested heavily in the development of this technology, it did not seek a patent, allowing competitors to gain access to it through Linde. The result was a transformation of both the cattle breeding and liquid nitrogen industries.[141]

It was not long before advertisements for special deep freezers began appearing in major scientific journals. In the early 1960s, the Linde division of Union Carbide ran a series of informational advertisements in *Science* claiming the techniques of preservation had come close to being perfected,

6 Prototype of the portable liquefied gas container produced by Linde in conjunction with American Breeders Service in 1956. The text on the container reads "Handle With Care." Reprinted with permission of ABS Global, Inc.

7 Advertisement for cryo equipment in *AIBS Bulletin*, 1962. The text accompanying the image reads, "Suspended animation—stopping and starting the biological clock at will—has been one of man's age-old dreams . . . including Jules Verne. Today, through Cryobiology, scientists are slowing down and even theoretically stopping the chemistry of life processes. Prolonging cell life is finding practical application in the use of frozen bull semen for artificial insemination of cattle. Ampules of frozen semen can be indefinitely stored at liquid nitrogen's −320°F temperature—actual tests have been successfully carried out over seven years. Successful programs are under way now for the long-term preservation of blood, bacteria, viruses and cancer cells in liquid nitrogen."

with "many of the inherent areas of risk in long-term experiments—such as chromosomal change or mutation, contamination of culture with bacteria or viruses with other cell lines, and loss of cultures" having been virtually eliminated.[142] The Denver-based company Cryenco captured the power of latent life at the dawn of biomedicine's ice age with a print advertisement featuring a tilted hourglass, signifying the ability to stop the biological clock.[143]

A cattle-breeding company—in partnership with corporations interested in the production of low temperature gases, like Linde—had become the source for knowledge on practical considerations of working with cryopreserved materials, as well as suppliers of equipment to those wishing to preserve and transport other kinds of biological materials on an industrial scale. Cryobiology, which had begun as the province of academic research in the United States and government-funded agricultural research in the United Kingdom, was consolidated by private industry, itself supported by funds from the United States National Science Foundation (NSF).[144] The brave new world of biology would be one that traded in fantasies of time travel and immortality but made its most dramatic impacts felt at the mundane level of biomedical infrastructure.

Archiving the Cold Frontier

By the mid-1960s, a number of forces—military, medical, and industrial—had converged to situate cold storage as an ideal technology for the preservation of latent life. Over the next decade, practices of storage and archiving would overtake efforts to access the fundamental properties of life at low temperature. Furthermore, as cryobiology began to gain status and legitimacy as a discipline, its practitioners became increasingly invested in policing its boundaries, ensuring that the promises that traveled along with practices of cryopreservation were ones that biomedicine might realistically be able to fulfill.

An application deemed suspect was cryonics—the project of freezing whole human bodies with a view toward their resurrection many years in the future. Though *Time* had seen fit to gesture to such an enterprise in its 1952 profile of Luyet, by the mid-1960s, cryonics was being pushed into the realm of pseudoscience. To be sure, the dream of freezing whole human bodies as a way of transcending normal life spans was not new. Even John Hunter, the famous eighteenth-century Scottish surgeon, had allowed himself to imagine that it might be possible to cheat death in this manner, only to later lament that "like other schemers, I thought I should make my fortune by it, but this experiment deceived me."[145] This did not stop others from pursuing the dream of cryobiological life extension.

In 1964, futurist Robert Ettinger published a trade book, *The Prospect of Immortality*, which laid out what he called the "freezer program," a strategy for medicalizing and postponing human death through the use of cold storage.[146] Jean Rostand, who had written about the protective properties of glycerol, legitimated Ettinger's enterprise by writing an introduction. In a remarkable chapter titled "Freezers and Religion," Ettinger eagerly confronted the questions of spirit that Luyet was careful to bracket off from his biological inquiries. Ettinger justified the need to freeze by explicitly invoking Judeo-Christian ideas about salvation, which he linked to the horizon of medical innovation.

At his boldest, Ettinger argued that if the future was filled with new medical technologies that could prolong life, failure to freeze oneself was tantamount to suicide, "a denial of life and therefore God." The freezer program was, in his view, "merely a medical means which will allow the present generation to share the longevity which our descendants will have in any case." Cryonics was predicated on the promise of salvation through science that asked potential "freezees" to make a leap of faith.[147]

Though some cryobiologists maintained a cautious curiosity about cryonics throughout the 1960s, cryonicists would later describe an aggressive campaign to derail the enterprise. Meryman reportedly found cryonics deeply threatening, going so far as to threaten to boycott Minnesota Valley Engineering, which manufactured the tanks used to store materials in liquid nitrogen, ranging from small specimens of blood to whole human bodies.[148] Perhaps he feared that if promises of cryonics were taken seriously and could not be fulfilled, funding for other experimental low-temperature enterprises would be frozen.

In the meantime, Luyet remained ensconced in his Madison laboratory and in 1962 the molecular biology division of the NSF awarded him a grant of $50,000 to support a project titled "The Freezing of Biological Material."[149] In 1964 he founded a new journal, *Cryobiology*, with Theodore Malinin, also a disciple of Alexis Carrel.[150] Two years later, in 1966, Malinin and Luyet would cofound the Biomedical Research Institute in Rockville, Maryland, and the Madison lab would be renamed the Cryobiology Research Institute.

In his introductory essay to the new journal *Cryobiology*, Luyet invoked a connection between organisms' (human and nonhuman) adaptation to cold in nature and the study of the control of cold in the laboratory, seeking once again to elevate the study of low temperature to an all-encompassing view about the order of nature.[151] This cosmological impulse was also seen in the breadth of articles commissioned for the first issue of the journal, including "Cryobiology as Viewed by the Botanist," "Cryobiology as Viewed by the Microbiologist," "Cryobiology as Viewed by the Engineer," "Cryobiology as Viewed by the Surgeon," as well as a review of the state of knowledge regarding the preservation of blood by freezing written by Harold Meryman.[152] Meryman, ever Luyet's loyal mentee, would later state that with cryobiology, "Man has altered the time dimension of his existence, since, for the red cell, the sperm . . . suspended at low temperature, time does in fact stand still."[153] The utility of cryopreservation had begun to overtake efforts to access fundamental principles of life at low temperature and, in the process, to reorganize the temporality of biomedical research.

Meryman had himself recently gotten into the business of biomedical archiving, having accepted a position as assistant research director of the blood program of the American Red Cross. (In 1974 Meryman would join the board of the AFBR BRI in Rockville.) There, blood intended for transfusion could also be used for basic biophysics and genetics research. The structure of the blood bank revealed its own latent potential to serve also as one among the many new frozen archives that were taking hold across the life and biomedical sciences.

As a strategy for managing such assemblages, cryopreservation offered an alternative to the "wet" anatomical collections that had begun in the mid-seventeenth century, "mirroring," as historian Hal Cook has argued, the "accumulation and warehousing of material objects valued by merchants."[154] The ability to stockpile otherwise ephemeral materials for indefinitely long periods of time made it possible for them to generate future use value for science as well as commerce. Use value, in particular the anticipation of future use value, emerged as an important metric for uniting the grand scale of scientific and spiritual inquiry into the essence of latent life with the practical ability to make biological matter mobile across space and time.

Ideas about salvage, the ability to maintain otherwise ephemeral materials from threatened life forms, thus became inextricably linked to hopes that such materials might provide salvation in the form of cures or strategies for preventing death. Take, for instance, the American Type Culture Collection (ATCC), founded in 1925 to provide supply lines of bacteria and fungi. While it had long stored most of its materials at 4°C, in 1960, ATCC and collaborators initiated a developmental research program to explore the effectiveness of preservation in the frozen state at liquid nitrogen temperatures as an alternative method for the long-term storage of a variety of life forms.[155] ATCC cooperated closely with the Linde Corporation, Cryenco, and others in the development of liquid nitrogen refrigerators, programmed freezing units, and associated equipment. With liquid nitrogen, ATCC began to maintain cell lines, emphasizing the particular utility of this ultralow-temperature storage for cells that were rare or hard to acquire, such as cells from certain wild animals or cells from patients with rare diseases.[156]

One of the most influential figures in the establishment of repositories for these kinds of human biomaterials—ones explicitly *not* intended for transplant or transfusion—was an immunologist and physician named Lewis Coriell.[157] With wartime experience in virology and biological weapons, he later became involved in efforts to combat polio. His polio research involved a good deal of tissue culture work, and he seized upon the potential for improving methods for mass production of cell lines. In 1959, while working as medical director of a hospital in Southern New Jersey, he called for the establishment of a central tissue culture bank and cell registry to certify and store cell cultures. His proposal gained the speedy endorsement of officials at the US National Institutes of Health (NIH), and his home institution in New Jersey was chosen as one of two national cell banks.[158]

Lewis Coriell was an important member of Luyet's network, contributing to the first issue of the journal *Cryobiology*, where he hailed the growing applications for cryopreserved materials as "the cold frontier," which included artificial insemination, blood preservation, banking of cells, and experimental banking of human organs. "It was obvious," he wrote, "that many cell cultures could be placed in a state of suspended animation by freezing."[159] Confidence in the reliability of liquid nitrogen led Coriell to create a bank of frozen cell cultures in concert with the US National Cancer Institute. By the early 1960s, a Cell Culture Collection Committee had been created, and a collection of "certified cell lines" was being distributed through ATCC.[160]

The original Biomedical Research Institute, which Luyet started around 1966 with a few rented rooms in an office building, was located on Rockville Pike in Maryland in order to take advantage of the proximity to a major funding source, the NIH.[161] The initial mandate for BRI was to study organ preservation by freezing and was funded both by the AFBR's endowment and through navy contracts.[162] From the beginning, the activities of the Rockville lab focused on cultivating techniques of cryopreservation, while the Madison laboratory continued to explore the fundamental properties of life at low temperature. When, in the early 1970s, the navy's interests shifted from cryopreservation to research on the development of vaccines against malaria and schistosomiasis, the AFBR BRI did as well. This was facilitated by a new low-temperature repository, formally established in the gray brick building in 1977, which was intended to serve as a centralized repository for researchers in need of space to archive medical specimens.[163]

The Rockville lab, beginning in the early 1970s, also established itself as an archive for information *about* cryobiology, including the membership records of the Society for Cryobiology. It became the home of *In Vitro*, the journal of the Tissue Culture Association and of the *Index of Tissue Culture*, a quarterly publication of papers representing techniques, methods, and procedures developed in the course of studies with vertebrate, invertebrate, and plant cultures. In 1975, arrangements were made to have the complete membership for the Society for Cryobiology added to the computer facilities operated by BRI. The same services were also extended to the American Association of Tissue Banks (which formed in 1975), for which BRI served as the business office. By the middle of the 1970s, cryopreservation-oriented services had supplanted cryobiology at BRI.

It is not a coincidence that historians' intellectual preoccupation with the archive began right around this time, during the late 1960s. Michel Foucault's *The Archeology of Knowledge* was concerned with the role of the archive not as a crypt of transcendent facts but as a system that establishes and legitimates what gets to count as an event or a thing.[164] The creation of an archive—a distinctive kind of collection—is a result of the desire to locate or possess the moment of origin.[165] With the rise of biospecimen storage, the archive also became central to constructing and deconstructing problems of knowledge in the life and biomedical sciences.[166]

In her history of the archive, Carolyn Steedman invoked the philosopher Gaston Bachelard to argue that "in the practice of History something has happened to time: it has been slowed down, and compressed. When the work of memory has been done, it is with the things into which this time has been pressed."[167] Archived frozen human biological matter began to be seen as valuable because it, too, could remember, and remember differently, than the persons to whom it once belonged. The ability of such material to persist in a state of latent life, or cryptobiosis, opened it up to new kinds of uses. Science, religion, mass media, morality, and commerce combined to inflate the archive as a record of the past, even, as we will see in the next chapter, as the potential of such a resource pointed to the future.[168] This compression of time and memory into material would come to apply to the use of both frozen tissue archives and paper ones, at BRI and elsewhere.

Conclusion: Life after Death

Basile Luyet died of Parkinson's in March of 1974. His protégé, Harold Meryman, immediately joined the board of AFBR, but financial constraints soon led to the closure of the Madison lab, and BRI became the sole operating entity of the AFBR. In the mid-1970s, it was the only research organization in the world dedicated to what they referred to as a "full range of study in low temperature biology." Yet there were signs that the center could not hold.

In a special addendum to the 1975 annual report, Vernon P. Perry—who had served as director of BRI after helping to found the navy's first tissue bank—observed with trepidation that the organization's focus had shifted away from attention to fundamental properties of life at low temperature. In Perry's eyes, new "applications to tissue banking," including areas related to histocompatibility and immunology, the technology of human leukocyte antigen testing, and organ perfusion had taken precedence.[169] One of the main areas of research at BRI had come to involve the use of schistosomes. The initial creation of cryopreserved colonies of schistosomes at various stages of the organism's life cycle at BRI was not to study their role as a problem in public health. Rather it had come about because the organism itself possessed immunological properties that made it an ideal model for dealing with tissue incompatibility, a major hindrance to organ transplants.[170]

The board concurred, acknowledging with some disarray that "[d]espite the obvious economic importance of cryobiology and the many ways by which its applications can improve the quality of life, curiously this scientific area has not received the emphasis that has been afforded to some more traditional areas of science." They agreed that, to a great extent, this was so because different aspects of cryobiology in the past had been incorporated into separate scientific specialties: plant hardiness in agriculture, hibernation in zoology, blood freezing and freeze-drying in industrial food sciences. "This fractionation on the basis of application has severely limited the kind of basic and interdisciplinary research that has been so fruitful in other areas during the past few decades."[171]

John G. Baust, a leading cryobiologist then at the University of Houston and, in the early 1980s, the vice president of the Society for Cryobiology, emerged as a potential savior for both BRI and the discipline of cryobiology. In the early 1980s, Baust wrote a report, appealing to Union Carbide for support in lobbying the US Congress, titled "Biotechnology and the Cryosciences: An Absence of National Awareness." The goal was to contribute to the "rebirth" of the American Foundation for Biomedical Research, which had "drifted away from cryobiology" in its research focus following Luyet's death.[172]

In his report, Baust proposed the creation of a federal program concerned with cryobiology, including a national repository for the cryogenic maintenance of biological samples. He emphasized that through the creation of a national repository of frozen tissues, "one of the major contributions that cryobiology can make to science and technology is that of *standardization* . . . . We can certainly envisage the day when NIH will require those biomedical investigators whose research does not require literally 'walking' experimentals [a euphemism for human research subjects] to secure 'standard' cells, tissues, and organs for study."[173]

Baust imagined the future of cryobiology in terms of its use value, determined by trends in biomedical research. When this funding plan failed to take hold, BRI's board found itself in the difficult position of adapting to a more modest role in maintaining tissues related to specific biomedical research projects, as opposed to setting the agenda for research about fundamental questions. The goal of widespread standardization of frozen tissues, which had also been pursued by the World Health Organization, as I will show in the next chapter, remained elusive at the national level.

Luyet, despite his gestures toward bracketing questions of the spiritual from those of the scientific, never got to the essence of life. The same man who promoted the study of life at low temperature with a religious fervor had little to say about the spiritual by-products of that endeavor. Meryman described him as a "naïve and gentle priest," recalling that Luyet "never really asked *why* the cells died. It was as though he avoided questions that might lead him to define life in physical terms and to a dilemma that the man of science and the man of God could not resolve."[174] In a publication created to honor Luyet after his death, one contributor likened him to "a kaleidoscope, revealing different things depending on the context."[175] We may never know for sure if Luyet realized the extent to which his career contributed to the transubstantiation of Christian ideas about messianic time into the future-oriented freezer time of his technoscientific age.

Examining the rise of biomedicine's ice age through the lens of latency and the peregrinations of Luyet shows surprising, unexpected properties of cryobiology itself. *Latency* came to refer both to liminality—an in-between state characteristic of suspended animation—and also to the ability of unknown or cryptic features of life or life forms to come into view.[176] As a historical actor moving between countries, disciplines, and cosmologies, Luyet functioned as a liminal subject, as anthropologist Victor Turner has observed, "at once no longer classified and not yet classified."[177] The same could be said to be true of liminal objects—blood samples extracted from people around the world, or even schistosomes, preserved indefinitely in repositories like BRI—which took on new roles as circumstances of funding, practice, and priority changed over time.

Liminal or latent objects, like the liminal lives described by literary theorist Susan Squier, "escape the fixity and regulation of clock time into a realm between what is and what may be" and "share with nonhuman life forms the possibility of being harvested for a use that transcends their own life."[178] The schistosome, for example, entered the freezer as a model organism for transplant immunology at one moment and was thawed in another as a resource for research on global public health. The technoscience of cryobiology also involved a gradual transformation of machines for producing low temperature into servants of human intention, even as efforts to study the science of life imbued those machines and their contents with new and unexpected potential.

Latency, when yoked to low temperature, would come to serve more broadly as a practical form of biotemporal reasoning. The practical applications cryobiology made possible—aided by new access to technologies of producing and maintaining low temperature—have become a powerful and often invisible dimension of a biomedical world that is itself cryopolitical: valuing the ability to make things live and not allow them to die.[179] When conceived within the confines of a linear or chronological approach to temporality, as we will see in the next two chapters, latency provided mid-twentieth-century experts in public health and genetics with a way of imagining the future that oriented action in the present. As practices of cryopreservation traveled, Luyet's fascination with latent life would reveal its own potential as an agent of untimely revelation.[180]

PART II

Temporalities of Salvage

TWO

"As Yet Unknown": Life for the Future

In August of 1949 a Yale epidemiologist named John Rodman Paul boarded an airplane in Connecticut, landed in Montana, took another flight to Fairbanks, and then finally a third to the very north of the continent. An expert on polio, Paul was in search of evidence of immunological resistance to a common strain of the infection. He would spend the next three weeks based at Point Barrow collecting blood from populations he referred to as North Alaskan Eskimos. This involved freezing and shipping small vials of the serum—the liquid component of this blood—back to his New Haven laboratory, where it would be analyzed and stored for future reuse.

A decade later, in 1959, the World Health Organization published a report titled "Immunological and Haematological Surveys." The report, which had grown out of Paul's and others' search for the etiology and distribution of polio, articulated a bold statement about the importance of acquiring and preserving tissue for a broad agenda of public health surveillance.[1] The WHO report's authors wrote: "The pattern of communicable disease through the world is constantly changing, owing to natural and man-made changes in the environment, and changes in man and his behavior. If samples of the [blood] sera collected in these surveys are stored in such a way as to preserve antibodies, it will be possible to examine them in the future and so to determine the past history of infections *as yet unknown* and to follow more clearly the changing pattern of communicable diseases all over the world."[2]

By 1959, the WHO had been in existence for nearly a decade, during which time its leadership had begun to establish its authority by promoting new laboratory-based forms of expertise about the relationship between bodies and the environments in which they lived. In particular, new techniques of molecular biology and new access to mobile cold storage—including innovations by industrial animal breeders—gave momentum to large-scale projects in taking and preserving biological stock. Cryobiology, the project of freezing life to study its fundamental properties at low temperature, began to be overshadowed by cryopreservation, practices that enabled biological material to be frozen or kept very cold to serve other kinds of knowledge projects. Access to tools of cryopreservation helped epidemiologists to engineer an experimental biomedical infrastructure from population-scale collections of frozen blood.

Paul and other experts affiliated with the WHO envisioned an explicit program of accumulating blood from populations living in distinctive environments in such a way that the various known and not-yet-known entities within the blood would be able to be accessed by multiple scientific constituencies, again and again, each time unfolding a new chapter in the epidemiological story of the sample.[3] The temporal disorientation made possible through the promise of low-temperature preservation—the ability to look forward to the past—was meant to provide a strategy for managing risk. Investments in the collection and preservation of blood from communities distributed across the globe emerged as a potentially powerful strategy for preserving fragments of a world that appeared to be increasingly in flux. Freezing blood was meant to enable the biological material to be studied in the present and especially in the future, when the individuals from whom it had been extracted were expected to have disappeared or changed beyond recognition. The phrase "as yet unknown" became a familiar refrain of those anxious about the environmental and human biological consequences of postwar industrialization and decolonization. If the world was changing too fast, then cryopreservation offered a technique for freezing it long enough to make the as yet unknown known.

The emergence of frozen blood as a resource for mid-twentieth-century epidemiology also involved new relationships between various communities of experts and between those they understood to be research subjects.[4] These relationships, forged initially through efforts to manage risks associated with the spread of infectious diseases, helped to shape ideas about expertise, value, obligation, and interdependence in the production of biomedical knowledge. In addition to human blood, these collections sometimes also included specimens taken from the nonhuman commensal species that were known to play important roles as reservoirs and vectors of diseases (including sheep and goats, swine, horses, domestic fowls, wild birds, rodents, primates, and bats).

In an effort to replicate warm-blooded ecological relationships in the artificial low-temperature milieu of the freezer, it was suggested that human and nonhuman materials be stored side by side.[5] Epidemiologists who drew and froze blood aimed to learn from nature's historical experiments by subjecting human-derived objects to further experimentation in the laboratory. The practical ability to accumulate blood samples and to freeze them in the service of an uncertain future, in this and other ways, involved fundamental reconfigurations of the relationship between the field, the clinic, and the lab. It also involved reconfigurations between researchers and their subjects. When physician-scientists traveled to the field, they had to confront circumstances about the lived experiences of their subjects that they would not have encountered had their subjects traveled to them. Freezing blood for the future also overshadowed forms of data that were more difficult to extract from specific places, including observations about the local environment.

At stake in these reconfigurations were not just new forms of knowledge production but also struggles for authority and resources amid the shifting boundaries of Cold War geopolitics.[6] Increasingly, ensuring the health, productivity, and political alliances of the so-called underdeveloped or developing nations was seen as the responsibility of the developed ones.[7] For the WHO, engineering a frozen-tissue based infrastructure was also a means of legitimating interventions in newly independent or transitioning states in the name of global public health.

Freezing blood accumulated from bodies of those who were least legible to state power would, in theory, allow those relationships to be preserved for restudy over time, even if the bodies themselves changed or died. It was a strategy for preserving ecological and political relationships as well as biospecimens.[8] In this chapter, I explore one strand of the initial adoption of practices of cryopreservation in the realm of epidemiology and public health. I begin by examining John Paul's efforts to prospect a Cold War human resource in Alaska in 1949, an initial step in what would become a global project of "serological epidemiology." The story then crosses the iron curtain to Prague, where the groundwork was laid for the international standardization and scaling up of a blood-based epidemiological infrastructure by the WHO. Back in New Haven, the temporal distortions built into this peculiar frozen resource become visible by looking at the WHO-sponsored serum bank at Yale, which served as a site for reproducing and circulating knowledge about the potential uses of life on ice. Speculative promises about the future value of cryopreserved blood helped to justify the technical and diplomatic effort involved in its collection, maintenance, and reuse.[9] The phrase "as yet unknown" referred directly to hopes of detecting previously concealed traces in the blood, which would be revealed at unexpected moments in time, a form of secular revelation called *kairos*.[10]

"Eskimo Sera Is Worth Its Weight in Uranium": Prospecting the Cold Frontier

Long before his trip to Alaska, John Rodman Paul had distinguished himself with his research on infectious disease. In the 1930s he had experimented with an approach to epidemiology based on the laboratory analysis of materials collected in the field.[11] During outbreaks of polio, Paul and his colleague James Trask had gone directly into towns around New Haven, collecting blood samples and analyzing them using lab-based immunological methods to understand how the disease spread.[12] This early work in what Paul took to calling "clinical" or "serological" epidemiology helped him appreciate the potential for using blood-based techniques to also identify traces of past polio infections in populations whose ecology—both social and biological—marked them as unique.[13]

On early serum collecting trips conducted during World War II, Paul and his colleagues had found high levels of antibodies in the blood of infants in Korea, China, Japan, the Philippines, and "some of the rather remote islands" in the Pacific.[14] His work in Alaska, then, was not his first effort to study blood he had collected for epidemiological purposes. Nor was he the first to conduct this kind of research. Earlier examples of similar kinds of blood serum surveys included results of Wassermann tests for syphilis in samples from normal populations of black and white people in urban areas.[15] In the 1930s, the International Health Division of the Rockefeller Foundation had conducted geographical surveys on the prevalence of antibodies against yellow fever in Brazil and Africa.[16] In the 1940s, the serum survey was used to track encephalitis. During the same period of time, a range of European researchers had begun to catalog the worldwide distribution of the few known blood groups and to identify new ones.[17]

By the late 1940s, however, Paul had become especially interested in understanding the extent of the presence of polio antibodies "in a remote population in arctic regions where, to my knowledge, little is known about the prevalence of clinical poliomyelitis."[18] Such populations, he explained to M. C. Shelesnyak, the director of the newly established Arctic Research

Laboratory in Barrow, Alaska, "are today rapidly coming more and more in contact with 'civilization' and its diseases. This impact of the *immigrant* populations on the *native* populations (and vice versa) is a matter to which the epidemiologist cannot be indifferent."[19] This was specially so when the epidemiologist himself was a new arrival. The idea of a virgin soil epidemic was emerging as a means of rationalizing the death of local peoples during contact with colonizers.[20] "[I]nsofar as the native Eskimos of north coastal Alaska are concerned," Paul elaborated, "a fundamental aspect of their medical ecology is their lack of resistance to our common infectious diseases—in other words the soil is fertile."[21]

That same year, at a Ciba Foundation Symposium on "The Eskimo," the eminent medical anthropologist Erwin Ackerknecht warned that "[t]he acuteness with which these diseases occur, having obviously encountered a population lacking any kind of immunity, and on the other hand, the fact that there still exist Eskimo communities which so far have never experienced these diseases, both speak clearly in favor of the recent introduction of these infections."[22] The perceived vulnerability of these communities, who were quickly also becoming a valued labor force for the newly established naval base at Point Barrow, contributed to their status as important subjects of biomedical research. Such populations appeared as natural and living laboratories for the study not only of infectious disease but of the impact of modernity.[23]

At the time the Arctic Research Laboratory was created, Shelesnyak, an experimental physiologist, had emphasized that partnerships between scientists and the navy should be "mutually beneficial," enabling "the scientist to conduct research in an atmosphere of scientific freedom and at the same time assuring the Navy that new knowledge is being created from which developments may later come to strengthen Naval power."[24] Such arguments situated scientists connected with the Arctic Research Laboratory as an elite reserve labor force, who in pursuing their own interests were also, implicitly, serving the state.[25] In addition to his position at Yale, Paul was director of the Virus Commission of the Armed Forces Epidemiological Board, a unit under the direction of the R&D Division of the Surgeon General's Office, US Army. In the earliest years of the Cold War, the government's activities at Point Barrow enrolled both itinerant scientists *and* local populations as critical in efforts to prospect the Arctic.[26]

In 1949 Alaska was still a US territory—statehood would not come until 1958—but the creation, in 1948, of the Arctic Research Laboratory at the military base at Point Barrow, along with the presence of contractors carrying out petroleum exploration for the navy, the completion of the Alaska-Canada military highway, and improvements in air travel made it possible for Paul to use it as a staging area for his fieldwork.[27] Ackerknecht had underscored the importance of the role to be played by medical and anthropological researchers in this political frontier region, declaring that the "[s]tudy of the Eskimo way of life is not only fascinating, but of every increasing practical importance, as the Arctic has suddenly shifted from the periphery into the center of this brave new air world; and it harbors great, still untapped, economic potentialities."[28] The shift had begun a decade earlier, in 1944, when the US military began exploring for oil, establishing five wells in the 37,000-square-mile area at an expense of $60 million and giving jobs to Eskimos—as tractor drivers, mechanics, and laborers.[29] Such workers came to Barrow from remote regions, creating a new community to serve military interests. Barrow was selected by Paul because it had *already* been selected by the military for its historical role in American arctic exploration.

"To our surprise," Paul had written to Shelesnyak in his 1948 proposal, "in a number of primitive and remote areas, high levels of antibodies for the Lansing strain of poliomyelitis virus have been found in infants within areas where the disease has been thought to be rare; whereas in other areas, low levels of these antibodies have been found in children within areas where recurrent and severe periodic epidemics have occurred."[30] The presence of antibodies to the Lansing strain would provide Paul with biomedical evidence that polio had existed in places where there were no written records of the disease. Such antibodies would also suggest that those who survived the disease had acquired immunity that protected them from subsequent infection. For Paul, such knowledge could serve as powerful demonstration that a vaccine might be effective in conferring immunity to polio without having to experience the illness.[31]

Paul's plan was to collect "[b]lood samples to be taken from native esquimaux of different age groups (0–4, 5–9, 10–20, etc)"—"otherwise the study will be practically worthless."[32] Access to the blood of differently aged subjects would allow Paul to reconstruct the history of exposure to various waves of potential epidemics. Paul believed himself to be working with extremely limited written knowledge about the history of polio in the region.[33] Since there was no access to paper records and Paul was skeptical of the verbal reports of those who might have lived through previous epidemics, blood from each cohort would serve as a biomedically verifiable—and therefore objective—living record of past infection and, potentially, future immunity.[34] As an added justification, Paul explained to Shelesnyak that "[f]rom a single blood sample one could attempt to determine a whole battery of antibodies . . . and so it would seem that such studies might be rewarding in more than one direction."[35] Paul repeatedly emphasized that the ultimate value of his research would not be limited to the relatively specific question of whether or not antibodies to the Lansing strain were present, but back at the lab could be analyzed for antibodies to other diseases, apart from polio.

Chief among the anxieties that preoccupied Paul prior to his trip was how he was going to preserve the blood he collected. He predicted that "[t]he preparation of these serum samples (namely its separation from blood) and their proper storage in the cold (preferably frozen) prior to their shipment to New Haven, will be a major problem," leading him to query Shelesnyak: "I would like to know if there is an electric ice box, available for storing serum."[36] This was his only special request for equipment. Upon learning that several refrigerators were indeed available to him at the Arctic Research Laboratory, Paul expressed relief that most of his needs could "be taken care of with a Bunsen burner and an electrical outlet."[37] He also brought several pounds of hard candy, which he was advised to do as an inducement to getting children to give blood. "Candy," another local official informed him, "is not easy to obtain in quantity here, so if you [were prepared] to give each child a half dozen pieces I am sure you would have no difficulty getting your samples."[38] Taking blood, Paul hoped, would be as easy as giving candy to a baby.

With the help of a local informant, Edith Tegoseak, described by an Arctic Research Lab official as "a motherly Eskimo lady," whom he paid two dollars per hour, Paul was ultimately able to recruit 250 volunteers from the 800 or so inhabitants of Barrow village and the smaller and more remote Wainwright village, a hundred miles southwest of Point Barrow, which had a population of about 225.[39] Tegoseak, who had grown up in a wood-framed sod house warmed with seal and whale oil, had a fifth-grade education and was praised by those at the Arctic Research Lab as being "intelligent and good with her hands."[40] Later in her life she reflected on a story her grandmother had told her in which there existed two worlds, which would someday come together. It was important, Tegoseak's grandmother said, that she "watch and listen . . . [for] when the two worlds are going to get together."[41] This she did, later assisting in developing genealogical and linguistic data that would become important in efforts to preserve knowledge of heredity as well as Inuit languages.[42]

At Barrow, Paul could only draw blood on a voluntary basis. The task was made somewhat more difficult by the fact that during the previous summer another investigator—Victor E. Levine, a dentist from Creighton Medical School in Nebraska and veteran of arctic research—had collected blood samples as part of a research project to study the "health of the na-

8 Edith Tegoseak and son, undated. Alaska and Polar Regions Collections. Reprinted courtesy of Rasmuson Library, University of Alaska Fairbanks.

tives."[43] The locals, Paul later remarked, were not wholly enthusiastic about "an annual session of blood letting."[44] Nevertheless, he deemed the overall response to be excellent. Tegoseak, who had assisted in that earlier blood sampling, notified "certain villagers, of her choice, that a Public Health Survey would be carried out in which blood tests would be done."[45] She even opened her home, on the east side of the village, as a site for blood collection.[46] Volunteers from the west side of the village reported to the local schoolhouse. Additional samples were collected from the workmen at the navy camp who, Paul noted, had "come into Barrow from all directions because of the opportunity for work with the [navy's] contractors."[47] Tegoseak was aware of Paul's desire to sample as many people as possible and, toward the end of his stay, made efforts to find out if he wanted to try and take blood from members of a family who lived in tents outside Barrow. Though it was she who had done the crucial work of recruiting volunteers, she remained deferential: "But Dr.," she wrote in a note, "it's up to you as I'm not the boss."[48]

Paul concluded a classified report on his time in Alaska—in which Tegoseak's efforts go unmentioned—by stating that he and his colleagues had come to "believe that this is a valuable series of samples and the most representative according to age group, to our knowledge as yet obtained from an 'isolated' population."[49] Regional public health officials were equally grateful for Paul's work. Dr. Jack Haldeman, the senior surgeon in charge at Barrow, explained, "As you probably know by this time information regarding disease incidence and transmission in Alaska is meager and most of the work needed to obtain baseline information regarding the disease picture is yet to be done."[50] In a 1950 lecture, "The Arctic as a Strategic Scientific Area," Shelesnyak warned that despite the relative lack of scientific knowledge about the region, researchers should "not . . . collect chaotic facts simply for the sake of collecting them. A basic structure of facts is essential; for the Arctic this has yet to be constructed."[51] The region itself was relatively unknown to the American government, as well as the immunological profile of its local populations.

Paul had already begun to broaden his Arctic inquiries beyond Alaska. In 1950, Paul learned of a Canadian team who was working with an Inuit community in the remote Chesterfield Inlet near Hudson Bay. There had recently been an outbreak of polio, and the ecology of the region—human built and otherwise—created opportunities for studying its impact.[52] Paul wrote to a colleague, C. E. Van Rooyan, a microbiologist based at the University of Toronto, that finding antibodies to the Lansing strain among the group in Chesterfield "would make me feel that Eskimo sera is worth its weight in uranium, if only to see what a *single* infection will do in a population presumably unexposed."[53] With this analogy, frozen Eskimo sera, the object of his biomedical prospecting efforts, assumed an equivalency with the search for nuclear fuels in similar sites of occupation in colonial Africa as well in North America with the establishment of Los Alamos laboratory on Pueblo lands in New Mexico, as well as the oil prospecting that was already under way adjacent to the Arctic Research Laboratory in Point Barrow.[54] This precious biomedical resource—frozen blood from as yet unstudied communities—would be the fuel that would propel his epidemiological experiment into the future.

During the epidemic in Chesterfield Inlet, Paul had heard that "several pathological specimens were obtained . . . shipped by aeroplane in the frozen state, and were received in good condition" by a Canadian laboratory.[55] Paul had been spoiled by the relative convenience of working at the Arctic Research Laboratory in Barrow, and he was not confident that the specimens collected at a more remote site would be up to his standards. It was "difficult . . . to imagine that there are any facilities for separating serum from blood in Chesterfield Inlet and . . . that whole clotted blood could be sent to a laboratory for separation without it being pretty sick on arrival."[56] Paul therefore proposed to Van Rooyan, "Perhaps next summer you and I could organize a trip up there with the proper equipment and collect the material ourselves. As an 'old Arctic investigator' I would be able to show you how to eat seal meat and like it."[57]

Later, Paul had second thoughts about the cavalier way in which he had flaunted his newly and rather hastily acquired arctic fieldwork experience. What the ambitious Yale scientist viewed as a self-evidently valuable project of transnational public health, requiring opportunistic knowledge of the impact of the Lansing strain, might be less appealing to those working in different countries and with different agendas. Paul subsequently wrote to Van Rooyan with the concession that "this is your epidemic, your Eskimos and your country. My interest is merely to find out whether a single exposure will result in the production of Lansing antibodies as it apparently did in 'my' eskimos in Alaska."[58]

Issues of ecology as well as territory, diplomacy, health security, and scientific prestige contributed to the transubstantiation of Inuit populations as members of communities undergoing transformation into objects of biomedical—and perhaps even political—value. This epistolary encounter with the Canadian microbiologist helped Paul to recognize the need to scale up his approach and make even bolder claims for the need to collect and preserve blood from groups living in remote regions.

Paul published findings from his Alaskan experience in 1951 in the *Journal of Immunology*, where he described how the laboratory analysis of serum samples revealed patterns of immunity that indicated polio had previously been present in the region.[59] This satisfied his interest in detecting the presence of the Lansing strain, but the paper also emphasized the value of focusing serological surveys on relatively isolated groups. "Ideally," he and his coauthors wrote, "the type of population which might best serve as a primary base line for work of this type would be a population completely free of infectious disease, but needless to say we do not have access to, or knowledge of, such a population; the nearest thing to a 'disease free' group, available to us, being the semi-isolated Eskimo population we have chosen."[60]

Paul and his coauthors acknowledged that the methods used for analyzing the serum were not new, but, again emphasizing the language of the baseline, noted, "We have believed that this population might therefore serve as a base line for future studies on other populations from different environments. Somewhat hopefully we have designated these as studies in 'serological epidemiology.'"[61] In the service of this future-oriented approach, he maintained the frozen sera with the explicit intention of reanalyzing them at a later date to answer questions he had not yet figured out how to ask, but that he believed would be revealed through time. Paul's faith in the generative possibilities of science made the possibility of using preserved blood to establish baselines and to ask new questions seem inevitable. Indeed, serological epidemiology was soon to become a major focus of the WHO's efforts to engage in global disease surveillance.

The Cold World: Scaling Up and Standardizing Serological Surveillance

The WHO is well known to historians for the uneven successes of its disease eradication campaigns.[62] However, there is much about the organization that remains obscure, including the multiple channels through which it cultivated its authority, as well as how far its influence reached across the life sciences in efforts to devise and promote new modes of conducting retrospective and prospective studies related to population health. The documentary trails of the WHO's engagement with serum banking between 1959 and 1970 reveal the standardization work required to transform the fleshy substrates of variable human bodies extracted from complex environments into the multipurpose building blocks of this emergent temporally and spatially distributed form of biomedical infrastructure.[63] During the Cold War, experts affiliated with the WHO advanced the argument that the ability to analyze samples of biologically varied tissues was a means of facilitating interdisciplinary as well as international diplomatic networks.

The tendency for a small group of powerful nations to dictate the appropriate means of participating in maintaining human health in a more intensely interconnected world had roots in nineteenth-century public health, specifically in the collection of national population statistics at home and in their colonies.[64] After World War I, new actors on the internationalizing health scene, like the League of Nations Health Organization (LNHO), began to coordinate efforts to monitor the spread of illness. According to historian Alison Bashford, during this time an older logic of quarantine "evolved into an increasingly global governance of subjects" as both national and international experts began to shape "vital statistics of the world."[65] By the 1920s, the LNHO had been authorized to act as a center for the coordination of standards on medical statistics and the conduct of international health work.[66]

The rise of the biomedical laboratory in the realm of population health over the course of the twentieth century involved efforts to ensure that laboratory-generated tests and therapeutics were consistently effective.[67] From the virology of influenza to the manufacture of reliable therapeutics and the testing of antibiotic resistance, the production of standards functioned to reduce variation in some biological agent that was either central to the work of the laboratory or to the products it generated.[68] An important impetus to develop international biological standards for this kind of laboratory work stemmed from the research of German scientist Paul Ehrlich. He had made huge strides in developing antisera—often produced through reactions with variable animal bodies—to afford protection against infectious diseases like typhus, diphtheria, and syphilis.[69] In order for materials to be used as therapies, the reduction of variability, in either the serum or the test animal, was imperative. The "standard" became known as the physical substance against which preparations of antiserum intended for introduction into a human body could be compared. The early twentieth-century project of biological standardization, then, focused on reducing any variation between samples of reference sera.[70]

At the same time, and in contrast to scientists like Ehrlich, immunogeneticists had also begun to use human blood sera to detect and map patterns of variability within and between populations. This mode of studying sera is associated with the work of Ludwik and Hanna Hirszfeld, who challenged Ehrlich by proposing that not everyone had the same "normal" sera. During and after World War I, the Hirszfelds established that the A and B blood groups were inherited along Mendelian lines.[71] They pursued a "biology of the blood," which sought to demonstrate that immune antibodies were a product of the unfolding and strengthening of genotypically conditioned cell potential.[72] A cascade of new blood groups was detected following the recognition of the Rh blood groups in 1940.[73] The first two serum group systems were identified in 1955, and many others soon followed, revealing the magnitude of variation within humans and in other species.[74] Obtaining the research materials for such a program of investigation involved transforming the world into a "lab of variability."[75]

The WHO's investments in creating and disseminating standards for making blood samples from diverse human communities appear useful for equally diverse purposes and would put a new twist on a project of biological standardization that had been under way since the turn of the twentieth century.[76] As they had done in their efforts to standardize serological substances, scientists would once again produce reliable technical knowledge in the "silence of the laboratory."[77] In the postwar period, however, they would rely on the application of standardized techniques and instruments to *preserve*, rather than eliminate, variability in natural substances.

Much as computers played a role in creating what Paul Edwards has called the "closed world"—characterized by global surveillance and control through the use of high-tech military power—frozen blood was being situated as part of the *cold* world of biomedical knowledge production for the purposes of global security in an age of cold war.[78] Serological epidemiological surveys could contribute to a nascent form of serum-based surveillance by providing biomedical information that could be used to justify and plan eradication campaigns. Prior to the establishment of the WHO in Geneva, the term *surveillance* was often used to refer to close observation of diseased individuals.[79] The idea of extending surveillance to the monitoring of populations was promoted by public health officials like Alexander Langmuir, who in 1951 founded the globetrotting Epidemiological Intelligence Service at the US Centers for Disease Control.[80]

Langmuir defined this new form of population-based disease surveillance, which did not necessarily involve analysis of blood, as involving "continued watchfulness over the distribution and trends of incidence through the systematic collection, consolidation and evaluation of morbidity and mortality reports and other relevant data. Intrinsic in the concept is the regular dissemination of the basic data and interpretations to all who have contributed and to all others who need to know."[81] It was an approach that Langmuir had worked out in dialogue with a Czech epidemiologist, Karel Raska.

In the 1950s through 1960s Raska was a leader in the nascent realm of epidemiological intelligence. A former colleague recalled that, at a time of high geopolitical anxiety, only scientists in Eastern countries seen as "reliable" by the West were allowed to work with colleagues from those countries, a demonstration of the uneven internationalism of agencies like the WHO. Raska was one such reliable figure, skilled at negotiating the complexities of Cold War politics by promoting his belief that "medical science had no boundaries and could only advance through collaboration."[82] He rose to become director of the Division of Communicable Diseases at the WHO, where he was an early supporter of smallpox eradication, often celebrated as one of the agency's great public health triumphs.[83]

Even Raska's skills had their limits. Just as his efforts were coming to fruition, during a WHO surveillance training course he was coteaching with Langmuir in Prague in August of 1968, the Soviets invaded. The course was abruptly ended, and Raska was summarily stripped of his honors and government support, though he remained affiliated with the WHO. It was not until the emergence of glasnost in the late 1980s that Raska's contributions were recognized and his reputation became rehabilitated in his home country.[84] Prior to this political freeze out, Raska had convinced the WHO to make epidemiological surveillance of populations a priority. It was widely recognized that under Raska's leadership the Czechs had developed an approach that was worthy of emulation. Through adoption of standardized practices of serological epidemiology—dependent upon laboratory analyses of large numbers of blood samples—Raska provided a more precise definition of surveillance, later adopted by the WHO at its 1968 World Health Assembly. This definition, as he later reflected, described

> the epidemiologic study of an infectious disease as a dynamic process that involves the biology of the infectious agent, the host, the reservoirs, the vectors, and the environment, as well as the complex mechanisms of the spread of infection and the extent to which this spread occurs. The concept of surveillance also implies the follow-up of specific diseases or infections in terms of morbidity and mortality in both time and place. Additionally, it tracks the circulation of etiologic agents in humans, and for some diseases, in animal populations and the environment as well.[85]

Those who supported the creation of a frozen-tissue-based infrastructure did so with the explicit intention that such preserved blood—human and nonhuman—could be used by different kinds of experts. WHO-affiliated experts emphasized that the many known *constituents* or elements of blood (in the 1950s this included particular blood groups, antibodies, serum proteins, and infectious agents) were already relevant to the work of multiple *constituencies*, including human population geneticists, biological anthropologists, immunologists, and epidemiologists. The potential of DNA remained largely in the realm of the "as yet unknown."

Serving multiple constituencies—who would in turn contribute to populating and characterizing serum banks—necessitated finding ways of standardizing practices of annotating and preserving samples of blood collected from variable human (and sometimes nonhuman) bodies, many of whom were indigenous or members of newly independent nations. The WHO accomplished this standardization largely through its administrative infrastructure, which included the convening of expert advisory committees—composed of certain representatives of these constituencies—and the publication of technical reports, which were circulated to members of other constituencies.

The approach that began to be developed by the WHO during this time was distinctive in that it was intended to preserve the natural variability between frozen serum specimens that came from real-world populations, such that their variability could be exploited in multiple biomedical contexts and across time.[86] In the most basic sense, such repositories, which the WHO referred to as "serum banks," would involve two things: (1) utilization of unique blood serum samples collected for one type of test for another type of test or tests and (2) the long-term cold storage of such specimens, along with information about the persons from whom they were collected.

Serum banking gained influence in the late 1950s, a time when a change in leadership led the WHO to devote an increasing amount of attention to biomedical interventions. Unlike his predecessor, Brock Chisholm, a psychiatrist and proponent of social medicine, Marcolino Candau, who served as director-general from 1953 until 1973, was a malariologist.[87] While Chisholm also supported vertically oriented disease eradication campaigns, Candau viewed the instability of new nations created following decolonization as a problem to be worked around rather than directly addressed. Candau and other experts affiliated with the WHO expressed concern that many of the newly independent nations—as well as other less biomedically engaged "developed" ones—did not possess reliable infrastructures to support adequate epidemiological monitoring.

This perspective was summarized in a 1970 report of the "WHO Scientific Group on Multipurpose Serological and Serum Reference Banks" that explained how, with a standardized system for accumulating such materials in place, "sophisticated laboratory investigations can be made outside the country of origin when national laboratory facilities are inadequate."[88] Cultivating a mode of epidemiological surveillance that relied upon the laboratory-based analysis of blood serum samples collected from unique populations in such locales would justify sending those materials to nations with robust biomedical facilities—nations that had often formerly been or still were—colonial authorities.

Such a view effectively situated those with access to unique populations and the ability to maintain and analyze their blood as strategic players in managing population health in the context of Cold War internationalism.[89] And as historian Sunil Amrith has argued, newly independent nations embraced the WHO precisely because of its role in promulgating universal norms for public health, even if they could not participate in shaping those norms.[90] The study of the variable properties of blood serum in the laboratory was supposed to provide a means of establishing knowledge about populations that was more objective and thorough than the traditional observation-based reporting of traits and illnesses still common in many "undeveloped" parts of the world. As one proponent,

Alfred Spring Evans, a protégé of John Rodman Paul, who would become a member of the 1970 WHO Scientific Group, would later put it, "By determining the prevalence in a population of antibodies to a specific disease at a point in time, a serological survey indicates the cumulative response of the population to the infectious agent at both the clinical and subclinical levels. In contrast, traditional surveys of the prevalence of disease indicate only those who show clinical evidence of the disease in an active or detectable form at the point in time at which the survey is made."[91]

Developing standards to make samples of blood serum from variable bodies usable by multiple potential constituencies served to maximize the epistemic and public health value of a given frozen collection. The same samples could and did serve as resources for producing knowledge about race, infectious diseases, drug resistance, and inherited patterns of susceptibility and resistance. Putting large-scale collections of blood serum samples into use as a multipurpose resource required acts of standardization that were rigid enough to enable such samples of human variation to be intercomparable across space (i.e., between labs in different locations) yet flexible enough that they could remain open to new and unanticipated uses. In the process, the effort to standardize the use of frozen blood led to its disarticulation from the ecological context from which it had originally been obtained.

"Ten to Twenty Years Hence": Techniques for Preserving the Present

In 1958 an international group of experts convened at the WHO headquarters in Geneva to discuss the viability of creating a standardized approach to serological surveillance. What legitimated qualified those invited to attend as experts was that each had acquired experience in working with bodily materials collected from real-world populations and were therefore aware of the potential challenges involved with efforts to develop a set of protocols intended to standardize the use of such materials. The study group consisted of five members, not including one Soviet scientist who was invited but unable to attend. The USSR had only recently rejoined the WHO, having removed itself between 1949 and 1957 over disagreement with the organization's orientation toward the provision of health supplies and criticism of its "swollen administrative machinery."[92]

Among those in attendance was Dorothy Horstmann, an influential voice promoting international strategies to contain existing and emerging infectious diseases and a close colleague of John Rodman Paul. The first woman to be appointed to full professor of medicine at Yale (in 1961), Horstmann was an epidemiologist and virologist and would in 1975 serve as president of the Infectious Disease Society of America. She is perhaps best known for discovering in the 1940s that polio reaches the brain via the blood, a finding that facilitated the creation of a vaccine in 1952.

Before she left for Geneva to attend the 1958 meeting, Horstmann received a letter from a comparative endocrinologist who urged her to persuade the WHO to standardize variation between laboratories located in disparate locations. He mused that "[i]f an organization such as WHO could set up a chain of Universities and Institutions throughout the world . . . then I for one would whoop with joy."[93] Horstmann was successful in communicating such priorities in Geneva, and they are reflected in the widely circulated proceedings of the meeting, published in 1959 as the WHO technical report titled "Immunological and Haematological Surveys."[94] It was in this report that the ultimate value of the samples—their "as yet unknown" potential—was articulated as crucial to the logic of long-term and large-scale serum banking.

The 1959 WHO report also stated that among the most important aspects of designing a serum survey involving human or, for that matter, nonhuman blood "is the proper storage of serum for future reference, conceivably ten to twenty years hence."[95] Such an enterprise—of collecting and organizing a frozen archive—relied upon a temporal strategy later referred to by those engaged in the maintenance of frozen collections as "planned hindsight."[96] The *Oxford English Dictionary* defines "hindsight" as seeing what has happened and what ought to have been done after the event, perception gained by looking backward. Planned hindsight, then, is an orientation to the future that attempts to prepare for the need to look back to the present, which will become the past.

The paradox of planned hindsight generates meaning in the desire to articulate a pragmatic compromise between the awareness that it is impossible to perfectly plan for the future and a belief that hindsight is twenty-twenty. It is a speculative approach to the production of knowledge that accepts its own blindness to the future as well as to what the future might reveal as having been latent in the past. This is a strategy that makes sense when the stated goal of an enterprise is focused on environmental change. However, a temporal orientation that sees the present primarily as a future past is one that may not accept the ways in which it might be blind to the conditions of the present.[97] It also facilitates and even justifies the deferral of problem solving in the present in favor of codifying standards that presuppose the shape of the future before it arrives.

"For example," the authors of the 1959 WHO report explained, "if [such collections] had been made before and after the 1918–1919 pandemic of influenza and stored for examination a decade later, when the influenza virus was first discovered, we should know much more than we do now about one of the great disasters in the history of the human race."[98] A logic echoed in later articulations of planned hindsight was being implicitly invoked to justify the collection of materials that could be used to answer questions that had, in 1918, yet to be asked. In other words, in imagining uses to which frozen blood might be put, they also imagined the present as a baseline against which to interpret future epidemics of flu and even of pathogens that had not yet been identified.

This was not an absurd thing to suggest, for in 1951 pathologist Johan Hultin had gained permission to dig up victims of the 1918 flu epidemic whose bodies remained frozen in the permafrost in Brevig Mission, Alaska.[99] He succeeded in excavating tissue from a mass grave but was not successful in analyzing it for traces of influenza. Fifty years later, however, scientists returned to Brevig Mission, and from the same vein of biological ore, preserved in "nature's freezer," they succeeded in gaining evidence of how the virus had arisen.[100] Hultin's work was not mentioned in the 1959 WHO report as a justification for storing frozen blood sera or other tissues. But in a very real sense, the embrace of artificial cold to preserve tissue was an effort to ensure that finding the answer to these kinds of epidemiological mysteries was not dependent upon such contingent circumstances. It was a temporal complement to Cold War efforts to achieve control over the unknown at a spatial distance.[101] Technologies of cold storage made it possible to engineer a standardized, tissue-based epidemiological infrastructure for an emergent regime of global health. Such technologies, however, were a necessary but not sufficient component of a larger system that included the ability to identify ideal subjects, to collect their blood, and to maintain it over time.

An entire section of the report focused on how exactly this might be undertaken. "Technical Considerations Regarding Methods of Collection, Shipment and Treatment of Blood for Multipurpose Examination" was devoted to the practical aspects of preservation. It provided exacting details—including the ideal amounts of blood to be collected (26 ml from each subject, separated into two vacutainers: 6 ml of oxalated blood and 20 ml of clotted blood, plus a glass smear for discerning the morphology of red cells).[102] Red blood cells were to be shipped by air immediately at 4°C, and serum was quick-frozen and stored at –70°C—the temperature of dry ice—or lower. At the very least, the report authors offered, these protocols could be field-tested by researchers; as it was, these initial guidelines were based on reports from those—namely members of the expert committee—who had begun to experiment with such approaches on their own.[103]

While smaller numbers of serum samples had, since the 1930s, been collected using older methods of cold storage such as dry ice or freeze-drying, the need to transport large numbers of serum samples from field to lab made it imperative to reappraise existing methods. Freeze-drying referred to any process where refrigeration was used as a means of reducing the moisture content in a material of interest without deterioration.[104] In the late 1930s, faculty at the University of Pennsylvania Department of Bacteriology began to recognize the epidemiological potential of accumulated human tissue, asserting that "human convalescent blood serum [was] being preserved by this method so that serum prepared at the close of one epidemic will be available in potent form when the next epidemic arrives, perhaps several years later."[105]

By 1970, Thorstein Guthe, chief of venereal disease and treponematoses at the WHO Division of Communicable Diseases, had argued in a widely circulated technical report on the subject that, under field conditions, dry ice was too expensive, as was freeze-drying.[106] Furthermore, he observed, "thermoses and wet ice can't provide necessary stabilization, especially if traveling by air-mail which can lead to repeated thawing and freezing."[107] Too often serum arrived in the lab contaminated or degraded, especially when it had been collected in tropical countries. Even when working in Alaska, Paul had been concerned about the integrity of his samples. Guthe concluded, "Modern cryogenic techniques involving low temperatures might also be usefully applied to the long-distance transport of human serum."[108] The reason he specified "human serum" was that such cryogenic techniques, by which he meant liquid nitrogen, had previously been the domain of cattle breeders, such as those at Rockefeller Prentice's American Breeders Service, described in chapter 1.

Rockefeller Prentice had resisted patenting the liquid nitrogen container, known as a dewar, he had developed in concert with Union Carbide's cryogenic division, which enabled Guthe to achieve "the successful adaptation for this purpose of liquid nitrogen transporter-refrigerator originally used for the preservation of bull sperm."[109] Together, he and Arthur Rinfret, the Union Carbide scientist who had worked with American Breeders Service, conducted a pilot study involving the transport of sera between a field site in Africa and laboratories in Denmark and France. Guthe reported that the liquid nitrogen canister allowed for three weeks of reliable ultracold storage. The "refrigerator-transporter" was filled in Europe and then sent to Lagos, where it was transported to the more remote rural areas where human blood serum was actually being collected. The entire trial took less than two weeks, leaving time to spare in the form of another week's worth of liquid nitrogen.[110] Though the use of the dewar was relatively simple, Guthe saw fit to caution novices to proceed slowly and with caution when inserting glass vials containing sera into the liquid nitrogen. Moving too quickly, faster than 10 cm per minute, could lead the nitrogen to boil and expose the glass to strain. As for the correct technique for removing the specimens from the liquid nitrogen upon return to the lab, Guthe stated only that it "must be learned by experience."[111]

What also needed to be learned by experience was just how long materials would last when they were archived in the freezer following initial analysis. Guthe summarized some of those results, noting that while preliminary research indicated that preserved serum becomes "inert" below –60°C, laboratories "around the world continue to store serum collections at temperatures of –20°C or higher in conventionally available commercial food refrigerators."[112] Guthe stressed that further research was needed, especially since the lower the temperature the longer materials could be preserved, perhaps far beyond the ten or twenty years predicted by the WHO experts. Even still, with the support of the WHO, Guthe had demonstrated the potential for using new practices of cryopreservation to further extend the enterprise Paul had begun decades earlier.

Universalizing the Local: Refining and Defining the Epidemiological Infrastructure at Yale

In a 1961 essay in the *Journal of the American Medical Association* titled "The Story to Be Learned from Blood Samples: Its Value to the Epidemiologist," John Rodman Paul extrapolated from his *own* 1949 experience in Alaska, noting first that "perhaps one of the best conditions for sampling exists within a small village, with a homogenous population of less than 500 of which everyone, young and old is willing to contribute sera." Of this now old serum, he went on to explain, "[It] was kept frozen for a number of subsequent years and, from time to time, as new tests for a variety of antigens become available . . . the sera yielded new findings. Indeed, the epidemiologic story which this work has gradually been unfolding is not yet finished."[113]

At the time of his *Journal of the American Medical Association* article, Yale had recently been designated as the site of an official WHO serum reference bank—the only one in the western hemisphere—and Paul named as its director.[114] There were also serum reference banks in Prague (initially under Raska's supervision), Tokyo, and Johannesburg.[115] These banks served

9 Yale Section of Epidemiology and Preventive Medicine, 1959. *First row:* J. T. Riordan, Dr. F. L. Black, Dr. Dorothy M. Horstmann, Dr. Robert M. Taylor, Dr. John R. Paul (chairman), Dr. R. H. Green, Dr. C.-D. Hsiung. *Second row:* Dr. B. H. Wilmer, Dr. James C. Niederman, Dr. Robert W. McCollum, Dr. J. R. Henderson, Dr. I. Yoshioka, Dr. M. G. Gudnadottir. *Third row:* Dr. S. R. Sheriden, Dr. W. H. Gaylord, Dr. E. P. Isacson, Dr. H. Sunaga, Dr. Edward M. Opton. Reprinted with permission of the Cushing/Whitney Medical History Library at Yale University.

less as centralized repositories than as demonstration sites for the production of knowledge about the care and use of preserved serum samples. The Yale bank became a critical node for generating information on how to store and maintain serum and facilitated the training of an international network of public health workers. Whether or not samples were literally stored in the same place would ultimately be less important than ensuring "local universality": that they were accumulated and managed in the same way.[116]

In order to become standardized, a variable human blood serum sample would have had to acquire the following qualities. It would need a provenance: it would have been tied, through its label, to a detailed genealogy, demographic profile, and basic medical data. This data would have been coded by technicians—generally female—and computerized via IBM punch-card technology.[117] These same technicians would perform other kinds of labor, including ensuring that a sample placed in the freezer could

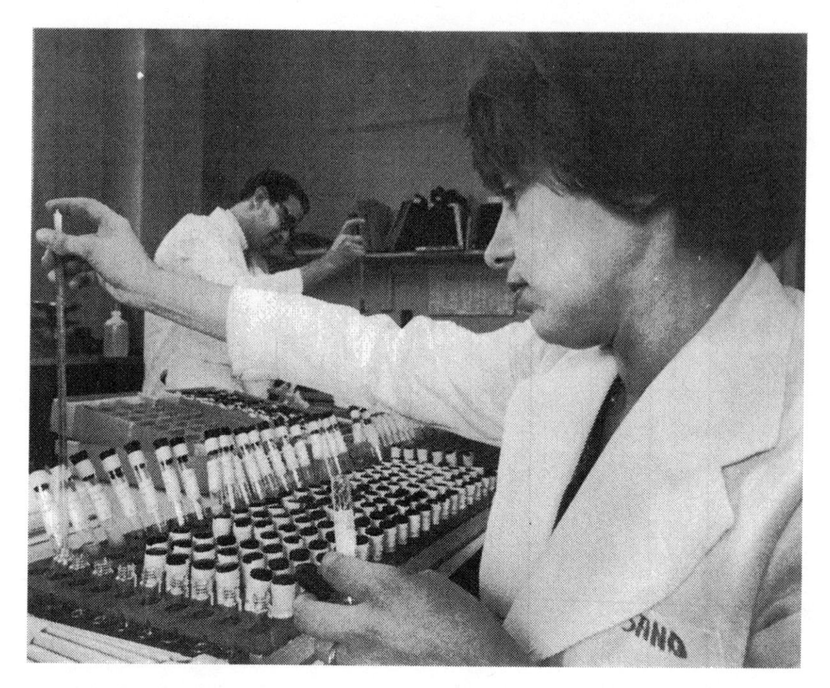

10 Technician Marie Pisano, shown in the foreground, and Dr. Raul Caudadro in the background at the Yale Reference Serum Bank. Reprinted with permission of the Cushing/Whitney Medical Library at Yale University.

be easily located and that it remained consistently frozen at a specific temperature.

The serum bank was a highly gendered workplace. Not only were scientists mainly older and male and the technicians mostly younger and female, much of the actual labor that took place was low status, repetitive, and highly standardized. Marie Pisano, the lab manager at Yale, oversaw a team of single and married women, including undergraduate interns.[118] The best technicians were ones who allowed themselves to be standardized, which meant conforming to a professional identity that minimized variation in affect and in technique.[119] Men like Dr. Raul Cuadadro could perform the work of organizing samples and preparing them for distribution but rarely stayed in such positions for long. Indeed, Cuadadro, already a certified medical doctor, departed the lab after a relatively short period of time and went on to pursue a PhD.

Due to the low-status routinization of this form of work, it can be difficult to reconstruct the experiences and specific contributions made by technicians.[120] However, as the Yale Reference Serum Bank became a resource and model for others seeking to establish similar repositories, it was seen as prudent to develop a technical manual. Each technician was encouraged to contribute tacit knowledge relevant to his or her particular area of expertise, which was then summarized by Paul. This manual became an artifact that could circulate to aid in the creation of standardized serum banks in other locations. In the foreword of an early draft, Paul explained that the countries that stood to benefit most from this approach were those "characterized by primitive health facilities and an existing dearth of vital statistics." Blood would fill the void made by the present absence of other kinds of information: "Especially this might apply to relatively inaccessible areas . . . [which] would be highly desirable . . . in that they could yield data which otherwise might not be easily acquired."[121] In 1965, for example, twenty copies of the manual were sent to the director of the Pan American Health Organization (PAHO), at his request, for distribution to Latin American countries that anticipated setting up serum banks in association with their local public health laboratories.[122] A version of the document was sent to University of Michigan human geneticist James Neel, who was also working to collect blood samples in the service of genetic baselines, the subject of chapter 3.[123]

By the mid-1960s, the Yale bank had also become recognized by its home institution as a key site for the reproduction of knowledge about epidemiology and the multiple potential uses of blood. This had occurred despite concern about the reductionism of biomedical forms of analysis required to make blood reveal its concealed secrets. For instance, in the mid-1960s Nobel laureate and disease ecologist Macfarlane Burnet sought to distinguish himself from biochemists and those fieldworkers he regarded as mere microbe hunters.[124] He feared that, if unchecked, molecular biology had the potential to crowd out the practice and values of interconnection that had come to define an ecologically oriented natural history.[125] With apparent skepticism toward investments in molecular biology that would only be realized in the future, Burnet argued that while medicine "must make use of all the sciences . . . it must also recognize the limitations that the process of evolution and the nature of man place on their utilization. It is a hard thing for an experimental scientist to accept, but it is becoming all too evident that there are dangers in knowing what should not be known."[126]

Paul, despite his enthusiasm for the techniques of the lab, was sympathetic to such views. At Yale, he openly promoted a version of epidemiology that was consistent with Burnet's view and influenced by British epidemiologist Jerry Morris, whose approach was focused first and foremost on attention to social inequality.[127] Paul's vision of medical or disease ecology resisted the purely statistical or rigidly scientistic approach, seeking instead to find ways of mobilizing the lab that held social data in equally high regard. Practitioners of disease ecology repurposed older techniques of bacteriology and virology in their emphasis on the importance of re-integrating humans into ecological thought.[128] In Paul's calls for a robust, even holistic, epidemiology, there was need to pay attention not only to the proverbial seed but also to the soil and the climate. "The entire flora and fauna of a place may be involved," he mused, "and this can represent a very complicated cosmos indeed."[129]

Nevertheless, the Reference Serum Bank—itself a complicated cosmos—gradually took on a life of its own, serving the purpose of coordinating activities at the global scale envisioned by the WHO and also playing an important role in the transformation of how the university trained students of public health. This shift was closely tied to the appointment in 1966 of Alfred Spring Evans as the second director of the Yale serum bank. Known for coining the term "the kissing disease" to describe mononucleosis, Evans credited his military service as a public health officer in Japan, where he had first met Paul, with stoking his interest in infectious diseases, their variable expression, and their distribution.[130]

Evans had previously worked for several decades in Madison at the Wisconsin Department of Public Health. There, inspired from afar by Paul, he became a vocal advocate for the use of serological epidemiological techniques.[131] At that time, many states required that individuals be tested for syphilis prior to being granted a marriage license. If the blood samples collected for testing of one kind of constituent—syphilis—were saved, he suggested, a given department of health could build a representative sample of the population of the region. These materials could then be used to ask different kinds of epidemiological questions, drawing on different constituents also maintained in the sample. Evans believed that public health labs had not taken advantage of the research materials that were, incidentally, at their disposal. Blood samples would often be discarded after being analyzed. In Evans's view, this was a lost opportunity. He argued that serum banking could give the public health lab "a new lease on life."[132] On this point, Evans cited a study that relied upon such preserved samples—which had been linked to basic demographic information—to make racial comparisons in serum cholesterol levels.[133]

In Evans's first year at Yale, he also taught its inaugural course on serological epidemiology (attended by five students, from Colombia, Panama, Venezuela, China, and the United States). Two years later, in 1968, the course was renamed "International Epidemiology" and broadened to include major diseases of worldwide import, both infectious and non-infectious.[134] The course remained rooted in the activities of the Yale WHO

Reference Serum Bank. Students used frozen variable serum samples to hone their skill in deploying serological methods, including the use of new electrophoretic techniques.

By 1970 the administrative apparatus and tissue-based infrastructure for global serum-based epidemiological surveillance was in place. Evans participated in the WHO scientific group that promoted multipurpose serum banking. The Yale collection represented over twenty-five thousand people. (In 1966 alone, 21,823 serum samples were distributed to scientists wishing to study them.) Its size required it to be moved off campus, to a new building of the New Haven Cold Storage Company, where Evans reported that "a corner of an enormous room has been partitioned off for the separate use."[135] The year before, at the twenty-second World Health Assembly, following the recommendations of Raska, the WHO had officially recommended serum-based epidemiological surveillance as a more "technically oriented" approach to the implementation of international health regulations.[136]

In terms of disease ecology, the 1970 WHO report on multipurpose serum banking acknowledged that the utilization of frozen specimens "had resulted in important contributions to research," including the recognition of a connection between high levels of measles antibodies and forms of encephalitis, that Epstein-Barr virus "may be the cause of infectious mononucleosis, and that an unusual circulating antigen called Australia antigen (HAA) is related to serum hepatitis."[137] This was an exciting demonstration of the ability for the standardized serum bank to generate new knowledge about infectious disease.

Moreover, the 1970 WHO report authors, while acknowledging important advances in the storage and management of samples and associated information, conceded that, while "practically all collections of sera stored in the last ten years in the WHO Reference Serum banks relate primarily to communicable diseases," the greatest impact of serological epidemiology thus far had been the creation of the frozen infrastructure itself. "From their inception," the authors added, serum banks "have systematically studied and made significant contributions to the processing and storing of blood and serum specimens, the standardizing of methods of testing sera, and the use of computers for the automatic data processing of population samples and results of investigations."[138] The frozen infrastructure appeared, in the decade between 1960 and 1970, to have been even more productive of knowledge relevant to its own generative capacities than in addressing public health dimensions of disease ecology. This was a demonstration of the "as yet unknown." Much as the cosmological investigations into the fundamentals of life and death in the realm of cryobiology (chapter 1) had created techniques of cryopreservation that could be used in cattle breeding and epidemiology, by 1970 a project of blood collection begun with the express intention of bettering human health produced an experimental system valued for its ability to reproduce and perpetuate itself.

Furthermore, in the intervening decade, a huge number of tests for new constituents had been developed, including those for the complement component C3, the Xm system, a1-acid glycoprotein, blood groups, hemoglobin variants, G6PD, 6-phosphogluconate dehydrogenase, red-cell acid phosphatase, phosphoglucomutase, adenylate kinase, and other red-cell enzymes. New tests for markers in the white blood cells, including human leukocyte antigens were also being developed, "although their stability on storage requires further evaluation."[139] The manifold entities concealed in frozen blood were subjected to a range of new kinds of analyses that were oriented toward an equally broad variety of epistemic ends and would justify the continued preservation of blood samples.

Multipurpose serum banking was also instrumentalized as a tool of public health diplomacy.[140] The authors of the 1970 WHO report repeatedly emphasized that serological surveys "can provide an objective basis for public health decisions," and here it is worth quoting from the introduction to the report at length: "Epidemiological surveillance is a basic concept for *modern* epidemiological services. It is a way of using existing scientific knowledge and facilities in an *integrated and logical* way for the effective *control* and prevention of communicable diseases. The advances in biological and medical sciences of the last three decades have increased the facilities for the study, *control*, and prevention of communicable diseases, so that the implementation of epidemiological *surveillance is fully justified* in all countries."[141]

In the first ten years of the Yale WHO Reference Serum Bank's existence, the words "modern," "integrated," "control," and "logical" were used to justify the practice of accumulating blood as part of a frozen tissue-based infrastructure for global public health. What changed in the intervening decade, and from the time of Paul's earliest expressions of the power of a holistic disease ecology, was that a laboratory-based approach to tracking population health on a worldwide scale—one that was dependent on collections of standardized tissue samples from variable bodies—had supplanted a previous system of aggregating observational reports of patterns of morbidity and mortality. The laboratory and the frozen blood upon which its experiments depended had become the gold standard for achieving epidemiological intelligence. It had also served to relocate the ecology of disease from the field to the freezer.

Historian Hans-Jörg Rheinberger has described laboratory experiments as "systems of manipulation designed to give unknown answers to questions that the experimenters themselves are not yet able clearly to ask. Such setups are, as Jacob once put it, 'machines for making the future.'"[142] The Jacob referred to was François Jacob, an eminent molecular biologist whose work would have been well known to those who advocated serum banking. Using cold storage to support the long-term and multipurpose study of serum samples from real-world populations, a form of natural historical collecting or biological prospecting, gave the lab new relevance in the realm of human biology and biomedicine.[143] Epidemiologists' anxieties about an uncertain and changing world produced its own unintended consequence: a new purpose for laboratories and for freezers filled with blood. The ultimate experiment would be seeing if blood that had already been prospected from bodies and preserved at low temperature could, when subsequently thawed, be made to reveal latent sources of value.

The idea of an unfolding epidemiological story expressed a future orientation that gave momentum to the long-term preservation of tissues, within and beyond the WHO. It was a mode of transforming otherwise ephemeral forms of living substance into enduring, albeit ambiguous, sources of value. Paul's early hunch that the freezer could add chapters—if not entire sequels and even *prequels*—to an epidemiologic narrative that might otherwise end after an initial set of analyses was realized in the WHO's reference serum banks, a distributed tissue-based machine for making many kinds of futures.

Whose Blood, Whose Futures?

By 1970 the Yale WHO Reference Serum Bank contained samples from communities Paul had sampled in Alaska as well as inhabitants of Easter Island. These places were understood to be natural epidemiological laboratories in which evolutionary processes had served to bring humans into a state of optimal equilibrium with their environments. This meant that their blood might possess unique and therefore valuable traces of the immunologic and hereditary impact of past infections. It could also contain infectious agents, "as yet unknown" to biomedicine and increasingly relevant as encroaching forces of industrialization and development disrupted these supposedly long-isolated communities.

In addition to the new tests mentioned above, the 1970 WHO report also indicated the emergence of those for identifying immunoglobulin allotypes, haptoglobins, transferrins, the Gc system, b-lipoprotein allotypes, pseudocholinesterase variants, alkaline phosphatase, a1-antitrypsin, and ceruoplasmin. Of the last of these, it was noted that, "variants are commoner in the American Negro than in the American white population, and appear to be very infrequent in American Indians and Orientals. This type of analysis should be extended to other populations."[144] Regarding tests for albumin, the authors observed that, "in western European populations variants are rare, but in some American Indian populations they are quite common."[145]

Attention to questions of race, in particular the comparison of tribal groups to those of European ancestry, echoed Paul and Ackerknecht's early interest in the study of geographically isolated groups, such as the so-called Eskimo. After all, disease ecology was part of a tradition that reached back to the nineteenth century. Then, practitioners of Alexander von Humboldt's scientific geography believed, in the spirit of Baconian empiricism, that if they could collect enough data they would be able to predict the response of human bodies to diverse environmental conditions. Naturalist Louis Agassiz of Harvard drew on Humboldt's geography to articulate his theory of zoological provinces and "natural racial zones," arguing that the various races of men could maintain and reproduce themselves only in distinct regions of the world.[146] As historian Linda Nash observed of nineteenth-century California, Euro-Americans, anxious about degeneration, held Indian bodies—seen as even more permeable than their own—as especially sensitive indicators of the region's salubriousness. They were "healthy and long-lived, but they were also ill adapted to progress and civilization."[147]

The authors of the 1959 WHO report had stated that the samples to be used for a serological survey should be selected in "some acceptable manner." At Yale, one initial approach involved collecting blood from military recruits around the world.[148] Paul noted the ease of recruitment to the survey, in this case, particularly in countries (unlike the United States) with "universal" military service. However, this only included males of a certain age, primarily between seventeen and twenty-four—hardly a representative sample of the total population of a given region. Another strategy involved sampling members of the same ethnic group living in two different locations. One such project sampled people of self-described Puerto Rican descent living in Puerto Rico as well as those living in New York City. Paul reported of this study, "Comparative serologic tests have detected a variety of differences in common and uncommon viral and other antibodies and blood components."[149] A third strategy was to track itinerant populations. Peace Corps volunteers were seen as ideal and would be sampled prior to leaving for their two-year stints in developing nations, and again upon return. A number of Yale undergrads were also sampled.

As the Yale bank got up and running, Paul redoubled his efforts to promote the practice of reusing preserved serum samples. One such venue for spreading the gospel of reuse was the 1960 Symposium on Genetic Polymorphisms and Geographic Variation in Diseases, organized by biochemist and self-described medical anthropologist Baruch Blumberg at NIH.[150] Blumberg, like Paul, had served overseas during WWII and was interested in the relationship between hereditary difference, ecology, and disease. In 1976, Blumberg would be awarded the Nobel Prize for identifying the viral origins of hepatitis B, based on the study of the Australia antigen—mentioned in the 1970 WHO report as an exciting by-product of serum banking—in large collections of frozen blood serum samples collected by himself and others in the Pacific.[151] He shared this award with biomedical researcher Carleton Gajdusek, who used large-scale collections of preserved human tissues collected from populations living in the highlands of Papua New Guinea in his prize-winning research on the etiology of Kuru.[152]

In his contribution to Blumberg's symposium, Paul emphasized the importance of developing protocols to facilitate a network of labs engaged in banking sera for prospective and retrospective studies. It was at this meeting that he made common cause with James Neel, one of the most influential figures in Cold War biomedicine.[153] Neel had gained early experience with the value and the challenges of collecting blood in the field through his work with the Atomic Bomb Casualty Commission in Japan and, as I will show in the next chapter, also worked through the WHO, in concert with Carleton Gajdusek and others, to intensify efforts to salvage the blood of so-called primitive groups in his efforts to develop human population genetics.

Over time, the Yale bank came to focus on materials from groups seen as geographically isolated, primitive, and endangered. Virologist Frank Black, also a collaborator of Neel's, while on the faculty at Yale, collected samples from members of the Tiriyo, a "remote and very isolated tribe on the Brazil-Suriname border" for Paul during a study of measles vaccination.[154] As early as 1959, experts at the WHO had asserted that the serological approach had demonstrated more value in "ascertaining the susceptibility of individuals . . . than in determining the immune pattern of a community." However, the authors continued, "the correlation of antibody patterns with the history of measles in remote areas such as Greenland and isolated communities such as the Pacific islands, would be a matter of considerable public health interest, particularly as vaccination against this disease [measles] may become possible in the not too distant future."[155]

If cryopreservation, through its purported ability to hold life in a state of suspended animation, promised to bring a slice of Paul's living present into the future, his embrace and promotion of a strategy of planned hindsight was also an attempt to guide how that slice should be made. This literal form of "primitive accumulation" effaced the local circumstances and conditions of labor that made certain bodies available to be sampled and their blood to be "banked."[156] Investing in the future value of these materials also required a certain faith in the progress of science, which would yield innovations that would enable stored sera to generate new knowledge, either through the identification of new constituents or through the comparison of new configurations of samples. In Paul's vision, these blood samples would also allow researchers to compare the blood sample, which would figure as a frozen baseline, to changes that occurred over time.[157]

The same year that Paul wrote his 1961 *Journal of the American Medical Association* article titled "The Story to Be Learned from Blood Samples," he received a clipping of a *New York Times* article about a community in Alaskan territory titled "Dying Language Being Recorded." The story observed that "the disappearance of a dialect [Eyak] appears to be part of a trend that may lead eventually to the consolidation of the native cultures in the far north and their assimilation into white culture." A handwritten note on the article said, "JRP, your friends?"[158]

Paul did regard those he had collected blood from as friends, or at least as humans in whose welfare he was invested as a group whose health was threatened by urbanization. Upon learning that Edith Tegoseak had left Barrow, Paul noted, "I hope she doesn't stay in Fairbanks too long as city life is none to good for Eskimos, I do believe."[159] At the same time, he was moved to "sound a note of warning to the effect that it would be unfortunate if the Eskimos of Point Barrow were to become an 'experimental' population." His own experience "was a case in point," having followed closely on the heels of the dentist's collection activities. "It was apparent to me," Paul wrote, "that one cannot always hope to fish the same pool twice with equal success . . . Nevertheless, it was surprising to me that so many of the Eskimo population responded voluntarily to the request that their blood be tested, even though they had been told that they themselves might not benefit individually. With tact and with the investigator's ability to give something to the community visited, as well as to take," Paul observed, "it should be possible as well as advisable, to continue in the study of these 'virgin' populations before, during, and after their inevitable visitation by our common contagious disease."[160]

Paul's experiences in Alaska—in particular in working with native communities—helped to give shape to his ideas about epidemiology and to WHO-supported practices of serum banking more broadly. Paul believed that it was important for any researcher following in his footsteps to "make the trip to the area and there collect the material himself, and obtain first-hand information as to the environmental circumstances and the character of the populations being tested . . . in other words to see the setting and thus get some appreciation of what the local ecology is."[161] The sample could not on its own yield the requisite data necessary for making sense of epidemics. Analysis, in Paul's view, was to be informed not only by what laboratory tests revealed but by what the individual seeking to draw conclusions had observed with all of his senses.

This process did not begin as a solely extractive, reductionist, or even paternalist enterprise. Those like Paul who first advocated the creation of frozen blood-based epidemiological resources were invested in understanding the health circumstances of research subjects living in specific environments. Ironically, though, the very practice of maintaining blood such that its epidemiological story could be unfolded—to be able to fish the same pool without appearing to exploit it—meant that blood serum would often be studied by scientists with very little connection to the circumstances from which it was initially extracted. While disease ecologists recognized that blood could potentially be used as a substrate to unfold stories about both the social and biological dimensions of illness, socioeconomic and sanitary explanations often remained latent and unexplored. As a result, and as will become more clear in the next chapter, practices of stockpiling blood serum from ecologically distinct groups facilitated the study of disease as the solution to a biological problem of adaptability rather than as a problem that would invite solutions to actively address the conditions in which people lived. It situated those who gave their blood not as members of dynamic and evolving communities but as frozen baselines, relics of the past.

THREE

"Before It's Too Late": Life from the Past

In 1962 human geneticist James Neel had recently returned from his first field trip to study the genetics of a tribe known as the Xavante, whose territory was deep within the borders of the Brazilian central plateau. This brief but intense experience had convinced him that mapping the genetics of Amerindian peoples would be a timely and productive research agenda. As he explained to an audience at Yale, "Our culture is changing with incredible rapidity. This must mean altered selective pressures. At the same time, agents, both clinical and physical, are being introduced into our daily living, which will almost certainly increase mutation rates, and mutations are one source of the raw material for selection. The perspective in which we view the genetic implications of these new agents and our changing culture is determined to a large extent by our knowledge—or lack of knowledge—of the manner of action of natural selection, both today and in the past."[1]

Neel was then curtailing his involvement with a long effort to study the effects of radiation on the survivors of the atomic bombing of Hiroshima and Nagasaki in Japan but was scaling up his international networks.[2] Whereas the Japanese survivors of nuclear war presaged a dismal future in which most members of the so-called modern world would live with mutations caused by radiation and other by-products of technoscience, the Xavante and other purportedly geographically isolated groups appeared to provide a portal into a less polluted human past.

Neel was not the only one who feared this contaminated future. In his 1963 polemic, *Science and Survival*, biologist and public intellectual Barry Commoner popularized concerns about human-induced environmental change that had been circulating for decades and attributed the problems facing humanity to the "wide disparity between the present state of the physical and the biological sciences."[3] The advance of what he referred to as the "physico-chemical" sciences—the by-products of which included radiation as well as chemical mutagens—"with incomplete knowledge" of their side effects on human bodies was tantamount to "acting like the sorcerer's apprentice" and, "in effect, conducting a huge experiment on ourselves."[4] Members of technoscientific societies, in their desire to master nature, had turned nature's wrath on themselves.[5] Moreover, these new risks undermined existing systems for assessing danger particularly because they often only revealed their harms over very long periods of time.

Commoner predicted that "[a] generation hence, too late to help, public health statistics may reveal what hazards are associated with these pollutants."[6] The human animal was the agent as well as the accidental subject of this inadvertent global experiment, and Commoner despaired that scientists might not have enough time to fully assess the damage let alone to develop a solution.[7] In the meantime, the governance of technoscience would need to be guided both "by what we know—and do not know—about life and its environment."[8] The challenge was *how* to live when life itself seemed imperiled. Scientists like Commoner grasped for ways to anchor their efforts to make sense of just how contaminated the bodies of Western society had become.

In his memoir, Neel would later explain the creed he had adopted during his first few weeks with the Xavante: "[T]o understand ourselves, and how the conditions regulating survival and reproduction had changed, we must understand the biology of pre-civilization much better."[9] The accelerating rate of change, the apotheosis of which was represented by the atomic bomb, would have major consequences for the evolutionary future, and perhaps even salvation, of the human species.[10] As Neel articulated it, the survival of his own society would depend upon the ability to make sense of the lives of peoples located differently in time, people he understood to have been insulated from modernity. Specifically, it would require understanding the adaptive purpose of mutations that had been selected for in human communities who were thought to have long lived in a state of equilibrium with their environments.[11] These mutations would be detected through studies of the blood. The catch, as Neel saw it, was that these particular communities were disappearing, making it crucial that their genetic signatures be preserved before it was too late. The indeterminacy of knowledge about humanity's evolutionary history became another problem for which frozen blood would promise future solutions.

Neel's ideas about the importance of focusing on so-called primitive peoples as a strategy for assessing the extent of this inadvertent experiment would become a cornerstone of the Human Adaptability arm of the decade-long International Biological Program (IBP). The IBP, which ran from 1964 to 1974, was a broad and literal effort to take stock of life on planet earth. Thousands of scientists from fifty-seven nations united behind the claim that "[t]he effects of man's modification of the environment should be studied. . . . Such changes are proceeding at accelerating rates and a global survey of present conditions . . . would form a base-line for all such future studies."[12] An acute awareness of time and its apparent quickening structured participants' sense of the stakes. If scientists did not make the effort to construct a freeze frame of the present, observed one booster, "society in the future will be the loser."[13] Another argued that "the sense of urgency is so great that some scientists call for a crash program, one saying, 'our greatest enemy is time.'"[14] Redemption could be achieved, if not via spiritual salvation, then through biological salvage.

The IBP consisted of seven sections, only one of which—Human Adaptability—required that biologists turn their gaze on themselves. The other six sections focused on what program organizers popularized as "biomes," including grassland, marine, and freshwater habitats.[15] The concept of a biome was a strategy for bounding the earth's variable land- and seascapes into manageable, intercomparable field sites or natural laboratories for the study of ecological processes. IBP ecologists' biomes, though they gave form to ecosystem ecology, excluded humans, choosing to conceptualize these spaces as pristine and unaltered by human activity.[16] According to the reflection of its scientific director, Edgar Barton Worthington, while the IBP made huge strides in ecosystem ecology, it did not succeed in bringing "all its component subjects together into a total ecology—the ecology of humans as well as the ecology of animals, plants and their environments."[17] For this reason, the IBP has often been remembered as a failure.[18]

The 232 projects of the Human Adaptability section, however—about 70 percent of which had been completed by the mid-1970s—brought new attention to the relevance of interactions between humans and habitat.[19] Collectively, these projects, which were conducted on nearly every continent, and in which genetics figured prominently, "aimed at elucidating the interaction of nature and nurture on the physiological, morphological, and developmental characters of human populations on a world scale" and, in this arena, it was agreed that "the IBP succeeded . . . beyond expectations."[20]

The Biology of Human Adaptability (1966) was produced to serve as an overview and introduction to the research projects of the Human Adaptability section of the IBP. The volume disseminated the proceedings of a 1964 research symposium, "The Biology of Populations of Anthropological Importance," held at Burg Wartenstein, the Austrian castle that served as the European conference center of the Wenner-Gren Foundation for Anthropological Research. The symposium was notable for the fact that it brought together population geneticists and biological anthropologists, two groups who shared an interest in the relationship between heredity and habitat but who had not previously undertaken significant collaborations.[21] Indeed, the salvage agenda of the Human Adaptability arm of the IBP would come to represent an important example of efforts to produce knowledge that transcended the bounds of particular academic disciplines.

In his preface to the edited volume, Sir Lindor Brown, a physiologist and president of the International Union of Physiological Sciences explained the sentiment that unified all the chapters: "If the adaptation of man to his environment is to be measured, if we are to predict and eventually forestall, we must have a baseline. . . . This book . . . reveals the really urgent need for man to look at and measure himself."[22] Crucially, the technical essays that made up the book encouraged human biologists to do so through the bodies of people they characterized variously as "primitive," "stone-aged," and about to disappear.[23]

Neel contributed one such chapter, "A Prospectus for Genetic Studies on the American Indians," coauthored with Brazilian population geneticist Francisco Salzano. They explained that from "intensive and extensive studies" of the frequency of the genes responsible for specific traits "will emerge a kind of genetic taxonomy."[24] This assertion was based on their initial review of 109 blood group surveys, 62 of which revealed a complete lack of the B blood group in a number of American Indian populations.[25] For Neel and Salzano, the absence of the B blood group was suggestive of some sort of fitness benefit, and similar studies would, as Neel later reflected, help "define a norm for humans that has obtained for millions of years, departures from which may create new evolutionary pressures, even as old pressures diminish."[26]

The broader justification for this work was summarized by another IBP-affiliated scientist, who asserted that "[e]ach such group, a unique baseline of human adaptation, genetically and culturally, to an unpolluted environment, is a priceless resource for understanding man."[27] The injunction to, whenever possible, collect blood from members of these groups—a practical strategy for fleshing out this genetic taxonomy—was a methodological red thread that linked Human Adaptability projects of the IBP to the serum banking initiatives of the WHO (chapter 2). The WHO, as will become clear, provided the administrative infrastructure to promote the study of so-called primitive people during the IBP, particularly through its promotion of collaboration between human population geneticists and biological anthropologists. In an ecological paradigm that sought to understand humans as functional agents in environmental systems—and which often had trouble accepting that humans were a part of those systems—scientists working under the agenda of the Human Adaptability section of the IBP situated their so-called primitive subjects as a part of nature, as biological baselines, while simultaneously locating themselves just beyond.

The ability to split living members of the human species into temporally and technologically distinctive populations—"stone age" and "atomic age," as Neel referred to them, was an ironic by-product of efforts of postwar biological humanists to construct a universal man.[28] In the early 1950s, physical anthropologists began to remodel understandings of human biology through studies of fossils from ancient humans and the behavior of hunting and gathering societies and primates.[29] This work, part of what anthropologist Sherwood Washburn called "the new physical anthropology," was meant to destabilize older essentializing typologies of race and replace them with antiracist concepts of population and plasticity.[30] Through the uptake of multivariate statistics, population genetics, and new techniques of molecular analysis, it was also meant to scientize and therefore relegitimate the biological study of human bodies. Over the course of the 1950s and 1960s, physical anthropology would become biological, and human population genetics would take on aspects of anthropology.

Donna Haraway has described how, through Washburn's efforts, "the discourses of Cold War, nuclear technology, global urbanization, ecological crisis, and sexual and racial politics . . . threatened now with intolerable rates of change and evolutionary and ideological obsolescence" were "[w]ritten into the bodies of early man and living primates."[31] In their contribution to *The Biology of Human Adaptability*, Neel and Salzano made clear their awareness that the uptake of an older "typological approach, so criticized by the proponents of the 'new physical anthropology,' when used to define morphological traits, is now being utilized by geneticists to define ethnic groups on the basis of gene frequencies."[32] What they hoped to bring to this agenda was a more scientifically defensible way of organizing the human species. This would involve replacing essentialist notions of race with evolutionary perspectives on time, mediated by technology and place. It would also require hybridizing human genetics with aspects of anthropology, such as the idea of the "primitive," which many anthropologists were actively discarding.

While the projects of the Human Adaptability section were not focused exclusively on so-called primitive groups, they were a dominant feature of the enterprise. As one of the leaders of the Human Adaptability section later recalled, "Such groups would provide object lessons of the actual adaptability achievable by man when relying largely on his biological endowment."[33] Blood from these societies, which could be preserved indefinitely at low temperature, became one of the most priceless objects of all. In the frozen state blood became a kind of living fossil evidence of human potential. The "concentration of effort on groups of this kind, still living their traditional existence often in extremely difficult habitats," it was declared at the end of the IBP, "has yielded information on man's response to situations soon to remain only in the folk memory of mankind."[34] It had also yielded collections of tens of thousands of blood samples accumulated from members of such human groups, the ultimate value of which would only be revealed in the laboratory as new techniques for its molecular analysis emerged.

Throughout the IBP, the engineering language of baselines provided a conceptual framework for transmuting ideas about racial difference into a language of neo-Darwinian adaptability while reorganizing expertise around biomedical problems that transcended individual disciplines. Blood, a protean fluid that could be partitioned in various ways and preserved, assuming that it could be gathered from disappearing "primitive" communities before it was too late, became an especially useful research material for population geneticists and biological anthropologists invested in determining how far humans living in the technoscientific present had strayed from their originary and purportedly optimally adapted past. Studies made from the preserved blood of so-called primitive peoples, situated as baselines, would also directly inform mid-twentieth-century medical theories that sought to redefine the relationship between the normal and the pathological.

Biology as a Weapon for Peace: Inventing the IBP

The IBP was promoted as a part of Cold War efforts to rehabilitate science from a tool of destruction into one of diagnosis.[35] In the United States it was also meant to destabilize the authority of physics as the dominant recipient of postwar funding from the federal government.[36] On May 8,

1959—the anniversary of the end of World War II in Europe—Roger D. Reid, director of the Biological Sciences Division at the US Office of Naval Research, addressed a group of specialized life scientists assembled for a conference at Notre Dame University.[37] As a keynote speaker, Reid declaimed his belief that biology could upend physics as "the new Queen of the Sciences." He made an impassioned proposal for an international, interdisciplinary program of collaboration.

Seemingly unaware of its cynical connotations, Reid quoted from the 1946 preface to Aldous Huxley's novel *Brave New World*, asserting, "It is only by means of the sciences of life that the quality of life can radically be changed."[38] To Reid and many others, the sciences of life would be instrumental for redeeming an existence made perilous by the physical and chemical ones.[39] Beyond his utopian misinterpretation of Huxley's dystopia—which cast survival as a choice between a totalitarian, bioengineered world and a wild, savage one—Reid's invocation of *Brave New World* conveyed a growing sense of the human condition as one that had become deeply imperiled. "I have therefore proposed," Reid announced, "an International Biological Year—or decade—because it will take many years to appreciate how fully biology is, in fact, *an effective weapon for peace*!"[40]

Reid was not the sole progenitor of the IBP; the idea had already been percolating through networks of biological humanists in Europe.[41] They had been inspired by the hugely successful 1957 International Geophysical Year (IGY). The IGY, itself modeled on a series of International Polar Years conducted in the 1880s and 1930s, coordinated investigations of a huge range of earth and atmospheric phenomena. This included calculations of the earth's ice content as well as rocket-driven explorations of the atmosphere and, eventually, outer space.[42] "In the process of preparing for the IGY," historian Jacob Darwin Hamblin has noted, "scientists came to a startling revelation that the 'normal' earth was retreating into an irretrievable historical past. With each nuclear test, the environment became a different place."[43] This vision of an increasingly pathological future made the establishment of biological baselines appear necessary for preserving a nostalgic notion of a more natural, less technologically contaminated past. In the success of IGY, scientists such as Reid saw value in rendering the biosphere itself as a laboratory for observing the effects of experiments run by both nature and the humans who wished to control it.[44]

Planning for the IBP began almost immediately following Reid's speech, led by British and American scientists. By 1961, as the newly elected president of the International Union of Biological Sciences (IUBS), left-leaning British biologist Conrad Waddington found himself in the position of helping to narrow the focus of the endeavor. Waddington was recognized also for his reputation as a biological humanist, having published a widely circulated 1941 pamphlet called *The Scientific Attitude* in which he described science as a state of mind that could serve also as a moral code. By 1948, when he published a revised second edition, he had become convinced that the problems facing society "lie far more in the sphere of ideals and values—in the spiritual sphere, if you are not afraid to use that term—than in the technological."[45] Embracing the "scientific attitude" would be a way of fulfilling biology's potential to serve as means of salvation for members of secular society.[46]

The cultivation of the scientific attitude of mind was also to be an inoculation against the emotional virulence of nationalism, which during World War II had, in Waddington's view, traded in myth and falsehoods—racism—to undermine the biological unity of the species.[47] Waddington had participated in crafting the revised United Nations Educational, Scientific, and Cultural Organization (UNESCO) Statements on Race, which sought to define inequality as a product of culture, not nature. He was wary of putting studies that could be construed as curtailing human freedom, such as those concerned with population control, at the center of any international program focused on the study of life.[48] In the spring of 1962, in Morges, Switzerland, Waddington presented a thematic approach for the IBP that attempted to background more politicized conversations about race by emphasizing "the biological basis of productivity and human welfare."[49]

American evolutionary biologist Ledyard Stebbins, an expert in plants, was tapped to publish programmatic articles announcing the urgent need for an IBP to address "threats to our society more serious than any which humanity has faced in its past history," which included communism, but more significantly, environmental change.[50] In a report on the Morges meeting published in *Science*, Stebbins wrote of the problems generated by "changes in the genetic constitution of human populations caused in part by the rapid advance of civilization."[51] As secretary-general of the IUBS, Stebbins was extremely effective in generating support. His efforts ultimately spurred the US National Academy of Sciences to establish an ad hoc committee to evaluate the merits of a potential US contribution to the IBP and, in turn, was perceived to have helped persuade the Soviets to participate.[52]

It was also at Morges that the importance of including a section devoted to the biology of humans was agreed upon.[53] Italian geneticist Giuseppe Montalenti, one of the meeting's convenors, was convinced that in the study of humans it would be "particularly important to make observations on human populations which were more or less still isolated on islands in valleys, etc., before civilization stirred these up beyond general recognition."[54] In a contemporaneous volume on natural selection in humans, another population geneticist, Theodosious Dobzhansky, offered the similar insight, based on his studies of drosophila, that "populations isolated on islands, or by some distributional barriers, are often very appreciably different from each other and from the continental populations."[55] This emphasis on island populations reflected the resurgence of biogeography—the study of the distribution of life forms—that harkened back to the nineteenth-century studies of Darwin and Wallace.[56]

The favoring of island populations for the study of human adaptability provided continuity with the biome approach of IBP-affiliated ecologists, as did the sense of urgency. The leadership of the Conservation Section, for instance, argued for the need to list threatened ecosystems for immediate study and preservation during and beyond the IBP. More focused efforts were made to do the same for specific varieties of crop plants. In Australia, plant geneticist Otto Frankel raised the importance of saving seeds and, specifically, "primitive cultigens or land races."[57] He promoted the need to take stock of "our genetic heritage," arguing in ways that paralleled human biologists' focus on so-called primitive people that "the virtue of these populations rests in their internal diversity as a store of variability."[58]

Frankel also mirrored the WHO's call for preserving variability for the "as yet unknown" future, declaring "material which is deemed to be potentially useful today may be even more so in the future when the growing depth of biological knowledge will doubtless enhance the precision and scope for identification and transfer of genetic elements." Frankel warned, however, that "by then, this material may have vanished unless urgent steps are taken now to safeguard its preservation . . . for future generations."[59] IBP organizers would recall that, "in an analogous way" to claims made in the other sections, in the Human Adaptability section, "emphasis was placed on the need to intensify the study of simple societies still living under 'natural' conditions."[60]

In addition to a sense of the urgent importance of establishing baselines, there were also political reasons for encouraging American and European scientists to conduct fieldwork among such societies. These groups often lived in territories where the Cold War was hot.[61] Frank Blair, a zoologist and leader of the US contribution to the initiative saw the IBP as "a chance to avoid a possible Vietnam, for example, in South America, simply by developing this kind of liaison with these people, this kind of cooperative research."[62] Blair later reflected, "IBP began during the years when the 'Wind of Change' was blowing vigorously through African and other former colonial regions."[63] IBP leadership hoped that the presence of international experts would allow the accumulation of resources for a biological science of the human and, perhaps, even contribute to the reshaping—or even adaptation—of local institutions in ways that would ally them with international organizations like the WHO and UNESCO.[64]

Cold War politics, environmental risk, evolutionary understandings of adaptability, the new molecular genetic sciences, and technologies of low-temperature storage converged to make the blood of members of so-called primitive groups appear as an especially vital research material during the IBP. That the groups in question were also perceived to be disappearing imbued the project of acquiring their blood with a sense of urgency. As the civil rights movement took root in the United States, accompanied by suspicion about the intentions of medical researchers, population geneticists and physical anthropologists recalibrated their studies to emphasize an interest in the unified evolutionary history of the species.[65] Yet, rather than leaving race behind, the emphasis on communities they situated as portals to the human past served to refract and redistribute race through time.

Adaptability and the Search for New Normals

Participants in the Human Adaptability section of the IBP, especially the British and the Americans, saw themselves as engaged in a virtuous project of sloughing off associations with the typological, racial thought of previous generations to fully embrace the evolutionary synthesis in biology.[66] Yet the resurgence of nineteenth-century ideas about natural selection presented challenges to an approach that was meant to move beyond Victorian ideas about race.[67] Scientists, in their efforts to define adaptability in the context of the biological study of human populations during the IBP, struggled to bring their understandings of biology into alignment with a social order being shaped by new global realities.

What adaptability or even adaptation should mean was not at all obvious to anyone involved in the IBP. In the foreword to the 1966 book *The Biology of Human Adaptability,* Sir Peter Medawar posited that a broad definition of adaptation was of a piece with the equally expansive goals of human biology. (Medawar, a Brazilian and British biologist, had won the Nobel Prize in 1960, along with Australian Macfarlane Burnet, for work on tissue grafting and acquired immune intolerance.) Channeling Waddington's humanistic claims about the scientific attitude, Medawar asserted that "human biology is not so much a discipline as a certain attitude of mind."[68] The phrase "attitude of mind" encoded its own morality; properly applied by virtuous scientists, it would bring about a new and improved social order.

Medawar held fast to the belief that adaptability was a flexible term that could contain seemingly disparate activities, even if he himself had long struggled to figure out how far the concept could be stretched. As early as 1951, he had reckoned with the multiple ways in which the related term *adaptation* could be used. He came up with three registers of definition: (1) adaptation as something possessed by an individual or a population, such as trait; (2) adaptation as a state of being, such as a population in equilibrium with a specific environment; and (3) adaptation as a process, such as in incremental adjustments to an environment conferring greater states of adaptation at each level.[69] Such a capacious and multitiered conceptualization of what could count as an adaptation helped to support the effort to bring an extremely diverse array of inquiries under the umbrella of the Human Adaptability section of the IBP.

Oxford University human biologist Geoffrey Harrison, in a thank you letter to Wenner-Gren director Paul Fejos following the 1964 meeting that led to the production of *The Biology of Human Adaptability*, declared that the gathering had been "absolutely crucial" for the successful articulation the Human Adaptability initiative of the IBP. Without the opportunity to meet face-to-face, it "would have been quite impossible to have crystallized the projects so precisely." He claimed it had been "a revelation to see how under the atmosphere of Burg Wartenstein [the European conference center of the Wenner-Gren Foundation] the apparently discrepant views of physiologists and geneticists could be so effectively integrated."[70]

These discrepancies centered largely on the variable definitions of adaptation and adaptability and the biological explanations implied. After the meeting, Harvard-trained biological anthropologist Paul Baker appealed to his fellow alum Gabriel Lasker to write an article that would publicize the multitiered approach agreed upon at Burg Wartenstein. In the resulting essay, published in *Science,* Lasker put forth a definition of adaptation that, following Medawar, was specific enough to satisfy detractors but capacious enough to bring the interests of physiologists, anthropologists, and geneticists together in a rubric of shared inquiry. His schema established adaptation as

> the change by which organisms surmount the challenges to life. In the broadest sense biological adaptation encompasses every necessary biological process: biochemical, physiological, and genetic. Adaptation can therefore be involved in (i) major evolutionary events, (ii) growth of the individual, (iii) behavioral and psychological changes lasting only hours or minutes. Human adaptation covers both functional processes and the structures on which they depend.[71]

Lasker's article provided a rationale and the Burg Wartenstein meeting provided a forum for organizing a diverse array of approaches to studying humans as biological organisms into three rough categories of practice within the Human Adaptability initiative: (1) environmental physiology, which emphasized themes of stress and homeostasis; (2) ontogeny, a concern with how human growth processes influence adult status; and (3) genetics/demography, which studied both microevolutionary variation and adaptations resulting from natural selection.[72]

This last category, the subject of natural selection in humans, was high on the agenda of many.[73] As the Darwin centennial approached, Geoffrey Harrison and his colleague, another British human biologist, Derek Roberts, inaugurated the Society for the Study of Human Biology with a conference on the subject.[74] They pointed to ecological geneticist E. B. Ford, who, in 1945, had suggested that variations in the polymorphic human ABO blood group system might be useful for studying natural selection in humans.[75] A study by Alice Brues had recently concluded that selection was likely at play in the ABO system.[76] Theodosius Dobzhansky, for one, maintained that this variation was due to random genetic drift, claiming that the "functional significance of blood groups in man is still full of uncertainty."[77] Nevertheless, at least one contributor to *Natural Selection in Man,* the 1959 volume that grew out of the conference, insisted that "many independent lines of enquiry have shown that human polymorphisms are subject to natural selection. The time has now come for a systematic attack."[78]

Those lines of enquiry included a spate of studies linking various blood group frequencies to infectious diseases including plague, syphilis, and smallpox.[79] In the late 1940s, Giuseppe Montalenti and J. B. S. Haldane had begun to consider the potential for infectious diseases to serve as agents of natural selection.[80] They proposed an adaptive evolutionary relationship between sickle-cell anemia and resistance to malaria.[81] Sickle-cell anemia was known to disproportionately affect African Americans, contributing to their stigmatization in the United States.[82] What if, in malarial regions in Africa, it had afforded a fitness benefit?

Neel had demonstrated in 1949 that sickle-cell anemia was transmitted from generation to generation as a heterozygous trait. This meant that if both parents were only carriers they would suffer no symptoms (a condition referred to as sicklemia) but if both passed on the gene, their child might suffer from sickle-cell anemia. Neel was initially able to trace the gene by looking at the actual sickled blood cells—which in about 8 percent of his sample assumed "various bizarre oat, sickle, or holly leaf shapes."[83] Soon he had successfully used new electrophoretic techniques to link the transmission of the trait to the inheritance of a molecule he and others referred to as an "abnormal" hemoglobin.[84] This finding verified biochemist Linus Pauling's study of sickle cell as the first "molecular disease."[85] The ability to study the sickle-cell allele by tracking differential electrical charges between the hemoglobin of carriers and noncarriers drove the hunt for other kinds of "abnormal" hemoglobin variants.[86]

Meanwhile, British geneticist Anthony Allison was studying blood groups as part of an anthropological expedition to Kenya, where he had been raised. This was one of the first efforts to conduct such study of the blood of members of communities known as Luo, Kikuyu, and Masai.[87] After collecting blood from hospital patients and laborers suffering from sickle cell anemia, Allison had specimens shipped back to London. There, they were subjected to agglutination tests for a small number of blood groups, adding to the store of blood and data derived from such materials being collated by serologist Arthur Mourant, the head of the Blood Group Reference Laboratory at the Galton Laboratory Serum Unit.[88] This study would lead Allison, in 1954, to prove Montalenti and Haldane's hypotheses that the sickle-cell trait had evolved to afford protection against malaria.[89] It also led many to seriously entertain the idea that infectious diseases could lead to genetic mutations, which would be selected over time for their adaptive benefits in endemic regions.

Building on Allison's work, and relying on blood samples collected by others in Léopoldville, then under Belgian colonial rule, Neel set out to demonstrate the role of the environment in selecting for the sickle-cell gene.[90] His student Frank Livingstone used blood collected in Liberia to determine the cline for the sickle-cell trait a few years later.[91] This was the kind of work that led to the collection of blood samples by other geneticists in 1959 that would later be used to discern the earliest traces of HIV-1.[92] By 1960, with financial support from the Rockefeller Foundation and the US Atomic Energy Commission, Livingstone and Neel had tested nearly three thousand West African blood samples for this "abnormal" hemoglobin and about a dozen blood groups.[93]

All of this research reinforced Neel's earlier assertion that "man is probably not as unfavorable an object for such studies [of selection] as is commonly believed."[94] For instance, "family histories, including data pointing to the probable time of origin and mean number of generations of survival of dominant mutations, can be obtained somewhat more readily from man than from Drosophila."[95] Yet tracing the inheritance of mutations within a given family would require finding communities that were relatively endogamous, bounded by high rates of intermarriage. Neel recognized that anthropologists, whose focus had long been on relatively isolated and so-called primitive communities, had a "start of many years on the geneticist in the study of populations." He indicated his interest in observing how the anthropologist "utilizes that experience as he becomes increasingly aware of the significance of the concepts of population genetics to anthropological research."[96]

In the mid-1950s Neel appealed to his colleagues about the need to "attract the interest of many not primarily trained as geneticists" as they attempted to prepare for the "Scyllas and Charybdises which lie in wait for us as we seek to extend our knowledge of inheritance in man."[97] Though he acknowledged "there is of course no key discipline in our complexly integrated science today," Neel argued that the study of human heredity had many ramifications, which "may lead us into strange fields."[98] "Pitfalls" to avoid in undertaking studies included focuses on populations who were inadequately homogenous and "failure to bear in mind the uniqueness of man," who, because he possesses culture, is not "an overgrown fruit fly" or an "enormous mouse."[99] Neel made a distinction between humans and the more classical model organisms for population genetics, mice and fruit flies. Yet, in his articulation, the uniqueness of man was not evenly distributed across the species. He emphasized the value of studying "primitive" groups because they were relatively homogenous, at least in comparison to himself and his cosmopolitan colleagues.

The French anthropologist Claude Lévi-Strauss would become one of the most enthusiastic supporters of population genetics, holding up Livingstone and Neel's work on sickle cell as an example of the demonstration of "optimal equilibrium" between humans and their environment.[100] Lévi-Strauss invoked the fitness advantage that the sickle-cell allele conferred to its heterozygous carriers in malaria-inflicted environments to make an evolutionary argument that coupled nature and culture, asserting "that the appearance of malaria and the subsequent spread of sicklemia must have followed the introduction of agriculture."[101] Sickle-cell anemia, in some places a mark of inferiority, would through the study of human biology, become an exemplar of adaptation through natural selection.[102]

Lévi-Strauss was not the only social theorist inspired by postwar research on the genetics of sickle cell. French philosopher and physician Georges Canguilhem had begun to revisit ideas about the nature of pathology he had first presented in 1943. Sections of *The Normal and the Pathological* written between 1963 and 1966 drew on recent studies in human biology and adaptability. Such cases helped him attempt to reconcile his earlier physiological-based insights with new knowledge about genetic equilibrium and natural selection. Canguilhem tested his earlier theories about the relationship between the normal and the pathological against the example of the apparent selective advantage of the "abnormal" sickle-cell trait in malarial zones.[103]

Canguilhem argued that, if the hemoglobin did indeed afford its carriers with a fitness advantage, it was only "abnormal" from the perspective of those who did not live in an environment in which it could reveal its beneficial potential. He emphasized this to caution against the essentialism that persisted within the adaptationist paradigm. Whether adaptation was considered to be the solution to a problem of the living being's demands on its environment or an expression of a state of equilibrium, in Canguilhem's view the environment itself or milieu was too often considered "as an already constituted fact and not as a fact to be constituted."[104] As a consequence, evidence of natural selection could be used to justify a particular kind of ecological determinism. Canguilhem's insistence on this point was meant to be a reminder of the fact that the relationship between organism and environment was always in flux because the organism and the environment were, themselves, always in flux and only capable of being valued relationally.

This nuance was not entirely embraced by human biologists. Their anxiety about their own society, which was seen to be in a "permanent state of crisis . . . [lacking] self-regulation," saw the identification of concealed traits that were specifically generated to solve problems of adaptation presented by their environments, as a means of defining the norm for the species.[105] "The abnormal, as ab-normal," Canguilhem insisted, always "*comes after the definition of the normal.* It is its logical negation."[106] In actuality, it had only been since the human biologists had categorized their own society as pathological that they set their sights on the "primitive" to serve as a benchmark of normalcy.

From the vantage point of practicing human population geneticists, like James Neel, who called himself a "physician to the gene pool," to let these latent potential solutions to *mal*-adaptation slip away would be tantamount to a form of malpractice.[107] Freezing the blood of "primitive" groups in the service of constructing biological baselines had the consequence of preserving a highly contingent relationship between the normal and the pathological. The result was that this frozen blood made the normal and the pathological appear as dichotomous and fundamental modes of existence rather than dynamic and relational categories of explanation. Freezing would allow them to be remade as such even as the role of machines and values in doing so was effaced (as will be discussed in chapter 5).[108]

The baseline, then, like the norm, was not a neutral or self-evident risk technology, but "a polemical concept" that disregarded the fact that what counted as the "environment" at any given moment was both perpetually in flux and a matter of perspective.[109] In this view, the normative project defined by human adaptability researchers—the effort to establish a baseline of human variation, the results of which would be a "genetic taxonomy" of adaptations produced from very specific local circumstances that could be deployed to ensure the survival of a universal man—contained its own critique. The emphasis on adaptation through natural selection redirected local concerns about race toward more global concerns about a curiously frozen or static geography of gene frequency. It also contributed to the valuing of blood as a research material that could travel and persist independently from reports of those suffering with the pain of sickle cell or other such genetic conditions.[110]

Accumulating "Primitives"

Through their connections to the WHO James Neel and his collaborator, the Brazilian geneticist Francisco Salzano, played an outsized role in focusing the priorities of the human biology on the salvage of genetic material from "primitive" peoples during the IBP. Neel's sickle-cell research, which sat at the intersection of human heredity and medical genetics, is what had led him in the 1950s to become involved in the Atomic Bomb Casualty Commission's work in Japan. There, he studied the bodies of survivors living in a natural laboratory of the human biological impact of radiation.[111] This experience made him acutely aware of how hard it was to detect genetic evidence of mutation from sources of risk—such as radiation—that might only be revealed in time.[112]

As Neel contemplated the challenges posed by the "as yet unknown" epidemiology of mutation, he became fixated by the possibilities of preserving blood from populations he saw as living human remnants of the planet's prenuclear past. Freezing blood would permit the search for evidence of natural selection to persist even if the societies who had once been in "optimal equilibrium" could no longer survive as such. A timely collaboration with Salzano, who had studied with Neel as a postdoc at the University of Michigan, led them both deep into the Amazon, where they conducted research that would provide experience to speak authoritatively about the importance of studying such groups, including in their contribution to *The Biology of Human Adaptability*.[113]

Salzano was a freshly minted PhD when he first wrote to Neel in 1955, requesting the privilege of spending some time in his University of Michigan lab. Dobzhansky, the drosophilist who helped found the field of population genetics, had made a strong impression on Salzano when the elder scientist visited Brazil in 1943. Neel responded enthusiastically, with the caveat that Salzano be prepared to adapt to a new pace of work. "When you shift your attention from Drosophila to man (as I have done)," Neel advised, "you must adopt a different time scale. Problems in human genetics move much more slowly than those utilizing fruit flies."[114]

The time Salzano spent in Michigan was productive and, upon returning to Brazil, he continued to write to Neel, seeking advice and proposing various possibilities for collaboration. Neel, claiming that his lab already had enough "negro" blood accumulated through his sickle-cell research, suggested that "Indian" blood would be a good complement.[115] Salzano, eager to pioneer the study of human population genetics in Brazil, was happy to comply. That year, Salzano sent Neel his first sample of "Indian" sera packaged on ice. The creation of this blood sample and its ability to be relocated from Brazil to Michigan would eventually lead to a reorientation of Neel's interest in human genetics toward studies of geographical variation and natural selection.[116] The desire to salvage blood from members of a supposedly disappearing biological baseline would lead him south, toward Brazil.[117]

By the late 1950s Salzano had begun to appeal to American funders with histories of investing in Latin American public health, including the Rockefeller Foundation.[118] He soon realized that, in order to make a good case for support, he would require the expertise of those who worked with the human species. Lacking suitable connections to anthropology, but authorized by the Rockefeller Foundation, Salzano once again appealed to Neel. This time, he asked for recommendations of a "good foreign physical anthropologist, in order to see the possibility of his coming to Brazil for one year."[119] Neel initially suggested William Laughlin, an anthropologist at the University of Wisconsin, Madison, who would soon become active in the Human Adaptability arm of the IBP. Laughlin, according to Neel, was one of the "new style" of physical anthropologists, who possessed a "sound training in human genetics."[120] Through his work with Alaskan populations, Laughlin was already very much engaged in the effort to identify blood groups, which he had cited as an important "new form of evidence" and a "useful addition to our tool kit."[121]

When Laughlin was unavailable, Neel encouraged Salzano to contact Sherwood Washburn, an architect of the "new physical anthropology," whom Neel described as having a "reputation for having his fingers on the pulse" of the field.[122] Several other candidates were entertained, including Frank Livingstone, a protégé of Neel's who worked on the sickle-cell studies and who would later famously declare "there are no races, only clines," based on his study of the global distribution of a different abnormal hemoglobin, thalassemia.[123] The position was ultimately offered to and accepted by Fredrick Hulse, a student of medical geneticist Arno Motulsky.[124]

All this activity motivated Neel to study up on the Indians of Brazil with a view to conducting a pilot study in the Amazon with Salzano. In March of 1962, Neel wrote to Salzano with a proposition. He began by reporting that his review of the literature had given him "a deep respect" for the ways in which the social structure of the Indian tribes of Brazil related to their population dynamics. "For instance," he casually noted, "have you by any chance seen the recent book by the famous French anthropologist, Lévi-Strauss, entitled, 'A World on the Wane'?" This, of course, was Lévi-Strauss's memoir and cri de coeur, *Tristes Tropiques*.[125] Neel observed that Lévi-Strauss's discussion of "several tribes of the Mato Grosso region brings out very nicely how we could go astray without a good understanding of the culture."[126]

Lévi-Strauss was working on a similar kind of comparative project, attempting to figure the relationship between his own society, which he would describe as "hot," and geographically isolated "primitive" ones, like the tribes of the Mato Grasso, which he valorized as "cold."[127] In thermodyamic terms, cold societies maintained equilibrium, whereas hot ones produced entropy. The crystallized social practices of "cold" societies were, therefore, present-day repositories of the human past. Such societies neutralized change—or, in the view of modernizers, resisted it—by integrating disruptions into preexisting symbolic systems.[128] What Lévi-Strauss articulated about social practices, Neel surmised, might also be read in terms of biological adaptability.

Though it is unclear if Neel and Lévi-Strauss ever corresponded directly, Neel *had* been corresponding with high-level officials at the WHO who were conducting studies about the health impacts of ionizing radiation. Neel had met with R. Lowry Dobson while both were in Brazil for a meeting of the United Nations Committee on the Scientific Effects of Atomic Radiation.[129] As Neel later relayed to Salzano, Dobson had revealed that the WHO was "possibly prepared to make the study of the surviving very primitive groups a dominant theme in their research program for the next 10 or 20 years."[130] The first step would be convening a working group to "survey the possibilities." As Neel reported it to Salzano, the WHO "regard our little pilot study as a possible model of how preliminary cooperative survey efforts can be performed."[131]

Here, "cooperative" referred to collaboration between scientists in different nations, not collaboration with members of the human communities intended for study. Neel concluded his letter to Salzano by stating,

"This little expedition of yours may be assuming more importance than we had realized initially." Salzano's efforts to cultivate human population genetics in Brazil was quickly becoming a partnership between himself and Neel and, even more so, an endeavor of international interest. Neel underscored the significance of the WHO's attention to this work by explaining to Salzano, "This places somewhat more responsibility on our shoulders than looked to be the case at the outset."[132]

Neel saw the WHO's interest in their "summer experience as a model" of the kind of study the agency was willing to support, as both urgent and doable.[133] On the one hand, Neel wrote to Salzano that having the continued attention of the WHO "would certainly help establish our position" in the nascent study of the population genetics of such groups. However, Neel also recognized how little they actually knew. He was keenly aware that "many of the participants will have had a very extensive field experience" and was nervous that their brief expedition would not be taken seriously as a blueprint on which to base future studies.[134]

It was with this in mind that, in the summer of 1962, Neel and a multidisciplinary team of researchers undertook a pilot study to investigate the feasibility of conducting human population genetic research on remote communities, beginning with one Xavante village, the males of which Neel described as "the most superb physical specimens I have ever seen."[135] With the knowledge that this trip was now serving a pilot project for WHO efforts to study the disease ecology of the "very primitive" people, Neel invited cultural anthropologist David Maybury-Lewis to join the team and bring his movie camera to capture information about cultural practice.[136] With a British cultural anthropologist, a Brazilian hematologist (Pedro Junquira, who headed the blood bank in Rio), Brazilian (Salzano) and American (Neel) human geneticists, and a German physical anthropologist (Friedrich Keiter), Neel led the team into the Brazilian bush, where they attempted to sample all members of a single village.[137] A significant portion of the work involved collecting blood samples, which would be transported back to Rio de Janiero and Michigan on ice, frozen, and subsequently analyzed for a range of serologic markers. Neel later admitted that, in addition to the fact that such populations were thought to have colonized extensive areas otherwise devoid of humans, a major reason for his interest in Amerindian populations was their "relative proximity to the lab in Ann Arbor," which was important because of his intention to be "sending a steady stream" of fragile blood samples back from the field.[138]

Ultimately, Neel and Salzano's relative lack of experience made them an ideal "model" team for demonstrating the ability to connect the lab

11 While a woman holding a child looks on, James Neel takes blood from a seated member of a Xavante community, circa 1962. Neel's left hand, which bears a wristwatch, holds the arm of his unnamed subject. A suitcase, marked fragile and filled with medical equipment, rests on the floor, underscoring the temporary nature of this intimate encounter. The faces of all figures are obscured as they observe the needle, though one eye of the child—who has turned away—is captured on film by one of Neel's collaborators. The shoulder of another collaborator is visible behind Neel's crisp white oxford shirt. From the James V. Neel Collection. Reprinted with permission of the American Philosophical Society.

and the clinic with the field: two denizens of the genetics laboratory from different countries—one "developed," one "developing"—working together under unfamiliar and physically demanding circumstances. That fall, having returned from Central Brazil both humbled and invigorated by the initial investigation—Neel described it as having been "encased in a temporary capsule of bygone time"—he gave a series of lectures in which he began to promote his new research program.[139] It was in this 1962 talk at Yale that Neel announced that an International Biological Program was on the horizon. Neel and Salzano's Amerindian studies would become both a cornerstone of the US Human Adaptability program and the model upon which many other similar projects in human biology would be designed.

CHAPTER THREE

Codifying and Contesting the "Primitive"

The WHO's interest in Neel and Salzano's summer with the Xavante was encouraging. Though the IBP provided an intellectual framework and political imperative for action, like the WHO, it was not in a position to directly fund this work. (Much of the actual expense was paid for by individual national governments.) Both the IBP and the WHO, nonetheless, played important roles in organizing the field of human biology and the tissue-based low-temperature infrastructure that transcended any one discipline.[140]

In November of 1962, shortly after Neel and Salzano had returned from their initial field trip to South America, they, along with a dozen others, assembled at the WHO headquarters in Geneva for a weeklong meeting of the Scientific Group on Research in Population Genetics of Primitive Groups.[141] The scientists in attendance had been invited by the director-general of the WHO on the basis of their quite varied experiences in human biological fieldwork. This research ranged from assessing the impact of radiation on Japanese survivors of the atomic bomb to seeking the causes of Kuru and establishing a guide to genetic polymorphisms in humans.[142] Neel was elected chairman.

The meeting yielded a report, "Research in Population Genetics of Primitive Groups" (hereafter referred to as the 1964 WHO Report), which was intended as a set of general guidelines for accumulating materials to support such research.[143] The 1964 WHO Report drew explicitly on previous WHO guidelines about serum banking, to the extent that it reprinted descriptions of cryopreservation techniques almost verbatim. It would also become an important document for disseminating ideas about the centrality of the "primitive" to humanity's evolutionary future and standardizing practices of collecting their blood along with relevant data relating to demography, genealogy, anthropometry, and environment. "Every attempt" the report's authors emphasized, "should be made to carry out such studies as intensively as possible so that valuable information can be recorded before it is too late."[144]

In addition to the urgency of studying populations characterized as "threatened with imminent cultural disintegration and in some instances loss of physical identity in the face of advancing civilization," the authors of the report cited a litany of justifications for focusing on "primitive" people. These included the availability of new molecular techniques for detecting evidence of adaptation. Such groups, which "due to the simplicity of their ecology" were also described as "more manageable" than larger societies and were situated as providing human biologists with the "closest approximation one can find to the conditions under which man has lived for the greater part of his existence." This last assertion was bolstered by the statement that "[i]t is probable that much of the genetic endowment of modern man has been shaped by the action of natural selection and other evolutionary processes at those cultural levels."[145]

The 1964 WHO Report devoted a special section to the "long-term storage of specimens." There it was emphasized that the likelihood that "the present rapid rate of discovery" of new constituents in blood "will continue," making it "important that red cells and serum or plasma be preserved in a biologically active form." Though further research was indicated as necessary in order to discover the best method, the temporal horizon for preservation was idefinite, making "the maintenance of the necessary low tempeartures without intermission for years . . . most important." The authors anticipated that so many people would be sampled that "the total volume of specimens is likely, also, to tax severely the storage facilities of most of the laboratories directly involved in the project" and suggested that cold-storage centers be specially established to achieve the "very high standard of maintenance" required.[146]

Among the concluding recommendations of the report were that WHO-affiliated researchers "explore the possibility" of collaborating with the IBP; "stimulate research on the production of blood-grouping antibodies" as well as the "development of new techniques for the storage and preservation of the biological activity in blood cells, serum, and biopsy material"; and "make available to the respective governments and appropriate international organizations concerned, any observations regarding the demographic decline of the populations studied."[147] Neel was, by this time, serving as a consultant to the WHO and noted that an additional "important byproduct" of the emerging connections between the WHO and the IBP was that where fieldwork was concerned, the international authority he was granted through his affiliation was a "great help in moving scientific equipment across national boundaries."[148]

Self-described medical anthropologist and virologist Carleton Gajdusek, who had already begun to assemble his own massive collection of frozen tissue at the NIH, was persuaded by Neel's assertion that "there is a certain urgency to the problem, so rapidly are primitive societies being disrupted."[149] After the 1962 WHO meeting in Geneva, in which he had participated, he wrote to Neel with a request: "Will you keep me informed . . . of any developments of our recommendations to the WHO or in the IBP interests in primitive groups?"[150] Gajdusek would himself salvage thousands of blood samples from communities in the Pacific as part of a capacious IBP-affiliated research program called "Child Growth, Development, Behavior, Learning and Disease Patterns in Primitive Cultures."[151]

Neel came to view Gajdusek as an ally in situating the nature, if not the culture, of "primitive man" as a uniquely valuable object of salvage. "As you know," Neel wrote to him toward the end of the 1960s, "not everyone is convinced that there are significant things to be learned from the study of the primitives, and I regard this as a bit of a missionary effort."[152] In a very real sense, participants in these projects of the Human Adaptability section of the IBP were salvaging blood as well as ideas about race, which many sociocultural anthropologists were actively trying to excise from the mainstream of the discipline.

Despite the consensus view communicated by the 1964 WHO Report, there were many anthropologists and geneticists who did not agree with the focus on such communities. In the mid-1960s, they argued, the "myth of the vanishing primitive" was already well past its prime.[153] Columbia-trained cultural anthropologist Stanley Diamond later charged salvage anthropologists like Claude Lévi-Strauss with "cold humanism." For Lévi-Strauss, Diamond argued, structuralism created facts as "a cyclotron creates subatomic particles . . . He thus presumes to reach into the minds of others," as human biologists reached into bodies "in order to grasp a truth beyond particularity" and in doing so, avoid "all mention of social processes such as exploitation, alienation, the extreme division of labor, modern war and the character of the state."[154] In the search for scientific universals, social realities were reduced to contingent details.

Biological and physical anthropologists advanced critiques of their own. In 1963, the International Committee of the IBP, of which Neel was also a member, had circulated a proposal to each of the national committees considering participation in the IBP for research on human adaptability. This proposal used language and assumptions similar to those outlined in the 1964 WHO Report. Gabriel Lasker was one of a dozen leading American biological anthropologists whose opinion was solicited. In his response, he dismantled the claim that "[t]here is now strong evidence that the present world distribution of genetically determined characters is in the main, the result of natural selection."[155] Lasker's sentiments would later be articulated in terms of the "neutral" theory of molecular evolution, which suggested that variation could be the result of mutations that were neither fundamentally beneficial nor deleterious, a perspective that resonated with Canguilhem's relativistic views.[156]

Lasker stressed that the proposal's animating emphasis on natural selection was not only scientifically dubious but "the political implications of these assumptions is that people are adapted where they are and will be healthier (or fitter) if they stay where they are. This may make the program unattractive in the have-not nations."[157] Here, Lasker highlighted the deterministic implication that supposedly disappearing "primitive" groups could not be expected to make the transition to modernity because they were biologically predisposed—adapted via a long process of natural selection—to thrive in an environment that could no longer support them. Nations, particularly those newly formed following the collapse of colonial empires, he surmised, would not warm to being told that they "had not" in either wealth or biology.

There was another related dimension to Lasker's critique in which he charged that the focus on "primitive" groups was geopolitically irresponsible. Pointing once again to what he felt was specious evidence that variation could be explained primarily via natural selection, he concluded that in the search for genetic variants, "[i]t would therefore be unfortunate to phrase the problem [so] that the socially pressing questions (the 'population bomb') are ignored while touchy questions (possible racial differences in adaptability to particular circumstances) are highlighted."[158] Lasker saw the proposed project as retrograde, particularly when there were many ways in which human biology could be applied to social problems in the present. For Lasker, the uncertainty surrounding what *might* be known was grounds for greater attention to tackling problems that had already been clearly defined.

At the root of Lasker's concerns was a fear that the focus on studying "primitives" in an effort to establish what Neel had earlier referred to as a "genetic taxonomy" in the service of salvation belied more quotidian desires to create a curiosity cabinet of rare human genetic polymorphisms, reminiscent of a distasteful tradition in physical anthropology with which he was all too familiar.[159] The legacy of his own teacher, Harvard physical anthropologist E. A. Hooton, had been tarnished by Hooton's inability to move beyond racial taxonomy.[160] Lasker later recalled that though he ultimately chose not to actively participate in any of the Human Adaptability projects of the IBP, "I tried to give the project a boost, because I saw the turning of attention to a different set of important issues as tending to leave the sterile topic of race in the dust of scientific progress. I regret to say that racist thinking dies hard."[161]

This criticism of the need to study "primitive" societies occured against the backdrop of economic development and modernization schemes that were taking root as part of the Cold War struggle for hearts and minds.[162] In the terms of the era, the "first" and "second" worlds represented the competing forces of capitalism and communism, respectively. The "third" world was the rural, farming peasantry, who some believed were poised to rise up and create an entirely new world order.[163] Members of the communities of greatest interest to Neel and his colleagues were not seen to have a future as part of the first, second, or third worlds; in the imaginations of human biologists, they were yoked to worlds that were on the wane.

While Neel and others who looked to the WHO to codify a set of practices for studying so-called primitive groups, another category had begun to circulate in transnational forums. In the postwar period, the identity "indigenous" was beginning to be forged by activists via the United Nations as a means of resisting assimilation as former colonies sought to establish themselves as independent nations.[164] In the 1950s, the International Labor Organization (ILO) had conducted its own global survey of "indigenous" peoples, making apparent that the category encompassed those who were poor and exploited. Over the coming decades, the ILO nonetheless created and reinforced a double bind in which participation in capitalist systems of production deauthenticated indigenous identity.

Such geopolitical classifications made indigenous peoples unworldly in economic as well as biological terms. In order to maintain the status of "indigenous" such communities would need to persist free from contamination by capitalism. The ILO's rationale could not accommodate an individual who was both indigenous *and* involved as a worker in a globally recognized system of capitalist production.[165] It created a circumstance in which such individuals would be excluded from the forms of enfranchisement, citizenship, and even political resistance afforded to all other laborers not classed as indigenous. If they were unable to be seen as workers or participants in an increasingly global economy, they could not be seen as capable of being exploited. As "unfree" laborers, indigenous peoples may have found participation in biomedical research a means of gaining access to resources and protection necessary for survival.[166] These were the circumstances that enabled "primitives" to be accumulated by an emergent system of biomedical research that did not view itself as engaged in a system of exchange nor did it understand itself as employing its research subjects as a labor force.[167]

Ultimately, concerns about the repercussions of focusing on these communities altered neither the trajectory nor the priorities of the research during the IBP. Those who supported the investigation of "primitive" groups believed that geographically isolated populations, by virtue of their insulation from modernity, lived differently in time and were about to disappear. Italian population geneticist Luigi Luca Cavalli-Sforza put a fine point on the matter. Of the group he would study under the auspices of the IBP, he stated: "It is thus hoped to obtain a picture of population structure of the Babinga Pygmies [a community living in an area that achieved independence from France in 1960] which may serve as an example of a population living in conditions very nearly as primitive or at least very

12 Almost a decade after the 1967 WHO meeting, many of those who participated—including Newton Morton—joined physical anthropologists at Burg Wartenstein, the European conference center of the Wenner-Gren Foundation for Anthropological Research for a symposium titled "The Role of Natural Selection in Human Evolution." *Front row:* P. Tobias, N. Morton, W. Fitch, J. Friedlaender, J. Huizinga, J. Gomilla, J. Neel, F. Vogel, D. Roberts, A. Motulsky, T. Edward Reed, L. Osmundsen, L. D. Sanghvi. *Back row:* F. Ayala, B. Clarke, P. Jacobs, N. Freire-Maia, H. Harris, F. Salzano, S. Ohno. From the Jonathan Friedlaender Collection. Reprinted with permission of the American Philosophical Society.

little different from those that must have prevailed for perhaps hundreds of thousands of years."[168] To Cavalli-Sforza, primitive communities were not necessarily people without history but people who *were* history.[169]

The Disappearing "Primitive"

In 1967 a new constellation of experts assembled in Geneva called "The Working Group on Research in Human Population Genetics."[170] In the span of six years, "primitive groups" had disappeared, but only in name.[171] The omission of the phrase did not mean that the assembly's focus had changed. Rather, concern about the particular ways in which "primitive" populations were being configured was now being registered from within the realm of human genetics.[172]

Geneticist Newton Morton had previously worked with Neel as part of the Atomic Bomb Casualty Commission and participated in a second iteration of the WHO working group, which met in 1967. He was not opposed to a focus on "primitive" groups, per se, but was wary about the approach proposed.[173] Despite having served as rapporteur for the 1967 meeting, his views were excised from the WHO report, published in 1968 as "Research on Population Genetics" (hereafter referred to as the 1968 WHO report).[174] Morton saw fit to publish his concerns elsewhere.[175] Writing in the *Journal of Physical Anthropology*, he acknowledged that "[i]n this generation we have a precious opportunity to study basic problems in human biology in rapidly-disappearing primitive populations." He cautioned, however, that this opportunity should be seized "vigorously and intelligently, but not with such uncritical enthusiasm that the enormous cost and dubious yield of a 'collect now, think later' philosophy will discredit our efforts."[176]

While he agreed with the view that "primitive" populations were endangered, Morton was displeased to find that anxiety and haste ("collect now, think later"), not what he saw as a technically defensible method, was driving the agenda of his peers. Cutting to the core of the salvage impulse that was guiding this work, Morton practically sneered, "The first fruits of a massive standardized program for research in population genetics of primitive groups will be a large number of Type I errors [false positives]. I prefer more sharply focused studies guided by some hypothesis."[177] He was deeply suspicious of any mode of knowledge production that drew on large-scale collections of data to generate hypotheses rather than to test them.[178] Contrary to the assertions of Neel and his supporters, Morton predicted that the "collect now, think later" salvage enterprise would eventually undercut the entire study of human variation, leading the field either to be discredited, or worse, to actively "obscure the important genetic parameters which, if not determined now, may forever remain unknown."[179] Without the appropriate kind of planning, the future might not ever arrive.

Morton proceeded to outline an alternative approach for estimating gene frequencies based on sampling an array of conveniently available local populations, as opposed to purportedly isolated ones. It would be better science, he argued, to increase sample size in "modern populations to detect small effects through such techniques as record linkage, rather than by a heroic but forlorn attack on primitive populations."[180] The biggest contribution the WHO could make to the study of human population genetics, Morton argued, was not a protocol for focusing on how to sample "primitive" people but greater investments in developing and funding "facilities for collection and long-term storage of biological material." This was where he and Neel agreed; such infrastructural resources were "essential" and "among the most valuable contributions of WHO."[181]

The disappearance of the phrase "primitive groups" from the 1968 WHO report signaled, then, that they had been naturalized as ideal populations upon which all human population genetics research could be modeled in a way that reflected a particular variety of Brazilian "primitive" preferred by Neel and his supporters.[182] Furthermore, the authors of the 1968 WHO report added to the list of "advantages" of focusing blood serum collections on so-called primitive groups including "the availability of immunologically 'virgin' populations, unexposed to infectious diseases" and "the likelihood of discovering unusual disease, abnormal frequencies of rare diseases, and sometimes new disorders."[183] This dimension of the research—the study of a "virgin soil" population—was one that would come back to haunt the legacy of the IBP when, in 2000, Neel was accused of knowingly exacerbating a measles epidemic in the Amazon in order to study the natural history of the disease (the controversy surrounding these accusations is described below and in chapter 5).[184]

With the IBP well underway, in June 1968 Neel shared his thoughts about what he had learned from studying the "American Indian"—he had begun work with a new group, the Yanomami, two years prior—at a workshop at the Pan-American Health Organization.[185] "One of the most provocative developments in the biomedicine of the past 20 years," he insisted, "has been the realization of the extent of man's concealed genetic variability, as revealed by modern biochemical techniques." Yet "understanding how all this variability is maintained is one of the great challenges to the human biologist."[186] It was no longer obvious to him that natural selection could account for the wide range of variation being uncovered through these molecular studies. New mathematical theories of population genetics and computer simulations would be necessary to "justify the effort expended" in collecting data about this variation. Nevertheless, Neel maintained that it was "[o]nly from the perspective of numerous studies in depth," that it would be possible to "begin with any assurance to factor out the significant common denominators that have guided human evolution."[187]

The effort would be worth it because, Neel explained, "[I]t can without exaggeration be said that ours is probably the last generation to be privileged to study such groups in a relatively pristine state. If there are important scientific lessons to be learned—and the WHO document [the revised 1968 WHO report] would seem to develop that case adequately—then there is an urgency to this problem not shared by many other areas of research." Neel agreed with those who felt "there has already been quite enough of surveying populations for gene frequencies in the vague hope that someday information will be valuable." However, he was optimistic that the problem had been defined clearly enough—the need to establish baselines grounded in population genetics—such that "considerable progress may be expected in the near future."[188] The ability to salvage evidence of life from the past and preserve it for the future would enable it to be retained to benefit the salvation of the denizens of Western civilization.

Neel concluded by connecting what he saw as "moral issues" involved with this agenda to the growing civil rights movement in the United States. "The world is watching my country as it agonizes over the Negro problem," he reported, and "it might equally well be watching the Americas for the signs of a belated, moral resurgence with respect to the Indians." With awareness that the forces contributing to the disappearance of the "primitive" stemmed from the twinned phenomena of decolonization and economic development, Neel implored the scientists in attendance to strive to "translate the results of [their] scientific investigation into concrete action programs, programs which must be carefully related to other governmental measures."[189]

He had given this talk on the heels of a harrowing experience in the winter and spring of 1968. Neel and his team had realized that a measles epidemic was unfolding among members of a Yanomami community living within Venezuelan borders.[190] Months earlier, Neel had begun attempting to source donations of vaccines from major pharmaceutical companies, including Roche and Merck, the latter of whom was apparently in a contractual relationship with the Venezuelan government to obtain a vaccine with a different strain. There is evidence that Merck was loath to provide vaccines freely to an American scientist when they could potentially get money for them from the Venezuelans.[191] Eventually, Neel was able to obtain two thousand doses of Edmonston B vaccine as well as permission from the Venezuelan government to use them. (Neel would later be accused of having chosen Edmonston B, a "primitive" measles vaccine, because it "provided a model much closer to real measles than other, safer vaccines in the attempt to resolve the great genetic question of selective adaptation."[192] Many have since argued that Neel's use of Edmonston was justified, as it had been approved for use by the US Food and Drug Administration).

Three doctors accompanied Neel to help administer the vaccinations—which needed to be brought into the Amazon on ice—but it was clear that the epidemic was more serious than anyone had appreciated. The team was frustrated and exhausted by their effort to vaccinate, while still attempting to collect blood (as well as stool, urine, and saliva). Though they succeeded in using all the vaccines they brought with them, their humanitarian intervention was far from adequate to address the extent to which other efforts to prospect the Amazon—for gold and timber—jeopardized the future of these portals to the past.[193]

Neel's experience in the midst of the measles epidemic left him even more convinced of his position, such that he sought to distill it into an essay for *Science*, in which he insisted, "In a world in which our heads are spinning under the impact of information overload, studies of primitive man provide, above everything else, perspective."[194] Anonymous peer reviewers of the paper "Lessons from a 'Primitive' People" articulated skepticism about what this perspective actually was. One wished to alert the editor that "a number of geneticists have taken issue with Dr. Neel's reasons for studying primitive groups," concerned that he was conflating populations living in 1970 with those who were evolving, say, fifteen thousand years ago.[195] Another reviewer charged Neel with basing his conclusions on "an unworkably simplistic set of anthropological assumptions," insisting that "Neel himself needs to learn more about his 'primitive' peoples before he aspires to teach their lessons to others."[196]

None of these criticisms prevented the paper from publication, which was remarkably well received. Anthropologist Ashley Montagu, an architect of the UNESCO Statements on Race, hailed the essay. Montagu had been involved with debates over race and genetics with Dobzhansky and Harvard anthropologist Carleton Coon.[197] The latter's unapologetic racism may have led Montagu to regard Neel's approach with more sympathy. "Lessons from a 'Primitive' People is magnificent!" he wrote to Neel. "This is not human genetics—it is what biological anthropology should be—a creative, living, constructive, and practical science, the applications of which will not only keep him in harmony with his world, but encourage his healthy development. . . . If anyone came to me and asked me what anthropology is really about I would recommend him your paper."[198] Indeed, "Lessons from a 'Primitive' People" circulated widely, its message of urgency, salvage, and salvation reaching across the life and social sciences.[199]

Too Late?

> "Before it is too late! Before it is too late!" The refrain runs through so much of the developing discipline that the needs of . . . salvage in the face of the impending extinction of peoples and their cultures dictated much that came to be anthropology both as a science and as a view of man.
>
> JACOB GRUBER, "ETHNOGRAPHIC SALVAGE AND THE SHAPING OF ANTHROPOLOGY"

With these words, anthropologist Jacob Gruber concluded his contribution to a symposium titled "The Vanishing Savage" held at the 1968 annual meeting of the American Anthropological Association. Gruber's talk described what he called the "salvage impulse," which he located in the nineteenth century. When theories of thermodynamic entropy and disorder initially began to circulate widely, "[p]eople began to sense the urgency of collection for the sake of preserving data whose extinction was feared."[200] Gruber, who had been closely following human biologists' romantic paeans to the "primitive," was disturbed that the return of Darwinian ideas about selection were accompanied by an emotional appeal to salvage. Anthropologist Adam Kuper would later write that the "idea of primitive society is perhaps even more potent when projected against an image of the future . . . in which we all inhabit a global village, set in a wasteland."[201] Cold War anxieties about the impact of planetary change made human biologists' claims for the urgency of studying the species' past seem self-evident. "Such a sense of urgency," which Gruber identified also in the writings of Claude Lévi-Strauss, "is not new."[202]

The only thing that was new was what was being salvaged. "[O]nly with the emergence of the population and its linkage to genetics," Gruber wrote, channeling Canguilhem, "does a pathological view of human variability and change for the meaning of races lose its importance."[203] The language of the baseline could reorient studies of human heredity away from issues of racial typology and toward epidemiology, survival, and even salvation, but not if it was rendered through the salvage of "primitive" blood.[204]

Gruber, speaking with convictions that presaged the coming reflexive turn in anthropology and the cultural turn in the history of science, condemned human biologists' choice to construct a baseline built from such bodies. His grasp of Canguilhem's notion of the context-dependent relationship between the normal and the pathological led Gruber to declare that "the notion . . . that data collection is itself value free, that facts are facts, is old-fashioned and naïve." The salvage impulse that was transmutating as it traveled from anthropology to human biology was not an a priori given, nor was the focus on "primitive groups." Rather, Gruber argued in 1968, it was a result of the "particular intellectual milieu that itself is a product of a time-centered sociocultural system."[205]

That Neel and Salzano were human geneticists by training, not anthropologists, may be important in accounting for their relative lack of attention to anthropology's long and vexed relationship to race and colonialism.[206] In a 1972 essay, one critic reflected on the courtship between anthropologists and population geneticists that, far from elevating themselves to the status of saviors, "a number of scientists are presently prostituting their mentalities to prove racist theory. They grasp at such findings as the . . . sickle-cell gene, predominately but not exclusively among blacks, as proof of the validity of the belief in race as an empirical object."[207] Even as these scientists claimed to reject racism, they studiously avoided the critiques that were emerging from within anthropology itself.[208]

In the ways they coordinated their efforts to salvage blood, human biologists were also salvaging ideas that were otherwise actively being discarded. As the blood of so-called primitive peoples was simultaneously venerated as containing latent archives of beneficial, albeit concealed adaptations to their environments, those environments were also cast as disappearing. A consequence of that disappearance would be the disruption of the purported stability of these "cold" societies. The hysterical cries of "Before it's too late! Before it's too late!" were loud enough to drown out those who suggested that perhaps such temporal anxieties could be resolved by a change of perspective.

Even a shift in perception that recognized that these groups might not be disappearing, but struggling to negotiate transition, still led efforts to provide help to be cast first and foremost as "urgent opportunistic" occasions to do research.[209] Overwhelmingly, the stated goal of human biological research on so-called stone age societies was to salvage information that might benefit "atomic age" communities' understandings of themselves. A pamphlet entitled "Man's Survival in a Changing World," produced to publicize the IBP in the United States, explained, "it can be truly said that this is the last generation of scientists to have the opportunity to study primitive cultures, and the first generation to have adequate tools in genetics, molecular biology and physiology to do it."[210] Above all, this included the ability to make bodies speak through the molecules that were latent in their blood. As the 1970s began, scientists, including Neel and Gajdusek, would use the resources at their disposal to intensify efforts to salvage blood from members of "cold" societies—whose purported stability was being thawed—such that it could be preserved in artificially frozen environments.

PART III

Collecting, Maintaining, Reusing, and Returning

FOUR

Managing the Cold Chain: Making Life Mobile

The hands of a clock on the wall read 2:45. The year is 1972. The location is a laboratory. On a ship anchored off the coast of an archipelago of islands near Papua New Guinea. The ship is called the *Alpha Helix*. It is a research vessel, designed to serve as a floating laboratory, funded by the National Science Foundation and administered by the Scripps Oceanographic Institute, based in La Jolla, California.

The shades on the porthole windows inside the ship are drawn shut, leaving beams from the fluorescent bulbs mounted on the ceiling to bounce around the stainless steel surfaces of a sink and backsplash. Even in this dim and fragmented light, caught on 16-mm color film, certain pigments stand out: black islanders, white scientists, and red blood.[1] The fourteen-minute silent film was created to document what members of the scientific team felt to be the most important elements of their expedition: namely their techniques for acquiring research materials—specifically, blood—to make knowledge about human biological variation.[2]

In the opening scene, a black man from the island of Rennell, a coral atoll and then a province in the Solomon Islands, is seated, shirtless, with a small child pressed against his left arm. They both stare at the vein that bulges in the bend of his flexed right arm, which is being held by a white man with rolled shirtsleeves. Another white man, who has abandoned his shirt to the heat of the tropics, bends down on his knees to insert a needle into the seated man's vein. For ten seconds everyone is focused on the blood spurting into a clear glass tube, known as a vacutainer. The kneeling scientist then removes the tube's cap from his clenched teeth and turns to the camera brandishing a proud smile and a vial filled with blood.

The camera cuts to the man from Rennell, now seated across from a third white man—this one in a striped, long-sleeved button-down shirt—who inscribes the islander's name, age, sex, and details about his health in an account book. The man from Rennell leans over to see what has been written, though it is unlikely he recognizes the English words. The child hangs on, like an extra appendage, against the arm that has not been punctured.

Off camera, as it cools from the temperature of his body to that of the room, the black man's extracted blood is already being allowed to separate into component parts—red cells on the bottom with a coating of white blood cells on top supporting a thicker layer of pale yellow serum. This blood sample, a few hours after it has been obtained, will be separated into several vials, each of which will be labeled and stowed along with thousands of others in one of the three freezers on the ship. Each component can be made to reveal different molecular markers of biological variation.

In 1972, no one is yet particularly interested in DNA, though it is known to reside in the nuclei of the white cells. The red cells, however, can be used to determine blood groups, a powerful means of establishing patterns of inheritance.[3] And the pale yellow serum can be subjected to a range of tests for antibodies and proteins linked to hereditary traits as well as acquired immunity, as described in chapters 2 and 3. At this time and for these scientists, the serum is deemed to contain the most valuable clues about human adaptability and evolution. These are the mysteries that have led teams of Euro-American scientists to gather blood from people living in regions far from their usual haunts in Bethesda or Cambridge, Michigan, or New York.

In a new scene, the camera zooms in on a single freezer, this particular one capable of maintaining temperatures as low as –80°C, and holds it in focus for a length of time that communicates its importance to the task at hand. Off the coast of Rennell, blood that had just hours ago circulated inside a warm human body is being chilled and rerouted, made newly able to circulate through the dense and globally distributed capillaries of biomedical science. The freezer, aboard a floating research vessel, is made to perform as a temporal and environmental prosthetic, a means of unmooring life and making it mobile across space and across time.[4]

Human Biology at Sea

With the IBP well underway, Euro-American human biologists ventured to regions remote from their home institutions. Their research included physiological studies of populations living at high altitude in the Andes and in regions of extreme cold, like Alaska, as well as research in human genetics and disease ecology.[5] Each of these forms of fieldwork created innumerable challenges, but none garnered more attention than the logistical problems involved with collecting and preserving blood at low temperature.

The ability to transform fleshy traces of human subjects into research objects was not seen as possible without the existence of a cold chain connecting the field to the lab. A cold chain is an infrastructure, sometimes made up of many different cold storage technologies, that enables low temperatures to be maintained as perishable materials move across long distances. As described in chapter 1, the first stage of the creation of the global cold chain that ultimately enabled blood to become mobile in the mid-twentieth century involved learning to generate artificial cold, a process that began nearly a century earlier. The second phase required figuring out how to manage products as they circulated, a practice innovated largely through the agricultural and food markets that developed into the first part of the twentieth century. Only then, as more kinds of increasingly fragile materials were put into circulation—such as bovine sperm and human biologics—did it become imperative to find ways to control temperature with precision. The final stage of the establishment of this cold chain was to extend its volume and reach.[6]

A transcontinental cold chain to support the circulation of blood had begun to form during World War II. Early in the war, when plasma (the liquid portion of blood) and albumin were thought to be sufficient for treating shock, a chain of manufacturing was established whereby the Red Cross would collect whole blood, chill it with ice, and seal it in large ice chests. It would then be shipped to pharmaceutical labs, where the plasma would be spun off in giant centrifuges. The plasma could then be freeze-dried, and the powder distributed for medics to carry in special kits.[7] The kits turned out to be awkward to use and army officials and naval researchers scrambled to make advances in the ability to transport, transfuse, and transplant tissues for injured soldiers.[8]

In one case, a procedure for shipping whole blood from California to the Pacific theater was developed whereby full bottles were gathered in Oakland and packed into insulated plywood boxes. They were then air-shipped to Hawaii. From Hawaii, they could be sent on to Guam, where they would be re-iced and forwarded to islands in the South and Central Pacific for distribution. In less than seven days, blood could reach soldiers stationed in the most remote theaters of war.[9] Systems like this for transporting blood helped establish protocols for shipping the perishable substance over long distances in nonmilitarized situations.[10] By the 1960s, biologists had begun to push for improvements in the cold chain to accommodate their worldwide stocktaking enterprise. When the infrastructure of the cold chain did not extend as far as their research ambitions did, they were able to take advantage of new instruments developed to serve the biosciences.

Commissioned by the Scripps Institute for Oceanography, designed by a physiologist, and named to honor the discovery of the helical configuration of proteins and genetic material, the *Alpha Helix* was built in 1966 with the intention of uniting lab-based molecular and natural historical field approaches in the name of "Big Biology."[11] To its administrators, the need for an *Alpha Helix* was continuous with the urgency that animated Cold War biomedical salvage projects. The ship's creation expressed the "accelerating and ramifying [trend] to carry problems from the artificial environment of the laboratory into the field, to see the organism as a whole, functioning in its natural habitat, [and] to look for the relatives of familiar laboratory species that are adapted to unfamiliar habitats."[12] The *Alpha Helix* was intended to extend the authority of the lab to the field.[13] In promotional materials, the ship's boosters had claimed, "It is this range plus the flexibility of location, type of organism and disciplinary approach that opens such vistas of opportunities and challenges biologists to learn how to make the best use of this instrument."[14] Here, the onus of maintaining and exploiting the authority of the floating lab, of making the most of the ship's instrumentality, was put on the scientist.

Like centuries of nautical explorers before them, the *Alpha Helix* provided scientists with a means of accessing regions seen as particularly remote from their home bases.[15] The ship might drop anchor, but it was not meant to put down roots. As a "truth spot," the floating lab was designed to provide a stable core of placelessness even when enveloped by the field.[16] No matter where in the world scientists ventured, with their lab in tow, they were meant to remain tethered to a familiar—albeit artificial—environment and worldview.[17]

This presumption of equivalence—that the lab at sea or docked off shore could or would function like the lab on land—was undermined by its very flexibility. Its primary instrumental value was not in making knowledge but in creating spatial and temporal distance between research subjects and the research objects extracted from their bodies. In fact, in ad-

13 The R/V *Alpha Helix*. From the Carleton Gajdusek Collection. Reprinted with permission of the American Philosophical Society.

dition to multiple wet laboratory spaces to support the practice of molecular science at sea, the ship had significant built-in refrigeration capacities: a forty-eight-square-foot walk-in freezer laboratory, held at a temperature of -20°C; a twenty-four-cubic-foot refrigerator in the main laboratory; and a twenty-one-cubic-foot freezer chest held at –18°C. There was also room and electrical power to support additional freezer space as well as liquid nitrogen tanks.[18] In practice, very little molecular work was actually conducted at sea. For human biologists, in particular, the *Alpha Helix* demonstrated its greatest value as a floating freezer, extending the global reach of a frozen-tissue-based infrastructure to support biomedical research.

The *Alpha Helix* was used in this capacity for three expeditions devoted to the study of human biology, two in Melanesia and one in the Amazon—each of which had connections to the IBP. The *Alpha Helix* provides a unique historical laboratory for fathoming contours of the intimate encounters that characterized a distributed Cold War enterprise of biological salvage on geopolitical and scientific frontiers. In both regions, those who made use of the *Alpha Helix* had to negotiate not only with members of the communities whose bodies would serve as biomedical baselines but others with necessary expertise and authority, including colonial administrators, missionaries, cultural anthropologists, and government agencies. Reading photos, film, field notes, letters, blueprints, maps, and ethnographies along and against the archival grain reveals the highly complex interactions involved with obtaining blood during each of these expeditions.[19] In the process, the otherwise concealed or previously silenced perspectives of many different actors who were involved in this enterprise become legible. Human biologists, as they attempted to deploy the administrative infrastructure created by the WHO and the IBP in order to create a frozen blood-based one, confronted practical challenges that were technical, ethical, and diplomatic. Though the ship was the same no matter where it traveled, those challenges were experienced differently in Melanesia than in the Amazon. Such frictions demonstrate the heated exchanges and uneven politics of postcoloniality that gave form to efforts to put life on ice.

Accessing the *Alpha Helix*: Damon and the Harvard Solomon Islands Project

In 1971, Albert Damon, a Harvard physician and biological anthropologist, was asked by the NSF to provide recommendations at the time of ship's fifth year review. He offered his opinion that, while the *Alpha Helix* had been used on a number of expeditions to study the biology of non-human organisms, "At a time when 'relevance' is a desideratum to justify research, human studies provide this, as well as scientific interest."[20] Damon's opinions had been forged through his experience in planning and executing several field seasons of the Harvard Solomon Islands Project, an IBP-affiliated initiative to study eight Melanesian populations, which began in 1966.[21] He had also participated in the WHO working group that produced the 1968 report "Research on Human Population Genetics."[22] In his feedback on the *Alpha Helix*, Damon reiterated the belief that human biologists had the most to learn from the bodies of those who were seen as least likely to endure. The primary goal of this research, he explained, was to undertake "multidisciplinary research on 'primitive' peoples while they are still primitive."[23] Damon's experience in the Pacific is an exemplary case study of what it looked like when human biologists with little experience in practices or preexisting connections to place attempted to put the agenda and protocols of the IBP and WHO into practice.

Damon did not have a lab at Harvard in which to analyze blood, so he worked hard to enroll a network of collaborators there and elsewhere. For scientists like Damon, international collaboration supported the compartmentalization, if not industrialization and outsourcing, of molecular forms of anthropological knowledge making. He informed his Harvard colleague, anthropologist William Howells, during the first season of the Harvard Solomon Islands Project that R. J. Walsh, the head of the Australian Red Cross, "is doing our blood work. . . . Hands across sea. All part of IBP."[24] To another colleague in Cambridge he described his plans to study a group in Ontong Java, "who have been a famous anthropological mystery: they live in Melanesia, have a Polynesian language, and look Micronesian! Your magical analyses [of amino acids on filter paper with blood and urine] should clear up the mystery."[25]

Damon had also sought collaborations with Carleton Gajdusek, who was well aware of the challenges in making use of samples accumulated en masse. Gajdusek agreed to conduct certain immunological and microbiological analyses of any blood collected, provided "we have an amply spacious timed schedule for its accomplishment," cautioning that "[o]n several such studies . . . we are not yet finished with the microbiological seroepidemiology of specimens over 10 years old, and the endless work will not be done a decade from now. I insert this only as a note of warning that the task is rarely finite."[26] By the mid-1960s it was clear that salvaged blood was being accumulated at a pace that outran scientists' ability, and sometimes desire, to ensure it was processed.

In the proposal he wrote to use the *Alpha Helix*, Damon explained how previous efforts to collect blood as part of the Harvard Solomon Islands Project had complicated the enterprise at the level of the cold chain. He explained to funders that the system he had relied upon involved a member of the team accompanying the local transport of samples, either by land or sea, each week and personally placing them on an airplane headed for an established laboratory. "When any link in the chain failed," he emphasized, "samples have deteriorated completely."[27] In spite of the fact that military bases and large mining enterprises had already colonized much of the Pacific, Damon insisted that "few places on earth are more isolated and devoid of natural resources than an oceanic atoll, hundreds of miles from the nearest land," making refrigeration his team's most significant "scientific problem."[28] For Damon, the *Alpha Helix* would perform its greatest contribution to science by serving as a means of overcoming gaps in the global cold chain.

Damon was successful in persuading funders to grant him the use of the *Alpha Helix*. Yet the record of his time on the ship is scant. By the time his trip had arrived, Damon had become stricken with cancer. He managed to conceal this information from the ship's crew, but it is easy to imagine how access to the ship—with its comforts and conveniences—made it possible for him to undertake otherwise very difficult fieldwork under personal duress. He died the following year, shortly after filing a brief report

14 The Harvard Solomon Islands Project team during its first field season, 1966. The man with the pipe and anorak is Lot Page, a physician and friend of Albert Damon, who stands beside him in white button-down shirt. Standing on the far right of the photo is Jonathan Friedlaender. The Solomon Islanders kneeling in front of the Harvard team served both as porters and as research subjects. The composition of the lineup reflects the kinds of colonial hierarchies that shaped the encounter between scientists and those they wished to study. From the Jonathan Friedlaender Collection. Reprinted with permission of the American Philosophical Society.

declaring the "enormous benefit the Alpha Helix Program has been for American Science" (but also expressing frustration about a "non-delivery of dry ice").[29]

Damon's 1972 *Alpha Helix* voyage focused on only two of the eight populations that made up the Harvard Solomon Islands Project, known as Ontong Java and Ulawa. However, the fieldwork he undertook prior to the voyage, fieldwork that helped him to justify the need for the *Alpha Helix*, reveals the extraordinary amount and variety of invisible labor required to undertake such a complicated expedition. The experiences of members of Damon's team in the years leading up to the first human biology voyage of the *Alpha Helix* demonstrates how these scientists sought to extend their efforts to the frontiers of Cold War biology.

In the early years of planning for the Harvard Solomon Islands Project, Damon was eager to experiment with new forms of cold storage. He was particularly keen to master the use of liquid nitrogen, as he believed it would allow his team to extend the cold chain deeper into the remote regions of the Pacific and to collect more kinds of tissue than just serum.

Damon's interest was piqued after corresponding with Richard Osborne, a professor of anthropology and medical genetics at the University of Wisconsin, who intended to set up a "Blood Genetics Laboratory," to "provide a more complete set of determinations on cells and sera than is done anywhere else."[30]

In 1966, before the first fieldwork season in the Solomons, Damon made a brief visit to Osborne at the McArdle Lab for Cancer Research in Madison, Wisconsin, to learn some basic blood-work techniques as well as to explore the relevance of then new practices of tissue culture for his project. Tissue culture required access to functioning cells, which needed to be preserved at lower temperatures than serum or even red blood cells. Osborne explained to Damon his intention to perform such analyses "only when our own equipment and specifications are followed for collecting and shipping specimens from the field." This equipment would involve the use of liquid nitrogen, "which we anticipate we will excel in," Osborne boasted, "as we have the decided advantage of having the world's specialists in the use of these units right here in Madison—namely the American Breeders [Service]," the site of Basile Luyet's low-temperature laboratory (discussed in chapter 1).[31] Osborne emphasized that a key aspect of the program was long-term storage of specimens on liquid nitrogen "for further studies as techniques develop."[32] Osborne was keen to recruit Damon into his venture as a means of testing his own system and helping to establish his lab as a central node for the preservation and maintenance of cold blood as a biological resource for the as yet unknown future.

Osborne suggested that Damon contact a representative at American Breeders Service to get him set up with the equipment he would need to use liquid nitrogen in the field. "I understand that . . . the equipment you have developed," Damon wrote to the cattle breeders, "seems to be just what we shall need for the collection, storage, and transportation of our blood specimens and possibly specimens of other tissues."[33] In return, Damon received ordering instructions and a copy of the March–April 1966 edition of the *American Breeders Service Newsletter.* The main feature of the newsletter was a centerfold on "Precision Processing" that demonstrated through a step-by-step series of pictures what it looked like to achieve industrial-scale processing of cryogenically frozen bull sperm. Visible in many of the pictures were the kinds of equipment—the racks, ampoules, and freezers—that Damon would need to support the collection and preservation of human blood.

Damon also wrote to Arthur Rinfret at the Linde Division of Union Carbide, in Tonawanda, New York, the American Breeders Service's main engineering affiliate. Rinfret, who had helped Rockefeller Prentice at American

Breeders Service to develop the first portable liquid nitrogen dewar, was eager to help. He shared knowledge acquired during wartime of places to acquire supplies of liquid nitrogen in the Pacific ("From World War II days I recall small naval installations at Santa Cruz, Florida Island and Bougainville"), inquired about the constituents of blood Damon wished to preserve (red blood cells or serum), and agreed to send a local representative to discuss the specific needs of the project. Rinfret indicated that Damon's biomedical research was of "great fundamental interest" to his colleagues in the realm of cryopreservation.[34] Rinfret was decidedly uninterested in what Damon was hoping to learn about human biological diversity. He *did* want to know what Damon might learn about transporting samples intended for long-term preservation. Damon's interest in putting a relatively new form of cryopreservation into practice in a new kind of setting would add to a store of cryobiological knowledge that might benefit a broad array of users.

When Damon first began planning for the Harvard Solomon Islands Project in the mid-1960s, not only did he have much to learn about cold storage, he had no prior experience working in the region. He was dependent upon those with longer-term relationships with the communities in question. Several of his Harvard colleagues already had such relationships, which is likely what kindled Damon's interest. Cultural anthropologist Douglas Oliver had been working with a number of communities in the Solomon Islands since the 1930s.[35] William Howells, a biological anthropologist who specialized in multivariate statistics, had published research on anthropometry and blood groups of Solomon Islanders, also in the 1930s.[36] This work had contributed to human biologists' conceptualization of the Melanesian Islands as highly diverse and therefore well suited for studies of human biological variation.[37]

In his proposal to request use of the *Alpha Helix*, Damon noted that "[s]ocial anthropologists resident for 1 ½ to 2 years will have learned the language and customs (including diet), obtained genealogies, estimated ages, and made practical arrangements for the biomedical survey."[38] These were mundane activities that included logging each day's subjects in and out, assigning and painting a number on the bodies of each, and supplying assistance to help each investigator as recorder, interpreter, and technician.[39] Among those he relied on were Pierre and Elli Maranda, sociocultural anthropologists who, over years, had established important bonds of trust with a community known as the Lau.[40] Damon dispatched the Marandas to ready the populations for study much as he might have ordered specimen containers from American Breeders Service, stating that "[a]s for Christian or pagan, we'd prefer the latter, if there's a choice. . . . Blood is the most important sample we take—I assume this will be no problem. As you know, we offer useful gifts for cooperation, as well as medical treatment."[41]

The presence of missionaries in virtually all of the communities studied by human biologists underscores the role of Christianity in preparing the way for science and ultimately contributing to its practice.[42] Christian missionaries, many of whom had lived for decades with the communities of interest to human biologists in the Solomon Islands and elsewhere, were a vital group of go-betweens who could make things either easier or more difficult for a potential outsider.[43] In certain instances, particularly in Melanesia, missionaries expressed paternalistic concern for those that scientists attempted to study, including those community members who sought work on plantations or in the new mines that were developing.[44] In others, they actively helped researchers build relationships with the communities.

Missionaries kept meticulous records of births, marriages, and deaths, which became invaluable—if not essential—resources for those intent on studying the inheritance of adaptive traits, in particular concealed heterozygous ones. Certain missionary groups even produced guides that would help fieldworkers define the boundaries of the population under study in terms of language. Human population geneticist Luigi Luca Cavalli-Sforza, who was also involved with the IBP, had recently demonstrated the potential for identifying genetically related communities on the basis of church parish records in the Parma valley of Italy and had augmented that study by claiming that language could also serve as a means of identifying and delimiting populations.[45] This form of data collection deemphasized indigenous practices of kinship that implied more dynamic relations between people and place in favor of biogenetic ones embraced by Euro-American societies.[46]

In the late 1960s, Jonathan Friedlaender, a Harvard biological anthropology graduate student who had earned his fieldwork stripes during the first season of the Solomon Islands Project, was eager to test Cavalli-Sforza's ideas. In order to identify the populations that would become the subjects of his own blood collection activities on the island of Bougainville, he relied on the work of a missionary organization known as the Summer Institute of Linguistics, a group dedicated to translating the Bible and spreading its gospel. The publications they generated helped make these linguistic groups understandable to other knowledge workers who wished to situate their speakers as natural laboratories.[47] Friedlaender made much use of the language guide produced by the Summer Institute of Linguistics for Bougainville. The annotated and tattered booklet became a treasured possession.[48]

Other graduate students of Damon's colleagues, including sociocultural ones who were already in the field in the midst of their dissertation research, were also enrolled in the service of the Solomon Islands Project. Cultural anthropologist Douglas Oliver explained the need for such help in a letter to graduate student Roger Keesing, who would later become a prominent researcher and advocate for the Kwaio of Malaita in the Solomon Islands.[49] Oliver stressed that "we shall be in a much better position to work up a good program if we know something about the place in advance—more than can possibly [be] learned from available documents."[50] Among his requests of Keesing were "population statistics, some knowledge about ethnic boundaries, degree of acculturation, logistical problems and such."[51] Oliver anticipated that Keesing might balk at the enormity of this request but offered only cold comfort, sardonically observing, "All this might delay your degree by a few months, but unless you are just dying to get out to Siwash Teachers College and begin that 18 hour a week teaching schedule . . . I suggest you look into the matter of chartering a small boat for long enough to get around."[52] With little sense of an alternative, Keesing agreed to help out and also enlisted his peer, another Harvard sociocultural anthropology graduate student, Gene Ogan.

The two graduate students set about planning for Damon to visit the Solomons in July and August of 1964. Keesing and Ogan dutifully absorbed the burden, which also exacerbated their own territorial anxieties about encroachments upon their field sites, including by the missionary Summer Institute of Linguistics. "These cats," wrote Ogan to Keesing, "have a big (140 people) operation in the New Guinea Highlands and spent last month studying Bougainville. The individuals I met were pleasant enough, but like most anthro types, I feel I've got a vested interest in the locals and don't want others fussing about."[53] The nature of Ogan's vested interest had as much to do with maintaining the fiction that he was studying a relatively undisturbed community as it did with becoming a source of labor for his human biologist colleagues. Years later, some of the sociocultural anthropologists who aided Damon and his team wrote to complain they had not been credited in publications stemming from the research.[54]

Ogan also expressed particular ambivalence about efforts of another Harvard graduate student, Gene Giles. Ogan was working intensively with a Solomon Islander group called the Nasioi when Giles approached him about collecting their blood. Ogan reluctantly agreed and began referring to Giles in letters to Keesing as "boy vampire."[55] He knew that Giles planned to spend a week, "main purpose being the collection of blood samples," but was skeptical as to whether Giles would be successful: "I fear they [the locals] will all be AWOL when it comes time to actually jab them with needles."[56] Ogan followed that statement, wistfully musing to Keesing, "It must be wonderful to have a lot of data to show for the time one has spent in the bush."[57]

Yet if biological anthropologists seemed positioned to get publications out of their data, sociocultural anthropologists had more access to happenings on the ground and greater insights into the ways in which the sting of these extractive encounters lingered, among colonial administrators as well as research subjects.[58] As he was preparing for the *Alpha Helix* voyage, Damon received a letter from the Marandas cautioning him not to repeat mistakes made on his first trip to the field. Pierre informed him that "we heard many not too favorable comments about the last visit of the team in this area [in 1966]." The criticisms broke down into two general complaints. Certain high-level administrators expressed "hurt feelings because [they] did not hear from" Damon or his team members personally after they had left the field. The other line of critique related to the politics of not involving local workers, stating that "a trained nurse could have done the same job for $500 in two months.'"[59]

When Friedlaender sought to commence his own research with members of the Solomon Island group known as the Kwaio in the mid-1970s, he reached out to Keesing, who warned him about the forms of resistance he might encounter. This shed light on the circumstances Damon would have been dealing with several years earlier. The year before, in 1976, the Solomon Islands had achieved self-governance. Agitation about the need to improve health-care infrastructure rippled into even the most supposedly "primitive" communities. Keesing observed,

> The Kwaio will want to know what's in it for them. . . . What is remembered about the 1966 expedition is not the measurements and data gathering, whose aims and benefits are and were meaningless to them, but the medical services by Lot and etc. [Damon's physician friends] that were provided during an epidemic that cost many lives. When I have raised the possibility of another study, all I can get them to talk about is wonderful American doctors who can cure people better than the ones at Atoifi [a small hospital established in 1965 by Seventh-day Adventist medical missionaries], and whom they want to stay on with them permanently. A few anthropometers and "$5 per family" may seem slim incentive.[60]

While there is no evidence that researchers affiliated with the IBP coerced their research subjects into participating, informed consent was not institutionalized in the way it would come to be in later decades (chapter 5).[61] Friedlaender recalled that the first group he intended to study refused him and he proceeded to the next. His own practice for obtaining permission to conduct research in the 1960s involved a short speech in pidgin to community members or leaders explaining that he wished to collect information to benefit his education.[62]

By the 1970s, this was no longer acceptable in the region. Keesing also warned Friedlaender of the impossibility of involving graduate students—"they are anathema at the moment"—who had done so much of the labor a few years earlier.[63] Any project that did not seek to actively enroll Solomon Islanders as collaborators in the actual conduct of research, and not merely as interpreters, would face resistance on the ground. While medical care was not the core of any visiting researcher's interest in Pacific populations, experiences with Western biomedicine had contributed to changes in Solomon Islanders' perceptions of the nature of exchange. Increasingly, they had become reluctant to serve as training grounds for those who would make their careers elsewhere. "A point of most serious political concern here," Keesing reported, "is that Solomon Islanders be trained by expatriate researchers so that this kind of work can be carried on within the country."[64] The Solomon Islands would declare independence from its British protectors the following year.

During the Harvard Solomon Islands Project, Damon's team had indeed made it a practice to treat acutely ill patients, but their objectives were by no means primarily humanitarian.[65] As in Neel's experience intervening in the measles epidemic among the Yanomami (chapter 3), humanitarianism was a by-product of what was fundamentally an uneven partnership between American researchers and local government.[66] The relative disinterest of the British protectorate in the Solomon Islands in providing acute medical care for its least accessible wards created opportunities for Damon's physician friends to get out of the clinic and into the field. Upon return, these physicians' efforts were cast in the newsletters of their home institutions—to a degree embarrassing even to them—as both charitable and heroic.[67] In his NSF proposal for the *Alpha Helix*, Damon stressed the multiple advantages to providing basic medical care, writing that "this activity, an ethical necessity, greatly enhances cooperation without interfering at all with the primary research objective."[68] Biomedicine was a resource to be exchanged—both with research subjects *and* colonial administrators.

In his 1971 NSF proposal, Damon was particularly attentive to problems of colonial administration. "[A]n integral part of the project," meaning the research to be conducted with the help the *Alpha Helix*, was that findings could be applied "to similar populations in the U.S. Pacific Trust Territory [now the Marshall Islands, Palau, and the Federated States of Micronesia] and to man-environment-disease problems in the U.S. Services to the Solomon Islanders and to their responsible administrators."[69] Whether or not Damon had a concrete idea of how his research might accomplish this—and there is little indication that he did—he understood how to leverage the Cold War landscape in which US territories, many of them islands, were coming to be increasingly geopolitically valuable as nodes in the expanding American empire.[70]

Aware also of the WHO's investments in serum banking, Damon emphasized the ancillary role he and his team could play in epidemiological surveillance. In his proposal to use the *Alpha Helix*, Damon explained how, during previous field seasons, as his team collected blood intended to be frozen, they also used small droplets to make malaria smears—a slide with a drop of blood enabling identification of the presence and/or level of infection—to be "read by Administration malariologists as a guide to eradication and control programs."[71] This was but a minor inconvenience given the extensive nature of Damon's research protocol.

In making his case for access to the *Alpha Helix* and for the Solomon Islands Project, more broadly, Damon emphasized the fact that his team sought to collect much more than just blood. Despite the fact that he had defined the purpose of the enterprise in terms outlined by geneticists like Neel, Damon wrote in his proposal for the *Alpha Helix* that "expeditions concentrating on genetic markers in the blood, like Neel's and Cavalli-Sforza's sacrifice other aspects of a biomedical survey, particularly the detailed anthropometry and the various special medical examinations. They draw blood and leave as quickly as possible."[72]

Indeed, blood draws were only one dimension of Damon's multipurpose agenda for the Solomon Islands Project. Participants were also subjected to an extensive medical protocol, including tests of vision, physician examination, dental impressions and X-rays. Additionally, physical anthropologists would take body, head, and feet measurements, fingerprints, and conduct tests on the ability to detect a bitter taste (a genetic marker). Additional procedures included assessments of the degree to which skin reflected light and photography, both full body ("somatotype") and details of head and face. They might have their ears swabbed (the texture of the of wax was an early genetic marker) and a few strands of their hair collected (the texture of hair was still another). Samples of urine, stools, and (for a smaller number) skin were also obtained and examined in the field.[73] The biopolitical dimensions of this work were later underscored by rumors that the data Jonathan Friedlaender was collecting for his dissertation research in Bougainville would be used to make his subjects American citizens, a prospect that some found desirable—despite the degree of poking and prodding that appeared to be involved.

In the mid- to late 1960s and early 1970s, it was only just becoming apparent that blood might be an especially generative research material. Nevertheless, members of the Solomon Islands Project research team were eagerly invested in its "as yet unknown" potential. "The blood is gold," Friedlaender recalled a colleague telling him as he learned to extract it during the first field season of the Solomon Islands Project, "and it's going to be important down the line."[74] By the time Damon had applied to use the *Alpha Helix* in 1971, he could already boast a number of laboratories, from New Zealand to Philadelphia, eager to analyze the blood he intended to collect.[75]

As will become clear from Carleton Gajdusek's experience on the *Alpha Helix*—its second human-biology focused voyage—it was the desire to accumulate blood, and the sociotechnical practices involved in obtaining it, that earned scientists the privilege of access to the ship. Indeed, the strongest case Damon could ultimately make for use of the *Alpha Helix* was as a crucial link in the cold chain necessary for transforming isolated human bodies into biological resources. More than as a molecular lab, he highlighted its value as a literal time-saving machine, advancing the seemingly urgent project of salvage and stemming the decay that menaced plans to preserve such rescued stock for future uses.

The Trouble With Freezers: Gajdusek in Melanesia

Immediately following the voyage undertaken by Damon's team, NIH virologist and self-described medical anthropologist Carleton Gajdusek charted an ambitious itinerary through the Banks and Torres Islands of the New Hebrides (administered through a British-French condominium which ended after a decade-long independence movement to become the Republic of Vanuatu in 1980); southern islands of the British Solomon Islands Protectorate; Pingelap Atoll (which was seized from the Japanese by the United States in 1945 and later became part of the Federated States of Micronesia, which gained independence in 1948); and the Caroline Islands (also US trust territories, divided today between the Federated States of Micronesia and Palau, which achieved independence in 1994). Gajdusek's meticulous field notes from the voyage emphasize how the *Alpha Helix* demonstrated its greatest utility as a floating freezer rather than as a mobile laboratory to facilitate the production of molecular knowledge about human evolution and variation in the field. This became most apparent when one of Gajdusek's freezers broke down at sea.

Reading his extensive notes for what they say as well as what they do not also permits a greater measure of insight into how certain Melanesian island residents understood a scientist who came to cast them as research subjects, research assistants, and intimate companions.[76] Gajdusek, arguably more than many other human biologists working at the time, found himself entangled in complex and enduring relationships with those he studied in the Pacific. He struggled with his scientific ambitions—which seemed to valorize the smooth cutting of networks—with his own desires to be transformed by his fieldwork encounters.[77] The blood he collected and froze would ultimately persist beyond his own life span as research objects entangled not only in incomplete relations of exchange but by ever-extending horizons of time.[78]

At the completion of his 1972 Solomon Islands Project voyage, Damon wrote to Gajdusek from Cambridge, Massachusetts, to extend his good wishes and welcome him aboard the *Alpha Helix*. Damon indicated how access to the ship had favorably augmented fieldwork, bragging to Gajdusek that it had been "by far the most successful of all our expeditions, in terms of numbers of subjects, (in)accessibility of locales, and particularly the quality of the laboratory preparations and analyses." Damon added, "Your itinerary makes us green with envy. We did our usual exhaustive (and exhausting) work-up on two populations, but you've picked some real exotica!"[79]

Gajdusek did indeed have an ambitious itinerary for his time aboard the *Alpha Helix*. In his journal he commented on the gratuitousness of the "luxurious stateroom" assigned to him as chief scientist and allowed himself a similarly outsized moment of anticipation, documenting his intention to "plan an attack which will make our expedition the most successful on record for 'new' observations in the Pacific area since La Perouse's [*sic*] and Captain Cook's voyages."[80] While La Pérouse and Cook had sailed to the frontier in an age of geographic mapping, Gajdusek understood that his accomplishments would be in using the *Alpha Helix* as an instrument to map genetic variation.[81]

His 1972 trip had been in the works for over a year and, in a letter to a scientist at the Kuru Research Center in Papua New Guinea, Gajdusek described what, specifically, this would mean. It was to be a first effort to create a "fibroblast cell line in the ship-board laboratory, which we shall then freeze down in liquid nitrogen, for cell revival later back in the laboratories at NIH." If he was able to do this successfully, it would set a precedent for all population genetic studies, for instead of preserving serum or hemoglobin, or red cells, "for future study of new pleomorphisms by newly evolving techniques, we plan to bank the *full genetic information of each* subject in viable cells. It is the working out of this technology under ideal field conditions that we hope to exploit on this trip."[82] The goal was to use the *Alpha Helix* to collect material that could be thawed and transformed into cell lines back in Bethesda, a more totalizing vision of a frozen-tissue based infrastructure than had been previously promoted by the WHO or the IBP.

Although the *Alpha Helix* came with built-in refrigeration capacities, Gajdusek's collection goals were ambitious enough that he wrote to inquire about the possibility of installing additional refrigeration capacity. In a perfect world he would have both a –70°C "REVCO-type freezer or a well-supplied liquid nitrogen or dry ice source." While a "liquid nitrogen tank plus small carrier reservoirs (which my laboratory can provide)" would be important, "ample 4C (or approx. that) refrigerator space for many things including blood, urine, cell specimens is essential as is FREEZER space." He explained that "temperatures of -20 can serve, but -70 (dry ice or REVCO-type box)" would be necessary for any virus work. While he had already made important breakthroughs in the etiology of kuru, virology was not a central feature of the work he planned to do aboard the *Alpha Helix*.[83] "Liquid nitrogen can cover this phase if the low temp. electric box is not available," he offered, "but the luxury of having both is great."[84]

Gajdusek wound up bringing both liquid nitrogen (at a cost of US$3,500) and a REVCO –70°C freezer (at a cost of US$1,500) onto the ship.[85] Both caused him a great deal of consternation. Before the ship even left port, Gajdusek realized the liquid nitrogen tanks he had shipped separately from his lab in Bethesda had failed to arrive. He repeatedly wired home to inquire about their whereabouts and to develop a contingency plan to get them on board. On September 14 he wrote to lab members in Bethesda: "Liquid nitrogen: NO SIGN OF IT AS YET . . . It is the most critical issue of the trip at the moment."[86] Liquid nitrogen was his preferred coolant for maintaining the metabolic function of the cell lines he intended to make to be recovered back in Bethesda. The next day, one 20-liter and four 9-liter containers of liquid nitrogen materialized, but Gajdusek worried that they were insufficient amounts to support the tasks ahead. He also fretted, "I have not yet seen the bleeding venules, nor the finger printing sets and am really worried, lest at the critical point of actually collecting data and specimens we lack the critical items."[87]

There is no indication as to whether or not more liquid nitrogen ever turned up, but on September 19, the team reached their first field site and was able to promptly get to work, bleeding, fingerprinting, and examining the children and adults they found. Gajdusek's research protocol was less exhaustive than Damon's, but it still included a range of anthropometric and medical tests. Gajdusek met personally with community leaders to explain, in the regional pidgin dialect, the nature of the research and to promise to provide acute medical care when necessary. In the film he made of the voyage, described at the beginning of this chapter, members of his team are

15 Gajdusek discussing the needs of his R/V *Alpha Helix* team with Chief Cedrik Tangata Teava upon arrival at Tikopia. This conversation, a contemporaneous form of obtaining "informed consent," occurred prior to Gajdusek's team commencing their biomedical research. Carleton Gajdusek Collection. Reprinted with permission of the American Philosophical Society.

shown working in the vicinity of a mission church, with a close up of a chest of medical supplies marked with the symbol of the Red Cross edited to bleed into a shot of the cross hanging outside the church. Inside the church, a clutch of villagers looked on as a team member examined an infant.

At each stop, the goal was to sample everyone—men, women, and children—("we have found 48 Merig residents and have left two babies unbled—a population of 50 on the island!"), and a routine for shuttling people and materials between ship and shore was quickly established.[88] A small skiff would transport the researchers between the *Alpha Helix*—which would be anchored offshore—and the island. Throughout the day certain team members would return to the ship with the blood collected. ("At 1:00 pm, Walter and Paul took the first 100 blood specimens back to the ship, where Paul managed to make the serum preparations of the whole series before supper!")[89]

Gajdusek preferred to remain on shore. He noted in his journal, "At night the ship is very uncomfortable and noisy everyone tells me . . . I have not been back on board since I landed!"[90] The ship's crew and Gajdusek cultivated a mutual distaste for each other as the trip progressed. He wrote in his journal of the "increasing friction between my scientists and our Navy-style captain, who wants tighter schedules, command, orders and discipline, not appreciating that we do not run our work by levels of authority."[91] Despite spending minimal time aboard, he still found room to criticize the ship and its crew's inflexibility, claiming that "[a]lthough excellently equipped for life at sea, the *Alpha Helix* is not equipped to provide shore parties with the optimum support, either for sleeping and eating, or for work."[92]

Other more serious frictions were generated by the very project of bringing the lab to the field. One of the most striking was the problem of waste. It was customary to dump certain forms of refuse overboard. Gajdusek was alarmed to discover that during the team's stay on Hu Island, a number of filled hemoglobin tubes turned up on the beach. He later surmised that they had been "foolishly" tossed into the ocean as garbage, close enough to the bay where the ship had been moored to reach the shore still intact. They had been subjected to some basic lab tests and then discarded to make room for the constituents of blood that were meant to be stored over the long term. Gajdusek found it "embarrassing to have them show up this way, and we have to assure the people that they have been studied and used and are now discardable."[93] The physical lab works by removing objects from their natural orders and resituating them in a social order that enables them to be manipulated and controlled in the pursuit of scientific knowledge.[94] At sea, the desire to cut relations could be undermined by shifting tides that reversed the flow of exchange.[95] The discardable "they" in Gajdusek's account was meant to refer to the hemoglobins and not the people themselves. Gajdusek's semantic conflation of research subjects with research materials betrayed the *Alpha Helix*'s ability to serve as a universal lab.

When people could be convinced to participate (though some simply refused), fieldworkers still struggled to combat attrition. Failure to adhere to protocol created problems. The most frequent culprit of this failure was fatigue of both researchers and subjects: collecting blood was hard work for everyone involved. Late into the voyage, Gajdusek chided himself for permitting too much flexibility in sampling protocol, which wreaked havoc on his otherwise carefully calibrated system. When members of a community on the island of Santa Cruz started fainting, Gajdusek was forced to reflect on what had gone wrong. "We had broken several of my tenets about taking blood specimens from villagers."[96] He listed these tenets without indicating which specific ones had been violated. However, at least two bear specific attention, for they reflect particular needs to control the temperature of both emotions and environment in the conduct of this form of knowledge production:

> Do not collect blood specimens and "release" any subject until all the population is assembled and "captured" and laughter and relaxation characterize the operation. If there are many bleeders then work at different parts of the assembled group and quickly shift to any part of the line where boredom, anxiety or too much discussion develops.
>
> Keep subjects out of the heat and sun before and after bleeding.[97]

Without proper attention, freezers, Gajdusek discovered, could also faint. One of the freezers, the REVCO described above, was being stored on *Alpha Helix*'s deck. This freezer contained many of the most fragile tissues collected during the expedition, including those he hoped to transform into cell lines back at the lab. About halfway through the voyage, members of the crew became concerned about the REVCO's ability to maintain the requisite low temperatures. They constructed a sunshade for the freezer but had to move the sensitive machine in order to secure it. From this point on the temperature gauge remained fixed at –38°C, much higher than it was supposed to be. Gajdusek noticed this, but since the freezer was already full, he later claimed that he was reluctant to investigate, lest any remaining cold be released upon opening.

In mid-November, with the voyage nearly over, Gajdusek's fear of freezer failure had become a reality. He and his team were docked and preparing to send several thousand samples on dry ice to Bob Kirk, an anthropological geneticist at the Australian National University in Canberra and member of the IBP, who had agreed to help with analysis. It was to Kirk that Gajdusek wrote, in all caps: "THEN THE TROUBLE STARTED." The REVCO freezer, though still registering –38°C on the gauge had stopped functioning. "Its contents," he was horrified to realize, "were not even frozen in the bottom. . . . The collection is unreplacable [*sic*] and represents an enormous effort. . . . Please, Bob," he urged, "do your best on them."[98]

The REVCO was easily repaired while the *Alpha Helix* was in port, though the temperature gauge remained stuck at –38°C. The next several days at sea, sailing toward the Caroline Islands for the last leg of the expedition, were spent trying to salvage two thousand samples—"a bloody mess"—from the failed freezer. Gajdusek surveyed the damage, of "tubes stuck together and frozen together as the clots refroze. . . . Only the sera are in excellent shape."[99] This is because the sera, generally hardier than cellular material, had been stored in the –20°C, walk-in freezer on the ship. Kirk eventually received the clots from the broken freezer but was ultimately unable to salvage them, describing their condition as "grossly unsuitable" for transformation.[100]

Gajdusek was wracked with guilt for not having investigated further when he first suspected something was amiss. He blamed himself for allowing the freezer to be moved to install the sunshade. The fact that he spent most of the trip socializing and sleeping on shore, away from the freezers may have aided him in the collection of blood, but it cost him control over its preservation. Maintaining influence over both human and machine regulatory responses required vigilance in both the field and the floating lab. Gajdusek chose the field and paid the price in sweaty samples. The surviving cold blood samples, including the sera, were shipped back to Gajdusek's NIH lab in Bethesda where they persisted for decades in the freezer as latent life, awaiting a future in which they could fulfill an as yet unknown potential.

"Out of Vacutainers and Knee Deep in Indians": Neel in the Amazon

On the other side of the world, in 1977, James Neel continued his studies using the *Alpha Helix* to travel the Amazon River as part of an expedition co-organized with Brazilian population geneticist Francisco Salzano—head of his nation's Human Adaptability arm of the IBP.[101] This was to be the third and final human-biology-focused expedition of the *Alpha Helix* before the program was ended and the ship was sold to the University of Alaska, which outfitted it for use in polar research.

While the development of new tests for analyzing blood continued apace, in the years since Damon and Gajdusek's voyages, research practices involving humans had come under scrutiny as the biomedical enterprise expanded to include all kinds of healthy or normal individuals as research subjects.[102] In December 1973, the United States Department of Health, Education, and Welfare instituted procedures that required all federally funded investigators conducting research on humans to submit a "questionnaire" describing potential risks to subjects as well as plans for obtaining informed consent.[103] Neel had long received funding from the National Science Foundation, and when he applied to renew his grant to cover the time he would spend on the *Alpha Helix*, he completed one such questionnaire. He characterized his proposed subjects as "primitive, illiterate Amerindians residing in the jungles of Brazil, Venezuela and Panama" who were "free agents" capable of determining whether or not to give blood. Consent would be obtained orally after the procedure of venipuncture had been "carefully explained" and demonstrated on a member of the research team.[104] Elsewhere on the application he repeated that "participation is voluntary; there is no effort at special persuasion," though he did add that the "great deal of medical care built into the program" would contribute to protecting the rights and welfare of individuals who chose to participate.[105]

The bureaucratic neutrality of the questionnaire was betrayed by an emphatic instruction to applicants, punctuated with four asterisks, which read, "****Please be certain that your 'informed consent' document contains the statement that the patient is free to refuse to participate (and to withdraw from experiments already begun) and that such refusal will not adversely influence his future treatment."[106] There was a clear awareness that subjects could change their minds about involvement in research over time, even if they had initially agreed. This clause, a reaction to revelations of coercion in medical research, signaled an awareness of how the coupling of research and care put patients in unique positions of vulnerability that they might only become attuned to as time unfolded. Refusal—the act of rejecting or resisting participation—which had been observed by researchers such as Friedlaender in his work in Bougainville, became codified as a legitimate position of agency for potential as well as active research subjects.[107]

Neel's "free agents" may well have been unperturbed by the practice of bloodletting. They may also have been appreciative of the healing properties of the medicines and vaccines his team provided. Yet what could not have been communicated through the demonstration of venipuncture or even a "careful explanation" of the procedure was the ultimate disposition of the blood—in a freezer in a faraway locale where it would persist for decades. In the mid-1970s, with more than a decade of fieldwork in the Amazon behind him, these new rules around informed consent were little more than a bureaucratic hoop through which Neel could easily jump. He dutifully completed the questionnaire and turned his attention to what he saw as more pressing matters. Neel was much less rattled by regulation of his work with research subjects than by the logistics of using the floating laboratory on the upper Amazon, an endeavor that required much greater diplomatic finesse than had been required of either Damon or Gajdusek in Melanesia.

For Neel and Salzano's expedition, Leticia, Colombia—one of two major ports on the Amazon—would serve as a base of operations, and the *Alpha Helix* would head into Brazilian waters from there. Control over the port of Leticia had been a source of violent conflict between Peru and Colombia since the 1930s. In fact it was the subject of the first dispute settled by the League of Nations.[108] Sensitivities persisted and were expressed in a desire to have Colombian scientists added to the roster of the *Alpha Helix* team or to otherwise risk being barred from basing activities in Leticia.[109]

Negotiating access to sail out of Leticia nearly led the trip to be cancelled several days before it was scheduled to begin. As a result of eleventh-hour interventions of ambassadors to Colombia, the team was ultimately permitted to proceed without the addition of Colombian scientists. Even beyond this encounter with South American postcolonial politics, the paper trail of Neel's voyage reveals how human biological research, when combined with use of the federally funded *Alpha Helix*, was interpreted by local officials as an expression of political power. In Latin America, where formal colonial rule was largely over by the mid-twentieth century, and practices of mining gold, rubber, and silver were well established, scientists found themselves harder pressed to justify their desire to access remote populations.[110]

Neel found himself in the reluctant role of diplomat, tasked with facilitating US-Latin America relationships, along with fulfilling his own research goals, which were focused on studies of mutation rates and natural selection. Neel's struggles to gain approval to study the "most primitive," to him the most epistemologically valuable communities, make the challenges he experienced aboard a floating laboratory an example of how the specificity of time and place mattered to his and his colleagues' efforts at salvage. His recurrent expressions of frustration underscore limitations and difficult realities of seeking out research subjects in circumstances far from the comforts of home. They also demonstrate his ambivalence about the political obligations that accompanied the use of the *Alpha Helix* in Latin America.

"The general research plan is quite simple," Neel wrote in his proposal to use the *Alpha Helix*, indicating that his goal was to use the ship as a base to "collect and process as many specimens as possible from members of the Indian tribes of the Amazon Basin."[111] Much as Gajdusek had proposed that access to the *Alpha Helix* would allow him to experiment with tissue culture in the field, Neel felt that he too should articulate some kind of novel scientific justification for using the floating lab. Neel described a "new and very exciting" approach to which he would be subjecting the blood he intended to collect. Electrophoresis, he explained, was providing geneticists with the ability to identify novel protein variants and had "set the stage for an entirely new approach to the problem of estimating mutation rates."[112]

When Neel's proposal was approved in February of 1976, Salzano came to Michigan to meet with him to discuss the complex logistics of the voyage. One problem was figuring out how, exactly, to locate their research subjects. Neel wrote to Scripps administrator Walter Garey, requesting extra time to allow him to meet with missionaries, whom he had come to recognize as essential to his research. "I must absolutely have two days for checking with the Mission Aviation Fellowship," he insisted, which was "the headquarters for the Summer Institute of Linguistics, and the headquarters for the New Tribes Missions." He believed that, more than any government agency, these religious groups were best situated to provide him with "the most up to date information they have on the whereabouts of all of the Indians we would hope to study."[113]

Later that month, a letter from a member of the *Alpha Helix* review committee revealed the explicitly diplomatic cast that this scientific expedition was acquiring—a dimension of the *Alpha Helix*'s instrumentality as a manipulator of international relations in the service of American political goals that had not been as pronounced in the Pacific. Scripps administrator and oceanographer Vera Alexander wrote to Neel of the importance of using the expedition "to further and improve cooperative research between the United States and the South American countries involved, and to generally assist in international relations." It was, in her view, "extremely important that local (e.g. Brazilian) scientists are given full participation status in the scientific effort and be offered adequate facilities, ship time, and other considerations." Much as US IBP director Frank Blair had observed that collaborations with colleagues in Latin America could be used to prevent another Vietnam (chapter 3), Alexander emphasized that, regardless of Neel's scientific goals, "[t]his must be considered an important responsibility and I am confident that you will carry out this diplomatic role with enthusiasm."[114]

In the Cold War contexts in which this work was being conducted, forming alliances with South American scientists was believed to be of great political significance. Neel, an experienced "bio-politician," accepted what seemed to be a straightforward request.[115] He had worked closely with Salzano for many years and was more than happy to conduct the expedition with him and his Brazilian colleagues. However, it was only in the coming months that he would begin to appreciate how the authority of the *Alpha Helix* as an instrument for "U.S. national needs" would get in the way of his efforts to use it as a floating freezer.

Shortly before Neel was ready to leave, he was confronted with the challenge from the Colombian government. Neel was alarmed that his obligations to include South American scientists would be extended beyond those from Brazil, whom he had already vetted. Toward the end of May 1976, Walter Garey at Scripps wrote to Neel to report that the Brazilian government had granted permission for the expedition and that a conditional clearance had been received from Colombia without any need of taking Colombian scientists on board. In keeping with the diplomatic dimension of the expedition, Garey reminded Neel of his obligations to ensure that collaboration would extend beyond the act of data extraction to include reports to be translated into Portuguese for publication in the journal *Acta Amazonia*.[116] Having assumed these responsibilities as a necessary by-product of the endeavor, Neel felt ready to proceed. And then, as Gajdusek wrote of his freezers, the trouble started.

On 8 June 1976, Neel called Walter Garey to express a new concern that threatened the voyage. Already in the 1970s Neel was encountering difficulty negotiating access with Brazilian representatives of indigenous people. Neel recounted how, in January, he had made a request to conduct research on isolated tribes to FUNAI—the National Indian Foundation, which is the Brazilian government body that establishes and still carries out policies relating to indigenous peoples.[117] FUNAI had been founded in 1969 as Survival International "by individuals appalled by the genocide of Amazon Indians."[118] In frustration, Neel told Garey that FUNAI "has waited until the last minute and has finally given permission for work with 5 tribes of low priority. Further, FUNAI has specifically stated that isolated or frontier tribes are not to be contacted. Since work with aboriginal peoples in isolated and frontier areas constitute the real objective of the proposed research," this was a development that could seriously undermine the expedition's agenda and goals.[119]

The colonial histories of South American nations like Colombia and Brazil experienced very different trajectories than those of the various islands of Melanesia. Colombia was settled by the Spanish in the early 1500s and achieved independence in 1819. Claimed as a Portuguese colony in 1500, Brazil became independent in 1822, before many regions of Melanesia even became colonies. During these centuries of colonial rule and mercantile relations, the numerous indigenous communities in the region experienced violent subjugation, if not outright murder, in the context of resource extraction—from rubber to silver and gold—as well as death from introduced infections.[120] This was the era of El Dorado, a time of rampant primitive accumulation that provided the infusions of wealth that spurred the rise of capitalism in Europe.[121] It has been argued that, unlike many Melanesians, Amazonian communities often deliberately avoided contact—they refused it—seeking to stave off the horrors that seemed inevitably to accompany European encounter.[122] Communities that did survive into the era of independence faced (and continue to face) persecution directly as they struggled to retain land rights sought by miners and loggers and indirectly as national governments struggled to address inequalities that exposed such groups to heightened risk of epidemic illness.[123]

Neel conceded to Garey that he and Salzano could, "appreciate the possible sensitivity and strained relations" that already existed due to the fact that the Brazilian government was attempting to build roads in the vicinity of indigenous groups' territory. Neel attempted to negotiate, conceding to Garey that he "would accept a denial from working with 2–3 tribes, but cannot operate under a blanket refusal for studies of any isolated or frontier tribes." While Neel had respected the need to accept refusal from potential research subjects themselves, he was resistant to the protocol of an organization tasked with serving as a protective intermediary. Studying "3–4 highly acculturated groups" could not be justified by Neel's research agenda. "Unless the Ambassador can convince FUNAI to give permission for this joint US-Brazilian study," Neel despaired, "the effort will likely have to be cancelled."[124]

Neel immediately also wrote to Salzano. He referenced a document that, in signing, he now confessed meant that "we made a blanket commitment not to work amongst any of the isolated border tribes. These were, after all, our primary objects. Those tribes to which they say we can have access are, as you are well aware, a very secondary interest."[125] His "primary objects," were potential human research subjects whose vulnerable status put them beyond his reach—even with the approval of the NSF's human subjects review board. Neel described six tribes for which "we have been assured the hospitality of the missionaries in contact with each of these tribes, and the full cooperation of MAF [the Brazilian Air Force] in flying us in and out." In contemplating the alternatives, Neel despaired, "I know absolutely nothing about how we would get to these other tribes (which we added to the list as 'insurance') and what we would find when we got there."[126] Without the permission of FUNAI or the assistance of local intermediaries, Neel and his team would truly be adrift in the Amazon.

In his letter to Salzano, Neel pointed out that under the usual circumstances of this kind of work (i.e., those not involving a high-profile NSF-funded research vessel) there would be greater flexibility in resolving these unexpected challenges. Now there was the additional burden of incorporating a big, expensive piece of instrumentation: "there is a ship waiting to lend support to a particular kind of investigative work and given the location of the tribes for which we do have some kind of permission, we will have great difficulty using either that ship or the MAF."[127] With no time to waste, Neel laid out a revised plan of action in several bullet points, which included: "1) Attempt through the NSF, the State Department and your Embassy to generate an 'official' request to FUNAI [and] 2) Attempt thru PAHO to convince FUNAI that we really are nice guys."[128] Neel's belief that the very "goodness" of his and Salzano's intentions could override a system of protections, however flawed they may have been, set in place to curtail just these kinds of extractive interventions, suggests that he was considerably more familiar with the machinations of international negotiations than with the lived experiences of the indigenous people he wished to contact before it was too late.

The trip, which began on July 14, 1977, eventually went ahead and was ultimately regarded by Neel and Salzano as having been a great success, even without being able to access the tribes forbidden by FUNAI. Upon arriving in Brazil, Neel's initial reconnaissance activities revealed that a group he called Ticuna had come to populate the areas along the shore. Within the past forty-five years, he learned, a majority of the population had been recruited by an ex-Catholic priest to join what he described as a syncretic religious movement.[129] This movement was unlike that of the more established Catholic and Protestant missions, preaching "strength in unity" and encouraging previously dispersed groups to band together in large communities, from five hundred to fifteen hundred, along the river banks, "where they would largely retain their own culture, but be able to take advantage of what civilization had to offer them." The practical implications of this migration for Neel's team was that, instead of taking smaller boats up the river's tributaries in search of more remote groups—those forbidden by FUNAI—they were able to "in fact anchor the Helix opposite these new concentrations of Ticuna, work ashore by day and return with our samples to the refrigeration facilities of the Alpha Helix each night."[130] In other words, the Ticuna were a social creation, which nevertheless made it possible for Neel to freeze their blood as evidence of an isolated Amerindian population.[131] The successes of an earlier program of spiritual salvation had corralled the Ticuna as a population seen as "primitive" enough to be salvaged as a biomedical resource by scientists on ships.[132]

Neel asserted that on the surface, members of the group he called Ticuna looked "very little admixed. We won't know the precise degree of admixture until we have done all of our genetic typings, but by now our eye for this sort of thing is pretty good, and I am ready to put high odds on there being less than 5% non-Indian genes in this very large tribe."[133] Neel invoked privileged access to a molecular gaze; he could see with his expert eyes what the lab would, inevitably, reveal. In keeping with the "new physical anthropology" that underscored the desire to salvage blood, racial claims were no longer to be made on the basis of phenotypic observations like skin color; they would have to be translated into the seemingly neutral language of molecular biology.[134]

The team collected 2,736 unique whole blood samples, which Neel intended to subject to a range of analyses including blood grouping, serum protein grouping, and red blood enzyme tests.[135] In his concluding report, Neel remarked that medical care was provided in the field, which "was not only considered appropriate under the circumstances but undoubtedly helped establish the rapport without which many of the medical studies . . . [i.e., the blood-based ones] would have been impossible."[136] He went on to predict that to get through the genetic typing of all the samples would take at least two years, a deferral of analytic potential made possible by the freezer.

With the fieldwork successfully completed, Neel was surprised to find that his diplomatic work was not over. Through unnamed channels, Neel had heard grumblings that he had inadequately used the *Alpha Helix*; indeed he returned it ahead of schedule. In disembarking from the ship early, Neel had violated the terms of use—he had failed to push the instrument to its extreme. He defended himself in a letter to program officer Mary Johrde at the Office for Oceanographic Facilities and Support at NSF and, in the process, offered a striking account of the factors that led to his substantial accumulation of blood.

His letter constituted a primer on the nature of the work he had undertaken, inadvertently admitting that the *Alpha Helix* best suited his needs in its role not as a mobile lab, but as a floating freezer:

> The backbone of our work involves the collection of blood samples from these Indian populations, [whose] samples undergo a wide variety of very sophisticated tests once they reach our base laboratory here in Ann Arbor. . . . I estimated that our team could count itself quite fortunate to collect as many as 2,000 such samples, which would

constitute some kind of a record. *However, because there is nothing more frustrating than to find oneself knee-deep in Indians and out of vacutainers,* I put in an extra 800 tubes as a contingency measure.[137]

The ship's ability to function as a mobile molecular biological lab was less important to Neel than its ability to serve as a freezer and a base camp—a scientific staging area from which a range of technical, political, and social activities could be coordinated. Neel saw no conflict of interest in making his case for maximal use of this instrument, as he explained to the NSF that, in balancing the needs of diplomacy and of lab work, he "solved the 'excess personnel' problem, by sleeping some of my people either in cots in the laboratories at night, or in hammocks on the deck."[138] In the conclusion to his letter, Neel argued that it was the fortuitous coastal migration of the Ticuna that enabled his team to complete their work—to use up all of the vacutainers—before they had used up their allotted time on the *Alpha Helix*. Slower than a jet, but faster than proceeding on foot, Neel had found the speed provided by the *Alpha Helix* to be just right, even if he was finished with it ahead of the allocated time. Yet, deploying the very criticism Damon had applied to Neel, he argued that other less scrupulous researchers "are jet setters . . . Because our work involved the delicate area of collecting biomedical specimens from Amerindians, I was confronted in planning this with a whole set of diplomatic and logistical problems that most of the persons using the Alpha Helix will not have to face."[139]

Neel resented the ways he was made differently accountable for his research activities and methods. His participation in a federally funded international research program brings historical specificity to challenges of collaboration in biomedical research—from the level of the research encounter to that of geopolitical diplomacy. At each of these registers, efforts to negotiate uneven power relations in the conduct of biomedical science, expressed themselves in the form of refusal. And at each of these registers, refusal was contested, subverted, or overcome through tactics of negotiation and improvisation. The temporal and spatial displacements involved with collecting and freezing blood for the future confounded and, in fact, foiled efforts to bring the accountability of researchers into alignment with the supposed agency of those whose lives they put on ice.

Fathoming the *Alpha Helix*

In concluding, it is instructive to turn to an earlier explorer, Captain Cook, to whom Gajdusek compared himself upon boarding the *Alpha Helix*. In her aptly titled article "Fathoming the Primitive," historian of anthropology Pauline Turner Strong emphasized the practices of self-measurement recorded in Cook's travel diaries.[140] The blood that Damon, Gajdusek, and Neel salvaged from supposedly disappearing "primitive" populations was intended primarily to help make sense of how far cosmopolitan humans like themselves had come from a state of ideal adaptation to their environments. Yet despite scientists' and science administrators' claims about the need to bring the lab to the field for analysis of exotic materials (claims that nonetheless performed very real political work), the immediate experimental subjects were the scientists who constructed stories about the landscapes and civilizations to which they traveled.

This was as true of fieldworkers who traveled on the *Alpha Helix* as those who did not. At a 1977 Ciba Foundation conference on "Health and Disease in Tribal Societies," physiologist and explorer Philip Hugh-Jones, who had collected blood from members of the Xingu in Brazil a decade earlier opened the event by expressing his hope that "in this symposium . . . we may formulate ideas about working with primitive people, where to look in order to find them, how to protect them when we go there, what information we ought to collect, how to collect and record it and why to preserve these people (not only ethically, because that is obvious), but because there is much to be gained for ourselves!"[141] Compare that sentiment with a contemporaneous one from ethnographer Roger Keesing, who observed that there "may be a chance for the Kwaio to exploit selectively the opportunities opened by Western education and medicine, without sacrificing their ties to tradition and their hold on self-sufficiency. Humans, everywhere . . . will need to transform their institutions and their consciousness in radical new ways if our species and our planet are to survive."[142]

Determining how to evaluate these different kinds of narratives also leads one back to Cook. The question of how Cook was perceived by those he encountered became a contentious topic in the 1980s as anthropologists reckoned with the violence of the field's colonial past and the latent racism that continually threatened to undermine its authority.[143] At the root of the debate—between anthropologists Marshall Sahlins and Gananath Obeyesekere—was the problematic nature of historical texts; taking fieldworkers' accounts of indigenous perceptions at face value risked confusing the attitudes of the researchers with those of the researched.[144] Unquestionably, the historical record of the *Alpha Helix* reveals more about the ideas of those who used the ship than those it was used to study. Yet, in working different perspectives of different kinds of fieldworkers against each other, it is possible to glimpse evidence of indigenous people innovating their own strategies for resistance and survival in the face of encounters with biomedical forms of modernity.

While the *Alpha Helix* may not have lived up to expectations as a radically "new" kind of instrument for life science, as a floating freezer the ship provided an ideal form of support for projects cast as urgent.[145] In extending an emerging cold chain, it was an instrument for facilitating the accumulation of a relatively new kind of biomedical resource: the frozen indigenous blood sample. The ship did not so much yield molecular knowledge as it supported the desire of human biologists—who feared the disappearance of these most remote "primitives"—to accumulate blood with speed while rendering them latent, both cold and cryptic, available to be mined by future scientists for novel purposes then as yet unknown.

The boat's ability to support a range of cold storage devices helped to bridge the gap between the existing cold chain and those human communities that were most valuable to science precisely because they were conceived of as living beyond its reach. The "floating" nature of *Alpha Helix* allowed scientists to extend their technical networks without having to build permanent facilities or concern themselves with enduring social relations, whereby local subjects might be recognized or come to recognize themselves as people with stakes in the future potential of their bodies as biomedical resources. As scientists sailed away from the field at the end of their collecting trips, they believed themselves to be unmooring the life of others—extracting and freezing body parts from communities in ways that foreclosed engagement with nonbiological approaches to vitality, kinship, and kind.[146] Even as they eagerly anticipated new uses for cold blood, they were unprepared to confront the many forms of latent life that would reveal themselves as the twenty-first century approached.

FIVE

When Futures Arrive: Lives after Time

In 1982, geneticists Ken Weiss and Ranajit Chakraborty lamented that the biological salvage work conducted by their mentors during the 1960s and 1970s had not resulted in more insights. Despite the urgent exhortation of human biologists affiliated with the International Biological Program and the World Health Organization to collect as much as possible before it was too late, and despite the promises of all the new laboratory tests that had emerged, Weiss and Chakraborty concluded, "Much, however, was not learned. Time and population were too limited, and as a consequence, much will remain forever a secret of the jungle."[1] The machine for making the future—the freezer filled with blood—was beginning to look like a relict of the past, an experimental system that had exhausted its potential.[2]

In fact, it was neither that this salvaged blood had failed to reveal new knowledge nor even that there was not enough available to analyze. Indeed, much of the blood, including that which had been acquired with the help of the *Alpha Helix* (described in chapter 4), persisted in freezers, never having even been unpacked. From the vantage point of anthropological genetics, Weiss and Chakraborty's early 1980s lament suggested that the spectacular and unending insights the blood was meant to generate seemed to have failed to materialize.[3] It sounded like a concession that the previous generation of scientists had not succeeded in generating the unending cascade of findings that would have subsequently justified the time and energy invested in salvage activities. The known techniques of immunology and biochemistry had not revealed the plethora of evolutionary selective mechanisms many had claimed just a decade earlier would be what made blood valuable.[4] The ultimate "as yet unknown" remained elusive.

Yet, even at its peak in the 1970s, the broader enterprise of collecting, preserving and analyzing blood had been far from a failure. Carleton Gajdusek used tissues accumulated from human populations in the Pacific and subsequently frozen to discern the etiology of kuru. Baruch Blumberg similarly mobilized frozen serum from the same region to reveal the viral origins of hepatitis B. For these achievements, the two men shared the Nobel Prize in Physiology or Medicine in 1976. Even earlier, John Rodman Paul, Dorothy Horstmann, and Alfred Spring Evans had used frozen blood to elucidate the nature and distribution of poliomyelitis (chapter 2). James Neel had even learned, much to his surprise, that the mutation rate among the supposedly "pristine" Yanomami was an order of magnitude higher than his early estimates had predicted, suggesting that they were not the ideal baseline for assessing the genetic risks of the postnuclear age.[5]

Before the end of the millennium—in the 1990s—this cold blood would acquire dramatic new futures, which enabled it to be repurposed as an archive of colonial and postcolonial Cold War science. Beginning in the mid-1980s, blood collected from the 1950s through 1970s was thawed to assess the origins of "emerging" infectious diseases, most notably HIV/AIDS. In a story that has been much celebrated, frozen blood that had been accumulated in 1959 by human biologists who had been seeking to detect genetic variation in African populations was thawed to reveal the presence of HIV-1 in a sample that had lain dormant in cold storage for decades.[6] In the early 1990s, fifty thousand frozen samples maintained at the Yale WHO Reference Serum Bank were also donated to the Viral Epidemiology Section of the US National Cancer Institute (NCI). The emergence of HIV and new interest in the relationship between cancer and viruses "contributed to the rebirth of seroepidemiological techniques and of the use of archival materials in epidemiological practice and research."[7] Around the same time, James Neel divided his collection of Amerindian cells and sera between several laboratories, including the NCI, "since," as he explained to his long-time collaborator Francisco Salzano "there was no one here [in Michigan] to carry on the tradition."[8]

Then new molecular techniques—most notably the ability to amplify and analyze tiny bits of DNA—would enable genomic revelations to be conjured from frozen archives distributed around the world but coagulated in the United States, Europe, and Australia. Through genomics, this blood would also be restored as a resource for studies of human migration and evolutionary history and, by the twenty-first century, as a crucial raw material for the commercial market for personal genomics and medicine.[9] Once again, it would be salvage that animated efforts to ensure that human blood could remain a source of technical knowledge. This time, though, blood would be salvaged not only from warm bodies but also from laboratory freezers that had hummed steadily for decades.

Historian Ann Stoler has described the paper traces in colonial archives as the "connective tissue that continues to bind . . . the relations severed between people and people, and between people and things."[10] In anthropology's genomic laboratory, frozen blood—a literal connective tissue—was thawed to serve a new genomic regime of historical knowledge production focused on questions of identity and kinship as well as epidemiology. Working individually and collectively, Cold War human biologists and epidemiologists had brought into being a new form of biomedical infrastructure, which—in time—had also become an archive, wrought from blood and other tissues of humans living in North America, Europe, Africa, Australia, Latin America, and Asia.

Meanwhile, members of a number of indigenous communities—who had not disappeared as feared—were finding common cause through an increasingly influential pan-indigenous rights movement.[11] They expressed their views via transnational institutions like the United Nations, which provided a forum for the drafting of international protocol to curtail the exploitative potential of biodiversity prospecting.[12] Scientists' efforts to extract the bioactive compounds of plants as well as human bodies without attention to indigenous lived experience, knowledge, or beliefs undermined the expertise of local healers but also gave them a reason to unite under the attenuated figure of what anthropologist Beth Conklin called the "generic shaman."[13]

In the United States, the passage of the Native American Graves Protection and Repatriation Act created legal mechanisms to support the return of bones and cultural objects to federally recognized tribes and nations whose lands were within the country's borders.[14] Activist groups, tapping into then new communication technologies like email and the World Wide Web, shared information about the persistence of blood that had previously been prospected and persisted in freezers in the global North. In some instances this led to demands that such blood be returned to donors and their descendants. This included blood salvaged by James Neel from members of the Yanomami, much of which, in addition the materials donated to NCI, would also come to be maintained in Ken Weiss's own freezers.[15]

In this final chapter, I examine twenty-first century practices and controversies surrounding the reuse of blood that, because it had been kept cold, augmented the life spans of those from whom it was acquired as it persisted beyond that of those who did the acquiring. I focus on a particular kind of reuse, known to practitioners as "freezer anthropology," which can be situated within the larger network of biomedical research enabled by an infrastructure of frozen human biospecimens. This is an infrastructure that encompasses the historical and present-day activities of a range of repositories, including the Biomedical Research Institute, the organization that emerged from the earliest efforts to institutionalize practices of cryopreservation (described in chapter 1).

Preserved in the latent state, a given biological substance may not appear to have changed over time, but its very persistence through time renders it as mutable, open to new uses *and* critiques in the present.[16] This includes mutations in biology as well as in ideas about the boundaries between life and death, matter and spirit, subject and object, human and nonhuman, nature and culture, individual and collective.[17] At the turn of the millennium, the practical challenges involved with the maintenance and reuse of cold blood were transformed as much by ethical as by economic regimes and provoked new efforts to recalibrate social and technical realms. Scientists who maintained cold blood were confronted with an unanticipated dimension of life on ice: their authority to claim the role of stewards of the scientific future was loudly contested. Along with scientific revelations, they were often also shocked to learn that frozen blood was also laden with spiritual, moral, and political essences. I call this surprise about the unexpected latent social potential of the freezer itself the "shock of the cold."

Historian of technology David Edgerton, in his book *Shock of the Old*, has argued that rather than focusing on moments of innovation or novelty, historians—especially those interested in telling global histories—would do well to focus on "technologies in use."[18] This means looking at what various collectives of people do with existing technologies as those technologies circulate to new places but also as they persist over time. Innovation comes from the ways in which old technologies become newly embedded in social circumstances, often with results that would "shock" those who first developed such technologies.

Edgerton even highlighted an instance of cold storage, the refrigerated railcar, as an exemplar of his approach. In the early twentieth century, the ability to make cold storage mobile enabled new ways of distributing life and its products in space and time. It justified the breeding and killing of animals so that the market for meat could expand. In the process, the problem of lag between the production of life and the realization of value, what Marx called "latent" capital, was solved by the creation of a stock market—which traded cattle at a remove from the farm—to manage futures.[19] The language of the "biobank," used to describe freezers filled with blood, is an intensification of this process, extending the financialization of life beyond the market proper by making the laboratory freezer a kind of safe-deposit box for bonds and promises that may, one day, be able to be cashed in.[20]

The "shock of the cold" is an approach to thinking about preserved substances that expands the framework of analysis at moments when blood is removed from the freezer to include those who might not otherwise be identified as central to innovation or knowledge production. It creates space for bringing new individuals and groups into view as well as for examining the difficulties, stutters, and resistances they may experience in accommodating new social and biological realities. In human biology and biomedicine, I argue, the freezer has performed "in use," not as the time capsule scientists were promised—providing the ability to stop and start life at will (chapter 1)—but as a machine for revealing the fundamental mutability and instability of life itself. As I will show, cold blood has been thawed to be reused, to be returned, and to reveal forms of biological and social relation that exist between people—as well as the microbial communities within them—which may have been, depending upon one's position, waiting to be revealed or actively suppressed.[21] The present-day management of blood frozen decades ago demonstrates that cold storage machines need not only be used and understood as technologies for fixing life in time. They also serve as technologies for becoming, for producing new and different kinds of relationships and knowledge about how to live.

Freezer Anthropology

One of the first human biologists to attempt to recover DNA from blood preserved in the freezer was anthropologist D. Andrew Merriwether, who earned his PhD in the genetics of human migratory history at the University of Michigan in the mid-1990s. A portion of his dissertation research, on the peopling of the New World, was based on the reuse of frozen blood serum samples collected from South American populations by James Neel and Francisco Salzano during the IBP. Merriwether had experimented with defrosting these particular old samples—stored for decades at −20°C—then searching within them for genetic material.

In 1999, Merriwether published an article promoting this novel form of biohistorical knowledge production. "Freezer Anthropology: New Uses for Old Blood," published as part of a special journal issue dealing with molecular genetics and history, emphasized the concealed latent potential of specimens kept viable in the freezer and reanimated using genomic techniques.[22] He did this using a then novel tool called polymerase chain reaction, more commonly known as PCR, which was developed to amplify genetic material that had been extracted from a living system. PCR is itself a protean and recursive phenomenon, having begun as a concept, which was made into an experimental system, which later became a technique. The technique generated concepts like "freezer anthropology."[23]

As a technique, PCR could be put to work using research materials from "the heyday of large-scale anthropological fieldwork in the 1960s and 1970s" that had lain dormant in freezers for decades.[24] In the freezer anthropology research program, PCR provided a new conceptual justification for transforming scientists' collections of old blood into archives for doing history. The ability to bring these archives into use through efforts to analyze the DNA contained within preserved specimens would generate still more concepts, giving new futures to decades-old blood.[25]

In the 1990s, PCR was such a novel technique—enough so that Merriwether cited the original paper describing it in his article "Freezer Anthropology"—that it was necessary for him to make explicit how latent life plus new techniques would transform frozen bodily extracts into priceless biomedical treasures. Merriwether emphasized that these "early blood samples" were of "tremendous value . . . represent[ing] a specific snapshot in time of the genetic variation present in the populations being sampled."[26] His use of the word "early" indexed them as among the oldest such blood available to his contemporaries, accumulated during a time when the potential long-term uses for frozen blood were unclear. Even if twenty-first-century scientists returned to sample members of the same communities—which they have in some cases done—they might find different things in the blood, making each sample biologically and temporally unique and therefore invaluable.

In "Freezer Anthropology," Merriwether named four specific collections of frozen blood that he saw as being of particular value to his peers. The first was the Harvard Solomon Islands Project, led by Albert Damon (about six thousand specimens); a portion of James Neel, Francisco Salzano, and colleagues' Amazon Indian samples from between 1967 and 1977 (about ninety-eight hundred unique specimens); Moses Schanfield's collection of serum for worldwide isozyme studies, culled from collections made by other researchers working on six continents; and Baruch Blumberg's collection of tens of thousands of serum samples from populations similarly far flung, but especially from Asia and the Pacific.[27]

Both the Harvard Solomon Islands Project and Neel's Amazonia studies were conducted under the auspices of the Human Adaptability arm of the IBP, discussed in chapters 3 and 4. Schanfield, professor of forensic sciences and anthropology at George Washington University, was not a fieldworker but received subportions of samples (also known as aliquots) from those engaged in fieldwork, including IBP-related projects. At one point, he claimed that his freezers held bodily extracts from .01 percent of the population of Papua New Guinea.[28]

Baruch Blumberg, who worked at the NIH and then at the Fox Chase Cancer Center in Philadelphia, did substantial blood collection himself but also relied upon others' fieldwork. This included members of the Harvard Solomon Islands Project such as Albert Damon and Jonathan Friedlaender. Prior to the ability to efficiently analyze snippets of DNA, Blumberg like many of his peers looked for the antibody signatures of past infections in the blood sera of members of Pacific populations. He also looked for antibodies in nonhumans, including mosquitoes—the primary vector for *Plasmodium falciparum*, the species of malaria most deadly to humans.[29]

The contingent technical circumstances surrounding the collection of such samples were of great consequence to Merriwether's enterprise. He knew that during these field trips, whole blood samples would have either been centrifuged or allowed to settle over a period of a few hours. Three distinct layers would emerge. Red blood cells would fall to the bottom. White blood cells would form a thinner layer on top, known as the "buffy coat." Pale yellow plasma or serum (which is plasma minus clotting factors) would rest on top of both layers and would be siphoned off from the other components. Because this work was being done under makeshift field conditions, and often in haste, the serum would frequently become "contaminated" by residual cellular material—the very thing that made it newly valuable to Merriwether decades later; DNA is only found in white blood cells. Blood that was initially made latent or frozen to preserve immunological signatures was being thawed, decades later, to reveal concealed human DNA and still other "as yet unknown" traces.

After graduating with his PhD from the University of Michigan, Merriwether become an assistant professor there and began teaching others how to search within serum for fragments of DNA that might have been accidentally preserved. A sample that was tinted red—indicating it contained red and white blood cells—could be assumed to have a greater likelihood of containing DNA. The approach relied both on field notes made at the time of collection and on new observations about the hue of the sample. Discerning and documenting the color of serum became a prospecting technique that allowed a given sample to be resituated as a vein of DNA to be mined.[30]

Although the samples were originally collected and used "for some specific purpose," Merriwether also recognized the value of the paperwork or metadata that accompanied the initial collection of the blood itself, including "pedigrees, morphological data, language, geographical location, isozyme data, etc.," which he explained could be "compared to the molecular genetic data typically being collected today."[31] He similarly acknowledged that human biologists had continued to work with and collect genetic material from populations all over the globe but that it was "rare to find a project that collects such large sample sizes, both in numbers of populations and numbers of individuals per population, making these early (mostly plasma [and serum]) collections a unique and extremely useful resource."[32] The planned hindsight of a previous generation of scientists had succeeded in enabling salvaged materials to serve a future then as yet unknown; it was time to begin planning again.

Ethical Unknowns

Merriwether's dream of reanalyzing these old materials soon became something of a nightmare. Frozen blood, despite compelling evidence of its extraordinary latent epistemological potential, was also turning out to be an ethically problematic resource. "Freezer anthropology" privileged frozen blood as a source of old human DNA. This aligned it with a new cultural awareness about the potential financial value of genetic material to the emerging global biotech economy. Mid-twentieth-century collectors had anticipated that future generations of scientists would find new uses for cold blood, but they had not imagined that such uses would be financially lucrative.

In the United States, the Bayh-Dole Act of 1980 enabled universities and their employers to retain rights in patented inventions developed with federal money and to license or sell them to private businesses. Also in 1980, a landmark court case, *Diamond v. Chakraborty*, upheld the patenting of modified organisms.[33] New techniques of assisted reproduction, made possible by access to cold storage, stoked a medical market for other human body parts, including sperm and eggs.[34] With the ability to extract human DNA from old samples came accusations, first from activists and then from indigenous people themselves, that their body parts would become sources of monetary value for industry.[35] Accumulations of blood from bodies cast in the previous decades as "primitive" had become the very so-called raw materials from which new fortunes could and would be made, a strangely literal and recursive form of what Karl Marx called primitive accumulation.[36]

Merriwether's innovations also coincided with the Human Genome Diversity Project (hereafter referred to as the Diversity Project), a direct descendent—in aims and actors—of the IBP.[37] The Diversity Project, first proposed in the early 1990s and led by Luigi Luca Cavalli-Sforza, was intended to be a large-scale effort to sample and archive human genetic variation.[38] The Diversity Project was not meant to be a commercial endeavor. Nor was it even conceived of as explicitly biomedical—it was meant to address concerns that the Human Genome Project, because it relied upon the genetic material from only a few individuals, did not adequately represent the fullest range of human variation.

Similarly to the IBP, the Diversity Project was also positioned as urgent. Organizers highlighted the importance of accumulating genetic material from indigenous populations that were, yet again, perceived to be vanishing due to, as one reporter noted, "war, famine, disease or simply what Cole Porter called the 'urge to merge.'"[39] However, the justifications given for the focus on indigenous, as opposed to other kinds of communities, situated such groups, once again, both as baselines and natural genetic laboratories for the detection of inherited diseases (see chapter 3).

Those who supported the Diversity Project were self-described antiracist scientists.[40] The fierce accusations of racism, colonialism, and biopiracy directed at them came as a shock.[41] Members of a younger generation of biological anthropologists were among those within the discipline who spoke out against the proposed Diversity Project. The fact that population geneticists were leading the initiative made these anthropologists mistrustful of an effort that sought to understand human diversity but which failed to explicitly include a significant number of those whose careers had been devoted to the genetic bases of human evolution and variation.[42]

Cultural anthropologists expressed mistrust as well. They condemned the continued categorization of indigenous peoples as relics of the past who were purportedly closer to nature than members of industrialized societies. They viewed this as a form of essentialism that perpetuated rather than overturned a long history of scientific racism.[43] Entangled with these racial critiques were also economic ones regarding the potential for indigenous peoples to be subjected, once again, to an extractive form of biocolonialism.[44]

The Diversity Project also received criticism from beyond academia. The first charges were leveraged not by indigenous community members themselves but by an activist group called the Rural Advancement Foundation International (RAFI). RAFI, made up of Canadian and American agricultural activists, had been vocal on issues of intellectual property and biodiversity for a number of years.[45] RAFI leadership contacted indigenous rights organizations, alerting them to the Diversity Project. Drawing connections between plant bioprospecting and sampling of human genetic material, RAFI charged, in a June 1993 memorandum, that Diversity Project organizers were focused on preservation of potentially valuable genetic material at the expense of conservation of indigenous lifeways.[46] The history of RAFI has yet to be told, but for a relatively small collective of agricultural activists, the group achieved astounding international attention as a voice of opposition to perceived exploitation of indigenous groups.[47]

RAFI also attempted to establish a link between the Diversity Project and the commodification of indigenous DNA.[48] At root of that specific accusation was the recognition of new US National Institutes of Health policies that encouraged patenting of innovations resulting from federally funded biomedical research. Though individual participants in the Diversity Project opposed patenting, they had difficulty diffusing suspicion about their relation to commercial interests. A few months later, the Third World Network, an important source of information on the needs and rights of people in the global South, echoed RAFI's call for an immediate halt to the Diversity Project.

It was at the end of 1993 that the World Council of Indigenous Peoples condemned the initiative as a "Vampire Project," in which Diversity Project organizers were accused of extracting the lifeblood of indigenous peoples for their own benefit.[49] Subsequent Congresses of Indigenous Peoples reinforced these claims. In 1995, the Pan American Health Organization, which in 1968 had actively promoted such activities in an article titled "Biomedical Challenges Presented by the American Indian" (discussed in chapter 3), reversed its position, declaring that "[t]his type of research will have a negative impact on future health programmes and projects in indigenous communities, by undermining indigenous peoples' trust in the medical and health professions."[50] Furthermore, in 1997, the "Heart of the Peoples" declaration from the North American Indigenous Peoples' Summit on Biological Diversity and Biomedical Ethics called for a moratorium on "all activities related to human genetic diversity involving indigenous peoples . . . and . . . for a return of all genetic samples taken from indigenous peoples without their full prior informed consent."[51] Times had changed.

Diversity Project organizers working in North America responded by developing a set of ethical practices intended to ensure that people understood the goals of the project and had enrolled of their own volition. This culminated in the drafting of a "Model Ethical Protocol," which yielded the idea of group consent.[52] Group consent was meant to address the fact that not all communities placed the bioethical value of autonomy above that of collective decision making. From the perspective of these Diversity Project organizers, group consent was an innovative extension of the biomedical practice of obtaining informed consent to participate in research from individuals. However, the assumptions embedded in the well-intentioned ideas of group consent and research partnership sparked new concerns that directed criticisms back to the epistemological foundations of the project and its focus on indigenous people.[53]

The Model Ethical Protocol attempted to recast indigenous peoples as research partners. This intervention was received differently both across and within various US communities. Some African American biological anthropologists felt that any effort to enhance the diversity and participation in the Human Genome Project should have also included attention to the genomes of those with African ancestry. In response, Diversity Project organizers sought to expand the list of populations to be studied beyond those seen as rare and endangered to include also African Americans and Mexican Americans. Other African American biological anthropologists resisted the call for participation, invoking a dark legacy of experimentation, as in the notorious Tuskegee syphilis experiments. Hopi geneticist Frank Dukepoo felt that while some American Indians would willingly participate, many others were likely to perceive an invitation to cooperate in genetic research as an extension of several centuries of colonization and assimilation.[54]

Two forces united all these responses: first and foremost, an organized indigenous rights movement, which made it clear that such communities had not vanished, and, second, the widespread sense within and beyond indigenous communities that efforts to make the Diversity Project more inclusive offered too little too late. Under the Model Ethical Protocol, organizers would retain the authority to define the key questions guiding the research. Those questions might not be those of greatest interest to various participants, and, more insidiously, their answers might produce new kinds of racially tinged harms. Ultimately, these criticisms caused the Diversity Project to falter. Jenny Reardon, a geneticist turned historian and sociologist of science, has interpreted the Diversity Project's derailment as resulting from its organizers' inability to coproduce or choreograph a program of research with social protections appropriate for a form of technoscience that relied on access to indigenous human bodies.[55]

Critical attention to the political authority and rights of indigenous peoples and the shifting legal and commercial status of human biologics within and beyond academia tempered Merriwether's initial enthusiasm for freezer anthropology research. Defrosting cold human blood required similar kinds of choreography of social and technical orders, which Merriwether, his mentors, and colleagues were unprepared to execute. He was unable to publish much of his dissertation research due to pressure from certain indigenous rights groups and communities who had since become more suspicious of extractive science.[56] Though he maintained an interest in the peopling of the Americas, his efforts turned to the study of migration patterns through an important commensal species, the alpaca. I spoke with him on several occasions about his previous work on human population genetics, but he was, for all of the reasons outlined above, much more keen to emphasize the exciting avenues provided by his work with nonhuman populations.[57] The sociopolitical circumstances that had transpired outside the freezer meant that old blood would be defrosted into a very different world—a world in which some descendants of the communities that had provided the blood were shocked by what scientists saw as biomedically innovative forms of reuse.

Immortalization or Dehumanization?

Biological anthropologist Jonathan Friedlaender, who had collected a large number of blood samples first in conjunction with the IBP and then as part of his own long-term research in Melanesia (chapter 3), became an important early collaborator with Merriwether in the project of freezer anthropology. As a senior scientist, he was in a different position to reflect on his experiences. By the late 1980s, then a professor of anthropology at Temple University in Philadelphia, Friedlaender recalled that he had pushed his research materials to what he thought was their limit and was at a loss as to how to proceed. At the time, he shared Weiss and Chakraborty's lament about the limitations of blood for studies of population genetics; he even contemplated quitting academia to become a high-end tour guide in the Solomon Islands.[58] Instead, he accepted an invitation to serve in a temporary position as US National Science Foundation physical anthropology program officer, where he learned the Diversity Project was on the horizon.

A major feature that distinguished the proposed Diversity Project from the IBP was that it would focus on collecting blood in a form that would enable it to be transformed into cell lines, a practice Gajdusek had attempted on the *Alpha Helix* before his freezers failed (chapter 4). Cell lines had recently become the basis for a robust biomedical infrastructure that supported both basic research and a huge range of commercial developments.[59] The creation of cell lines was made possible through techniques of tissue culture, which enabled unique somatic cells to be transformed into "immortal" resources.[60] These cell lines, which needed to be stored on liquid nitrogen in order to maintain metabolic function, broadened the possibilities for studying the activity of the cell as well as gene-environment relations. They also promised an essentially infinite source of DNA to be harvested at any point in time.

Friedlaender had maintained favorable relations with a number of communities in the Pacific and would return in the 1990s to a community in the Solomon Islands he called the Nasioi to collect blood that could be transformed into cell lines for emerging genomic studies connected to the Diversity Project. Before the Diversity Project was derailed, blood to create 1,064 cell lines had been collected, including the cells of the Nasioi. It persists as part of a panel at the Fondation Jean Dausset Centre pour l'Étude du Polymorphisme Humain (CEPH) in Paris. The cell lines were gathered from various laboratories by the Diversity Project and CEPH and frozen on liquid nitrogen "to provide unlimited supplies of DNA for studies of sequence diversity and history of modern human populations."[61]

By the end of the first decade of the twenty-first century, DNA extracted from these samples had been circulated to over a hundred researchers, leading to more than two hundred scientific papers. Another significant collection of indigenous cell lines persisted in freezers maintained by Kenneth Kidd and Judith Kidd, human biological variation researchers at Yale University School of Medicine who were intimately involved with the Diversity Project. Friedlaender had also sent portions of the Nasioi blood to the Kidds in New Haven and to Cavalli-Sforza, who was then at Stanford.

It was around the time that Friedlaender had returned to the Nasioi that his colleague, an anthropologist named Carol Jenkins, made international headlines. Jenkins had worked for years with a community in the New Guinea highlands known as the Hagahai. In the early 1990s, in collaboration with a range of biomedical researchers at the NIH, Jenkins found that a member of this group, as well as others in Colombia and the Solomon Islands, possessed a viral gene associated with HIV called HTLV-1.[62] The analysis of the gene was conducted in partnership with several researchers, including Carleton Gajdusek and his younger colleague Ralph Garruto. HTLV-1 appeared to provide clues about the origin of HIV, but since HTLV had been found in the blood of those in the Pacific as well as in the Amazon, there was an additional possibility, as Neel explained of his decision to donate a portion of his samples to a special retrovirus unit at NCI, of "using viral epidemiology to reconstruct some rather obscure aspects of human prehistory."[63]

In the case of the Hagahai, it was actual viruses that existed within human cells—not the immunological traces of viruses—that had become potentially valuable. Under NIH director Bernadette Healy's direction, funded researchers were required to apply for patents on any discoveries made by government scientists that could conceivably put to commercial use, even if that use had not yet been defined. This mandate was reinforced by the Supreme Court of California, which ruled in 1990 that a person had no right to any share of the financial profits of anything developed using his or her "discarded" body parts.[64] In accordance with regulations recently put in place by the US federal government, Jenkins applied for a patent for the HTLV-1 gene. She reportedly told the Hagahai about the patent at the time she made the application and agreed to donate her own portion of any royalties to the community.[65]

Shortly after filing the patent application, RAFI sued Jenkins, along with Friedlaender, for patenting indigenous genes.[66] RAFI's efforts catapulted the Hagahai along with the scientists who worked with them to the center of nascent discontent over issues of biocolonialism and biopiracy.[67] Such concerns also involved the purported illegal sale by the low-temperature repository at Coriell in Camden, New Jersey, of cell lines from indigenous communities living in Brazilian territory known as Karatiana and Surui.[68]

The immediate furor over HTLV-1 led NIH to drop the patent.[69] Friedlaender, who had not, in fact, been engaged in patenting, defended himself over the newly influential NATIVE-L electronic bulletin board, with the aid of testimonials from health officials in the Solomon Islands and Papua New Guinea. Soon after, Friedlaender was invited to guest-edit an issue of *Cultural Survival Quarterly*, a journal founded by David Maybury-Lewis, the cultural anthropologist turned activist who had accompanied James Neel on some of his earliest trips to collect blood in the Amazon.[70] Maybury-Lewis's own initial field trip to the region had been at the behest of the Brazilian Indian Protection Service, which had commissioned a report on the status of such groups.[71] The *Cultural Survival Quarterly* issue, published in 1996, dealt explicitly with the subject of human gene patenting and featured commentary from indigenous activists, scientists affiliated with the Diversity Project, and a representative of RAFI.[72]

The commentaries made clear that concerns about patenting were inextricable from understandings of the negative, enduring impact of colonialism and capitalism on indigenous lifeways. Solomon Islander and indigenous activist Ruth Liloqula's critique centered on the ways that new genomic techniques of human resource extraction not only were oriented toward benefiting the biotech industry but also privileged certain forms of ancestry making over others. There was, in her view, a fundamental lack of alignment in the interests of those who wished to collect and use blood to answer questions that did not appear to serve those who had provided the substance. She declared: "Indigenous people know who they are, where they come from, what genetic traits are in their families, and have always known where they are going. The problem at the present time is having enough knowledge to cope with the newly introduced ways of life, sickness, and social problems, problems associated with new way of economic and political development, and from being exposed to the wider world."[73] Aroha Te Pareake Mead, a Maori activist, pushed the critique further to encompass the epistemological foundations of the enterprise, arguing that "Western science goes to great lengths to de-humanize the humanness or life-force of human genes; hence, terms such as 'specimens,' 'materials,' 'properties,' and 'collections' are adopted as a means to ignore the essence of life contained within. . . . [T]he same perspective is carried over to issues of replication, immortalization, transgenic engineering, and cloning."[74]

In the 1990s, scientists found human DNA *and* indigenous peoples' anger about past and future exploitation to be latent forms of biosocial life in frozen blood. Mead's argument about dehumanization, in particular, pointed to the spiritual dimensions of deanimation that early cryobiologists like the Catholic priest Basile Luyet had shied away from addressing as they pursued indefinite horizons for extending life through time (chapter 1). These vigorous frictions did not bring an immediate or universal end to the practice of using blood for purposes other than that for which it was initially accumulated. However, they did begin to reverse the flow of some of these valuable and newly precarious frozen assets, which began to move out of biomedicine's freezers and back to members of the communities from whom they had been extracted.

Millennial Revelations

The arrival of the millennium brought with it conflicting revelations about the potential of blood in biomedicine's freezers. This included scientists' beliefs about its untapped latent potential but also certain indigenous peoples' claims that the persistence of this blood amounted to an incomplete form of death.[75] In 2000, Oxford professor of biological anthropology Ryk Ward agreed to give back the 883 blood samples he had collected from the Nuu-Chah-Nulth, a First Nations community in

Canada. He had made the collection in the 1980s in conjunction with research on rheumatic diseases, which disproportionally affected the group. When members learned that Ward had also used the blood for research unrelated to the health questions but to questions of lineage, they demanded that it be returned.[76]

Ward died unexpectedly before the transfer could occur and it took until 2004 for the blood to be returned to researchers at the University of British Columbia, where a Nuu-Chah-Nulth research committee was created to regulate the use of the samples. Larry Baird, the head of the research committee said that the point of oversight was to ensure that "[w]e'll know who's coming into our communities and we'll have the ability to make sure they follow our protocol and rules." Commenting on the case, Laura Arbour, a professor of genetics at the University of British Columbia, concurred, explaining that outside researchers "have to understand that DNA is not just DNA; in many aboriginal communities, DNA has a hugely spiritual importance."[77]

The return of the Nuu-Chah-Nulth samples, perhaps the first instance of the international repatriation of indigenous biospecimens, was overshadowed by another episode. In 2000, journalist Patrick Tierney published a book that leveraged scathing accusations at scientists who had engaged in research in the Amazon during the IBP. *Darkness in El Dorado: How Scientists and Journalists Devastated the Amazon* accused James Neel and his collaborator, cultural anthropologist Napoleon Chagnon, of exacerbating the 1968 measles epidemic that took place during their studies so that they could study the natural history of the disease in a virgin-soil population.[78] Tierney also wrote about approximately three thousand blood samples accumulated by Neel, which persisted in Weiss and Merriwether's freezers as well as at the NCI. By 2001, as anthropologists and human geneticists began to investigate the charges about the measles epidemic, Yanomami leaders and allies, including certain anthropologists, began to demand that the blood be returned.[79]

Robert Borofsky, a cultural anthropologist who has written one of the most comprehensive accounts of the controversy that ensued, emphasized that the Yanomami did and do not necessarily speak with a collective voice but with many voices, many perspectives.[80] What became clear, though, was that Yanomami who did speak were, in general, concerned about different kinds of issues than the anthropologists and geneticists who sought to respond to Tierney's charges in *Darkness in El Dorado*. Borofsky's anthropologist colleagues, who had been authorized by the American Anthropological Association to create a task force, were primarily engaged with assessing the validity of Tierney's accusations. The Yanomami representatives appeared more troubled by the knowledge that the blood Neel had extracted decades earlier persisted in the frozen state. It is in their statements that references to "spirit" or "essences" reveal themselves to encompass much more than what scientists often heard as claims about religion.

Davi Kopenawa, a Yanomami shaman and philosopher who has since become an internationally recognized advocate for indigenous causes, was quoted as saying, "The white man didn't tell us . . . 'We're going to store your blood in the cold, and even if a long times goes by, even if you die, this blood is going to remain here'—he didn't tell us that! Nothing was said."[81] This was, in the first place, an accusation that members of the community had not been properly or fully informed about what the researchers intended to do with their blood and therefore they had obtained permission to extract but not to retain it.

Another member of the community, Toto Yanomami, asked for a reversal of the flow of old blood, arguing that "Yanomami don't . . . take blood to study and later keep [it] in the refrigerator. . . . The doctors have already examined this blood; they've already researched this blood. Doctors already took from this blood that which is good—for their children, for the future. . . . So we want to take all of this Yanomami blood that's left over."[82] This was a critique of the open-ended horizon of a biomedical research program that preserved human remains without specific ideas about who would benefit or how to distribute such potential benefits.

In 2001, Kopenawa traveled to the United States to participate in the annual meeting of the American Anthropological Association. There, he affirmed his distress about the persistence of cold blood, beginning his comments by stating: "I would like to speak again about the book [*Darkness in El Dorado*] and the blood which was taken from my kin and taken from there and today is stored in a refrigerator. I would like to know what they want to do with this blood and why do they keep it. I want them to give the blood back to me so that I can take it back to Brazil and spill it into the river to make the shaman's spirit joyful."[83]

Neel could not reply. He had died in February of 2000. The papers he left in the manuscript archive, including a folder titled "Insurance," which meticulously documented his intention to stem the epidemic, were mobilized by American historians to address a future Neel had anticipated but hoped would never arrive, one in which he would be accused of harming rather than helping the Yanomami.[84] This folder facilitated Neel's posthumous exoneration by professional anthropologists and human geneticists.[85] The persistence of the cold blood, however, made it difficult to dismiss the specific Yanomami claims that it did not belong in North

American freezers, especially in view of the human subjects review questionnaire Neel had completed at the behest of the NSF in the early 1970s, wherein no mention was made of explaining to subjects that their blood would be stored indefinitely at low temperature.[86] At the core of such assertions was the injunction that the blood had never been Neel's to keep; moreover, they suggested that those whose body parts fueled the search for the "as yet unknown" should be the ones who got to decide when enough was known.

In 2009, the prosecutor of the State of Roraima in Brazil formally requested the return of the blood, and by 2010 several US institutions had begun the complicated and protracted process of repatriation. The possibility that some of the blood might contain pathogens that had, since the time of their preservation, become known and classified as biohazardous, produced uniquely sticky red tape.[87] Calls for the return of the blood to the Yanomami deeply unsettled members of the scientific community. One biological anthropologist articulated his feelings about the repatriation in terms of loss. Invoking a longer tradition of salvage, he argued that "very important collections (bones, blood) have been saved by biological anthropologists. It can be argued that these collections now belong to mankind's patrimony, destroying them would be an absolute loss and would impoverish not only museums, but also future generations. Also, offering one's blood for epidemiological or genetic studies, not only helps fundamental and applied research, but can be seen as an affirmation of solidarity which connects *ego* to the rest of mankind."[88]

Such claims participated in a narrative of human relations that perpetuated assumptions that indigenous blood provided a baseline for the history of universal Western man without seeming to consider that indigenous forms of knowledge provide different means for constructing cosmologies about the order of nature and kinship.[89] Based on her own experiences in negotiating the repatriation of indigenous bones, biological anthropologist and historian of science Ann Kakaliouras has argued that what has been at stake in debates over repatriation is not that "one way of seeing a thing is more true than another, but that the things themselves are produced, maintained, conceived of, and operate in different worlds."[90] Freezing had contributed to transforming blood extracted from the fluid meaning and uses of one world into a hard and ossified source of knowledge in another. In circumstances where blood has actually been returned, it is something different still, having absorbed and expressing the values of both worlds. In late 2015, the Yanomami samples were returned not to Brazilian freezers but to the earth, under circumstances I will return to in the epilogue.

The Frozen Archive

There was another substantial set of blood not listed in Merriwether's "Freezer Anthropology" article, but with which he was soon to become familiar. In the late 1990s, the tens of thousands of blood samples that Carleton Gajdusek had accumulated over his career had become the responsibility of Ralph Garruto, the biological anthropologist and long-term member of Gajdusek's lab who had participated in the HTLV-1 research. At NIH, Garruto had worked closely with Gajdusek, performing much of the initial analysis of the cold blood the senior scientist had shipped back from the field. Garruto was invited to sail on the 1972 *Alpha Helix* voyage but stayed behind in Bethesda to receive blood as he finished his dissertation.

Garruto inherited Gajdusek's collection in 1996, when the elder scientist pled guilty to having sexually molested boys he had adopted during his time in the islands of the Pacific. Gajdusek had to abandon his lab to serve one year in prison. Upon his release, Gajdusek exiled himself to northern Europe, where he remained until his death in 2008. During this tumultuous time, Garruto undertook the responsibility for the vital legacies of Gajdusek's life's work, thousands of glass vials of blood that his mentor had either collected himself or acquired through personal intellectual networks over the course of his career and the accompanying data that imbued this blood with scientific provenance.[91] Securing this inheritance required incalculable hours of administrative, technical, and emotional effort.

In 2003, Garruto relocated from Bethesda to his hometown in upstate New York. He obtained a substantial grant from the NIH to take the frozen blood with him in his new role as professor of biomedical anthropology and neurosciences at the State University of New York (SUNY) at Binghamton. The recruitment of Garruto and his research materials was part of an effort to prepare students to pursue postgraduate degrees that could lead to careers in biomedicine and biotechnology, a new potential hub of industry in the region, which had become economically depressed since IBM's facilities closed at the end of the Cold War.[92]

Garruto designed a facility to house this old blood and support its reuse in an explicitly biomedical, genomic paradigm. He called this facility a "serum archive." The blood came to Binghamton accompanied by labels and numbers that allowed it to be linked to the "investigator's original handwritten field notes and bleeding lists," which Garruto also maintained in a tall filing cabinet and a volume of leather-bound maroon notebooks.[93] By the time he arrived at the Anthropology Department at SUNY Binghamton, Merriwether had also recently been hired. When Merriwether relocated from Michigan, he too brought freezers full of human blood serum from Latin America as well as Melanesia, including material collected by Friedlaender and by Neel. He rolled the freezers into the back of a truck where they could be plugged in and drove them to their new environs.

Binghamton was becoming a major locus for the conduct of freezer anthropology, cemented by Garruto's creation of a master's program in biomedical anthropology. Every student who passed through the program would be required to participate in the work of organizing the frozen blood as part of training as a "serum archivist." According to the program's website, "The work of archiving serum samples is a long-term multi-step process, carefully designed to preserve the integrity and value of the samples for future use by researchers at this and other institutions."[94] Serum archivists undertook meticulous and sometimes risky work to process the frozen samples—a good number of which had never been removed from the original packaging in which they arrived from the field—into resources for the genomic era.

Between 2009 and 2010, I spent more than a month at the Binghamton serum archive to learn about the forms of labor involved in running such an enterprise. Much like the work of technicians in John Rodman Paul's WHO Reference Serum Bank in the 1960s (chapter 2), the labor of technicians—most of whom were female—was uniquely embodied. Day after day technicians used pipettes to transfer serum from glass tubes, which sometimes crumbled in their hands, into plastic cryovials. They then scrutinized the pigment of each sample, carefully registering what they saw in a lab notebook, which would later be transferred into an electronic database. The risk of boredom and fatigue loomed large. Becoming distracted could mean more than losing track of one's progress; it could mean contaminating the samples or, worse, exposing oneself to the unknown and potentially infectious life forms that might be latent in the sample.

The repetitive nature of the work led some technicians to find ways of bringing their own forms of meaning to the enterprise. Over the course of my visits, I noticed a small wipe-off board that bore the evocative insight, "Blood will tell, but often it tells too much." No one claimed to know where the sign had come from, but everyone agreed that it communicated the need to regard this bodily substance with cautious reverence. For one technician, the sign was a reminder that kinship could be challenged through the blood; one could learn that her father was not who she had been told. Another invoked forensics, that blood could lead someone to be convicted of a crime, or, more hopefully, to be exonerated.

Upon graduation many serum archivists would go to work in the emerging genomic facilities popping up in public health departments, crime

16 "Blood will tell, but often it tells too much," one of several anonymous messages written on a wipe-off board in the Binghamton Serum Archive, circa 2010. Photo by author.

labs, museums, and industry. The knowledge and experience technicians gained working with the uniquely fragile materials in the serum archive was a particularly valuable form of training to support an increasingly vast biomedical infrastructure dependent upon access to high-quality biospecimens. Denizens of the serum archive were also aware that the skills they were acquiring, in addition to creating job prospects, had the power to resolve or to disrupt social relations.

In the winter of 2010, science journalist Rebecca Skloot's book *The Immortal Life of Henrietta Lacks*, a biography of the woman and the cell line produced from her cervical cancer tissue, had just been published. The publicity surrounding the book directed unprecedented levels of attention to practical and moral questions surrounding reuse of human blood and other tissues in biomedicine. News outlets interviewed the author on TV and radio. Before long Henrietta Lacks's ghost as well as Skloot's conscience was making appearances within the serum archive. As one technician mused that February, "The 50th anniversary of the HeLa cells recently passed and that's really interesting. Sometimes I think, 'did this person ever have any idea that they'd be outlived?' I think it's really weird. I think that when you die you die and we have some other part of us that lives on that's not our body, so it's hard for me to think of anyone caring about their body being in an archive, but I do think about it."

Cell lines aside, the sheer number and diversity of samples frozen in even one serum archive, such as that at Binghamton, makes it difficult to fathom how and whether personhood could or should be restored to each of these fragments, as Skloot did for Henrietta Lacks's cancer cells.[95] However, it has been argued that indigenous groups are united by a "pervasive belief that bodily substances continue to retain the essence of the individual, even after removal from the body."[96] Such arguments are augmented by claims that recognize the vulnerability of bodies to commodification when they are alienated from their persons.[97]

A review of Aboriginal Australian's beliefs demonstrates that blood performs important roles in a variety of rituals and practices, including functioning as a medium of exchange, a form of power, a substance for healing and regeneration.[98] Among some Northern Australian Aboriginal groups, the body is often equated with the environment. Bodily secretions, including blood, are regarded as being capable of metamorphosing into geographical sites.[99] In this view, any use of blood is also a form of action upon both spirit and land.[100] Yet, the heterogeneity of beliefs among Aboriginal peoples means that such attitudes cannot be taken for granted or assumed. Therefore, the authors of the review conclude, "researchers have a responsibility" to engage with community members to see how particular beliefs may impact specific studies involving the use and reuse of blood.[101]

Ethnographic studies suggest that similar caveats apply to the use of blood from North American indigenous communities.[102] A study conducted with fifty-three members of one Native American group found that indigenous opinions about the appropriate use of frozen blood samples are varied. Individuals who identified as more "traditional" were more likely to believe that specimens should be returned at the end of the study for which they were collected, that they were the property of the person who provided them, and that they should not be used after a participant's death.[103] A similar study conducted regarding the Alaska Area Specimen Bank, a collection of indigenous tissues managed through a collaboration between the Centers for Disease Control and Alaskan natives, affirmed the need for researchers to consult with communities to understand cultural beliefs, specifically relating to the handling of body parts at the time of death.[104]

Garruto was keenly aware of calls for the return of cold blood. In addition to the Nuu-Chah-Nulth and Yanomami repatriation claims, in 2003 members of an indigenous community in the American Southwest, known as Havasupai, who had earlier agreed to provide blood to researchers interested in questions about diabetes and hypertension, learned they were being used to ask other kinds of questions. These were questions about migratory history and mental illness that community leaders had no interest in having answered. Furthermore, they were concerned about the system of reward that allowed their blood to benefit people who asked those upsetting questions. They responded by demanding that the blood be returned. In 2010 members of the community reclaimed the blood from the freezer in the presence of the steward of the collection and a reporter from the *New York Times*.[105]

Garruto saw the material in his care as sacred in a different way, unable to be discarded or allowed to die, dormant until some moment at which its salvific, "as yet unknown" potential could be realized.[106] As such, he devoted significant time and energy to maintaining these old materials to ensure their transfer to the next generation of scientists.[107] Looking around his own freezer room, Garruto described his reluctance to relinquish the vital legacy of his mentor and his own career. If he had not maintained this blood, his comments implied, it would not even be available to be returned. He reflected: "This would all have been trash. I don't even know how to value it. How can we know the future scientific potential? It could wind up saving the world! I'd love to put a value on it, but how do you a put a value on it[?] . . . A sample has a history, but you don't know what its scientific potential is. . . . And it's not enough to just archive it, you have to be willing to put time in if you're truly invested in making it available."[108]

Indeed, Garruto invested an enormous amount of labor in taking on the stewardship of the serum archive. This was labor that was also invisible, meaning that it will never be compensated or recognized in academic publications. Garruto—by spending time and effort in rehabilitating frozen blood for use in genome science—took on a huge responsibility that required technical expertise and skill in navigating a bioethical regime that had not existed when the blood was first collected.[109] Garruto approached the challenges involved with securing the future of the legacy of his mentor with the same degree of care he brought to his own research.

By the time Garruto sought to salvage this blood from Gajdusek's freezers in the late 1990s, the practical enterprise of archiving old human serum samples had come to require adherence to the ethical protocols of institutional review boards (IRBs). Such IRBs were formalized long after the samples were initially gathered and, as such, needed to be applied retroactively. Garruto obtained approval from at least two IRBs—NIH and Binghamton.[110] While these IRBs granted him permission from these two institutions to archive old sera, they also constrained how they could be used. The NIH insisted, in keeping with the bioethical norm of confidentiality, that samples be deidentified before use. This frustrated Garruto. He explained that while anonymization would be acceptable in some circumstances, in others it would be akin to showing a historian a letter without telling her who it was from or when it was written. It would disavow the paper trails—many volumes of field notes that included genealogies and medical records—which imbued the bodily substance with special epistemological and historical value.[111]

Garruto also appreciated that IRB approval did not resolve larger questions about the moral or ethical dilemmas posed by finding new uses for latent life.[112] The very procedures designed to certify research as "ethical"—which include the use of consent forms and practices of anonymization—often merely formalized the alienation of bodily substance from persons. It is in this sense that IRBs have been critiqued as "regulatory filters" that function to separate bodily substance from donors, allowing it to be treated as a biomedical resource.[113]

Although it may be passed through increasingly refined systems of scientific analysis, cold blood preserves and accretes meanings and intentions that cannot be contained by the freezer or by science. Discourses of bioethics and property have shaped the terms upon which claims can be made about the appropriate use of parts that have outlasted the bodies from which they were extracted as well as those who extracted them.[114] Yet practices of informed consent or strategies like "DNA on loan"—in which communities agree to time-delimited uses of their genetic material—do not capture the full range of problems posed by efforts to maintain life on ice.[115] The kin of those who gave these samples often do not share scientists' ideas about when the future has arrived, or about when it makes sense to thaw old blood for new uses, or when the loan period has expired.

Unsettling Relations: "Flashback to the 1960s"

As calls for the decolonization of the freezer intensified in the beginning of the twenty-first century, so did efforts to demonstrate the biomedical reuse value of these old blood samples. One of the new biomedical futures scientists granted to the human blood samples from the Binghamton sera archive involved permitting some of them to be searched for the genomes of malaria parasites preserved within, with substantial financial support from the US National Institutes of Health.

At Binghamton, biological anthropologist Koji Lum led an NIH-funded project focused on human diversity and its imbrication with that of *Plasmodium falciparum*, the form of malaria most deadly to humans. As a graduate student, Lum had participated in some of the earliest research using mitochondrial DNA to trace human migration in the Pacific.[116] His doctoral students, who had trained with Garruto as "serum archivists" on their way to pursuing their PhDs, were actively looking within human blood samples from the archive to locate malaria DNA that *also* persisted within. Certain of these specimens were those collected by Albert Damon and Carleton Gajdusek during their respective 1972 voyages on the *Alpha Helix*.

Just as the messiness of separating plasma from blood samples had facilitated Merriwether's ability to find human DNA in old, frozen blood serum, it also enabled the recovery of nonhuman DNA and even entire microbes that had persisted in a latent state. In some cases, Lum and his colleagues found multiple different strains of malaria infection in a single human blood sample. The technical challenge facing Lum's lab was to first demonstrate that malaria DNA could be recovered from frozen specimens and then to find forms of the plasmodia in situ, against which it could be compared. This had led Lum, Garruto, and their students back to Vanauatu, known during the colonial period as New Hebrides, to collect mosquitoes and still more human blood.

This research program involved prospecting blood for several kinds of genetic material: that of human DNA and malaria DNA within archived cold blood *and* DNA of contemporary malaria found in twenty-first-century humans and mosquitoes. This was an enterprise that one graduate student informally dubbed "mosquito anthropology." The biomedical future animating the research was the hope of applying pharmacogenomics to malaria, a disease that is both biological and social in etiology.[117] With their proximity to the "serum archive," Lum and his students had unique access to blood samples containing DNA from strains of *P. falciparum* collected before the microorganism had evolved resistance to the inexpensive antimalarial drug chloroquine.[118] Lum believed that understanding the biology of drug resistance could have an important public health impact, most significantly in helping to manipulate the chemical structure of chloroquine to restore its effectiveness.[119]

This approach was described in a 2006 publication entitled "Flashback to the 1960s: Utility of Archived Sera to Explore the Origin and Evolution of *Plasmodium falciparum* Chloroquine Resistance in the Pacific."[120] The paper built on Merriwether's "Freezer Anthropology" article to demonstrate the concept of using PCR and other genomic techniques to find novel uses for old blood at a new register of anthropological investigation. This time, the focus was on not only humans but their relationships with other species. Whereas the blood sera collected in the 1950s through 1970s was understood to be valuable for its ability to reveal how malaria had impacted the human genome—through the case of sickle-cell anemia (chapter 3)—in the twenty-first century, this same blood, now decades old, was being used to reveal how humans had impacted the malaria genome.[121]

Accordingly, Garruto began to instruct his serum archivists to salvage any mosquitoes inadvertently frozen along with blood or accidentally squished into the boxes in which samples were shipped back from the field. What made such mosquitoes and the plasmodia in frozen old blood serum appear valuable was the sense that they were relatively uninfected by modernity, or at least by technoscientific efforts at disease control.[122] They were, in a quite literal sense, a new kind of indigenous subject that could be salvaged from the blood of the so-called primitive initially accumulated during the Cold War projects of the WHO and IBP. In the broader project of freezer anthropology, these premodern mosquitoes and malaria also became anthropological, understood as part of what makes up humans, biologically and ecologically.[123]

This was the kind of use—to overcome antibiotic resistance—to which Garruto was referring when he spoke of archived blood saving the world. His strong belief in this blood's potential to help solve public health problems activated a new sense of urgency in getting what was in his head into written records so that scientists of the future could make it useful. The ideal was to transfer his institutional knowledge and every document associated with each sample into a computerized database. His engagement in the work of archiving linked efforts to preserve the past with efforts to anticipate and care about the future, the moral dimension of planned hindsight. This archiving was also about the labor of social reproduction, of maintaining the conditions of the present so that they could endure beyond any specific individual's life span. This is a process in which donor-descendants of cold blood—who might aid in decolonizing the archive and reconstituting the circumstances under which science is done—are only beginning to be enrolled.[124] Yet even the technicians who worked so intimately with cold blood in the serum archive were only partially aware of the circumstances that had led to its creation.

One slow day in Lum's lab, a PhD student who had begun his training as a technician and had been involved in the search for malaria in archived human blood, asked me if I had heard about the 2009 Bosse Lindquist documentary *The Genius and the Boys*, which sensationalized the circumstances surrounding Gajdusek's arrest.[125] I knew that a number of eminent historians and medical anthropologists had been interviewed, but I had not seen the final product. The PhD student, who had obtained a copy, loaded it on one of the lab computers, and we watched the documentary together.

Featured in the film were interviews with an ailing and mentally unstable Gajdusek, as well as clips from the films he had made in the field decades earlier. Our shared discomfort over Gajdusek's attitude toward his criminal conviction was disrupted by one serum archivist's exclamation that she may have worked with some of the very same blood being showed in the old film footage. She recognized the name of the village and claimed to have recently processed those samples in the serum archive. The shock of connecting the cold blood, which she believed she had held in her hands, with the unfamiliar warm bodies she saw on the screen sparked a heated conversation about why and how the blood had initially been accumulated.

This was where my forms of historical practice met theirs and I experienced my own shock of the cold; the revelation that much of the research I had been doing to understand the history of efforts to freeze life before it was taken up by human biologists was continuing to unfold in different directions, including at Binghamton. As I shared what I had learned about early cryobiology and the uptake of cryopreservation by human biologists, I was stunned to learn that John G. Baust, the cryobiologist who had been involved with the Rockville-based BRI during its early years, had been the previous occupant of Garruto's Binghamton lab. Baust was nearing retirement but still kept an office just down the hall.

When I met with Baust soon after, he recounted his career, which he had begun as a physiologist and biochemist.[126] Gradually, he became interested in life at low temperature, completing his PhD at the Institute of Arctic Biology in Fairbanks, Alaska. He had traveled a great deal to the Soviet Union under the auspices of knowledge transfer in the realm of cryobiology. Baust spent time in Novosibirsk doing research on cold of all types. During the Cold War, he explained, the Communist Party invested major money into storing the bone marrow from politburo members in case of nuclear war. The Russians asked Baust about the work of important scientists in the United States. Baust became a valuable contact for the FBI and CIA on matters of bioterrorism, alleging that they wanted to know what would be involved in harboring microbes for biological weapons, "because freezers would be necessary to keep the stuff cold until it would be used."[127] By the twenty-first century, the largest cryobiology program in the world was in Ukraine, and Baust had succeeded in maintaining a collaborative agreement that had outlived the thawing of Cold War tensions.

In his Binghamton office, the walls of which were decorated with images of polar bears and icebergs, he recalled that he had also been active in the effort to change the bylaws of the Society for Cryobiology to exclude cryonics, the practice of freezing whole human bodies described by Robert Ettinger in his popular book *The Prospect of Immortality.* In retrospect, Baust called his actions "a defensive move, not wanting to be in the public limelight over something so out there. They were arrogant to think about immortality in that way. I think all those bodies will just become a protein source." His belief was that human flesh that was placed in the freezer after death, as it had been with cryonics, could never be reanimated. The only plausible use for the dead, it seemed to him, was to be consumed. It was a joke about cannibalism that belied the emergence of a biotech economy that depended upon the metabolism of life itself.[128] I left with hard copies of full runs of Basile Luyet's journal *Biodynamica* as well the more recent incarnation of *Cryobiology*, published as *Cell Preservation Technology*, for which Baust had served as editor.

Later that week, Baust introduced me to his son, also a cryobiologist, named John, with whom he had begun a for-profit cryopreservation business called Cell Preservation Services in nearby Owego, a small town about thirty minutes by car from Binghamton. Owego was built up around the 1860s but had been in a state of economic depression for decades. Yet, at the end of the first decade of the twenty-first century there were signs of new forms of life, cell preservation services among them. Front Street, one of the main drags that ran along the Susquehanna River, was filled with quaint storefronts, and Cell Preservation Services, Inc., was located right on the corner of Front and Court Streets, the major intersection in the town. The old brick building was freshly painted, and in the first-floor window the company's name was written in delicate script. The warmth of the office interior, with exposed brick and glass doors on every office, each with its own window, was disarming. Cell preservation technology had become part of the new lifeblood of the main artery of late industrial America, developing applications of low-temperature technology to improve knowledge about cancer and cardiovascular disease. The "as yet unknown" futures of life at low temperature revealed themselves to be increasingly multiple, cloaked in the appearance of the past but presenting new challenges in the present.

Rescaling Relations

To many twenty-first-century human biologists and biomedical scientists, frozen blood has become an invaluable, vital legacy of their careers and those of their mentors. Maintained in the latent state, it continues to serve as a source of concealed potential.[129] To members of various indigenous communities, the persistence of body parts from their ancestors has been experienced as a form of incomplete death, a mode of alienation that does not reflect their worldviews.[130] Beyond this rhetorical binary is even more variation and more possibility for alternative trajectories, including the fact that, in some cases, indigenous people have become human biologists, and in others, human biologists have begun to incorporate concerns about the subjectivities of those whose body parts had previously mattered primarily as research objects.[131]

In the anthropological lab, through the use of genomic, epidemiological, and cryobiological techniques and their attendant forms of labor, multiple species have also been called upon or identified *as part of* humans. This shift parallels feminist cultural anthropologists' practices of thinking of humans as multispecies beings, a move informed by indigenous philosophy.[132] Idealized and actual uses of the freezer have played an important role in reconfiguring the relationship between human and nonhuman, insides and outsides, and producing hybrids that appear not to fit in existing schemes.[133]

The practice of thawing archived human blood to explore new genomic expressions of human-environment relationships began at a moment when biomedicine, more broadly, began to call upon genomics to reconfigure what it meant to be a human organism. The 2007 *Nature* article that announced the Human Microbiome Project—a NIH-funded effort to characterize microbial communities found at several different sites on the body—claimed "dramatic implications for how we think of ourselves because it challenges the view of ourselves as atomistic individual organisms. This new concept may recast us as an amalgam of us and them . . . we are co-existent rather than independent beings."[134] The effort to use frozen blood to recover the evolutionary relationships among species living within the human body could be a means of fulfilling the unmet promise of holistic disease ecology articulated in the early and mid-twentieth century, even as it may also be a mode of recognizing that to be human is to contain a microbial cosmos within.

The emphasis on coexistence and interdependence has ironic resonances with forms of indigenous attitudes about the relationship between bodies and environments—ironic because the outcomes of the Human

Microbiome Project emphasize the benefits to the body of an individual person, "a new age of pharmacogenomics, personalized medicine, and personalized nutrition," not the soul, let alone society, or even the environment.[135] The Human Microbiome Project (HMP), despite its claims to interdependence, participates in recapitulating the reductive holism that characterized mid-twentieth-century disease ecology. During the IBP, the human was seen as a functional and optimally adapted agent in the environment. One of the most enduring concepts to emerge from that endeavor was that of the "biome."[136] In the twenty-first century, the human is still figured as such an agent but *also* as a biome, a distinctive milieu or environment for the support of latent molecular forms of life.[137]

"To study the human microbiome," its boosters have suggested, "a few specific islands (humans) could be characterized in depth." Not unlike the emphasis on the "primitive" as a special kind of sentinel body for making knowledge about the human during the Cold War, "an early outcome of the HMP will be the establishment of 'human observatories' to monitor the microbial ecology of humans in different settings."[138] Even as blood collected to examine genetic diversity in a previous era is rerouted from North American freezers back to members of some Yanomami communities, new groups of researchers are collecting new kinds of specimens—albeit with new provisions for collaboration—but still using them in ways that continue to reinscribe indigenous people as fecund portals to the past. For instance, in 2015, research on samples collected more recently from members of an "isolated Yanomami Amerindian village with no documented previous contact with Western people" celebrated the unprecedented levels of microbial diversity found in this Amazonian tribe.[139]

As biomedical researchers locate new frontiers and new futures inside human bodies, the blood that persists in the setting of the freezer has been assigned new kinds of value. Innovative techniques and epistemological frameworks generate promises that may or may not be realized but carry forth the structures of thought and practice from the past. In the meantime, frozen blood floats between worlds where it is differently and ambivalently situated between latent life and incomplete death. Nick Tipon, the vice-chairman of the Sacred Sites Committee of the Federated Indians of Graton Rancheria, an organization that represents people of Coast Miwok and Southern Pomo descent, has argued that many tribes struggle to reconcile the two. "If someone could come to us and say 'yes, if we destroy this ancestor of yours, maybe we'd find a cure to cancer,' would we still have the same feeling?" Answering his own question, Tipon continued, "Our traditional cultural feeling is you're buried, that's where you rest in peace, but all societies change. We talk about it. We wonder where the right answers are."[140] Through time, and understood in these different registers, a freezer does more than collect and preserve. It provides material conditions for informing and even transforming ideas about what it means to be human, which humans get to participate in innovation and exchange, and how in the warm present.

By following blood as it has been conceived of and collected as a research material, and then as it has been cared for and reused, I too have been continually surprised, and often shocked, by the liveliness of freezing. Freezing, as both a material practice and metaphorical perspective on the limits of life, is not something that machines do without the aid of humans. People build, use, and inhabit machines that mutate their relationship to space and time. These mutations, like the genetic ones scientists initially sought to preserve—and like technology itself—are neither good, nor bad, nor neutral. They take on value through the ways that people imagine and experience them. Freezing, as well as thawing, relies on many spatially and temporally distributed forms of visible and invisible labor. Recognizing a broader variety of forms of life as well as these forms of labor—technical, ethical, legal, emotional, and even that which is often regarded as spiritual—is a first step toward making biomedical knowledge that serves the widest array of lives.

Epilogue: Thawing Spirits

To be in transit is to be active presence in a world of relational movements and countermovements. To be in transit is to exist relationally, multiply.

JODI A. BYRD, *THE TRANSIT OF EMPIRE*

In September 2015, the US National Cancer Institute returned 474 frozen blood samples to members of the Yanomami, whose present-day territory is in Northern Brazil and Venezuela. Earlier that year, in April, 2,693 blood samples collected by James Neel and his colleagues—maintained by Ken Weiss at Penn State and Andrew Merriwether at SUNY Binghamton—were returned to members of the Yanomami living within the bounds of the Brazilian state.[1] According to reports, over one hundred Yanomami, including fifteen of the individuals whose blood had been collected in the 1960s, attended a ceremony to honor those who in the meantime had died. Officials from various agencies of the Brazilian government were present as traditional dances and songs were performed to mark the event.[2]

Davi Kopenawa, the Yanomami shaman, philosopher, and spokesperson for members of the community, made a statement to the press prior to the event in which he explained that the blood was to be thawed, buried in a hole in the ground, and capped. "That place will become sacred to the Yanomami people."[3] He has said, "Nobody imagined that the blood would be kept in their freezers. . . . We were all very sad when we realized that our blood and the blood of our deceased ancestors was being preserved. . . . Science is not a god who knows what is best for everybody. It is we Yanomami who know whether or not research is good for our people."[4] The return of this blood, which had been frozen for decades, was a bittersweet victory for Kopenawa and his supporters. For the scientists who had inherited and maintained the blood, none of whom had been involved in its initial extraction, it was experienced as a loss of part of the vital legacy of science.

The specific number of human specimens that continue to be maintained in American freezers is difficult to pin down. At the turn of the millennium, the NIH commissioned a study that estimated that there were more than 300 million frozen human tissue specimens being preserved for research purposes in the United States, accumulating at a rate of 20 million per year.[5] This means that in 2015, there might have been as many as 600 million preserved human tissue specimens in the nation's freezers. Likely there are many more.

In the United States, as frozen human blood samples extracted from a huge range of bodies have been able to be used again and again for new purposes, arguments have been made about the urgency of acquiring and preserving "high-quality specimens from as wide a swath of the country's population as possible."[6] In testimony before the US House of Representatives, one official argued that recent calls for a national repository were "triggered in part by the acknowledgement that the value of biospecimens and other scientific research collections is not always recognized." The same official also stated that it was crucial to develop best practices for maintaining them "since many such collections are priceless and irreplaceable."[7] Consolidating these stored tissues—both undervalued and invaluable—has allowed them to become legible as populations to be cared for and a resource to be exploited.[8]

Concerns about privacy and property have dominated discussions about the ethical, legal, and social issues of what has been referred to as "the stored tissue issue."[9] The US Food and Drug Administration has struggled to regulate commercial services that rely on access to such specimens to provide individuals with genomic information about ancestry and health risk.[10] Indigenous peoples, whose DNA often forms a basis for such tests, have expressed concern that questions about ancestry could jeopardize their land claims.[11] The Department of Justice has supported the reuse of tissues initially collected for anthropological and biomedical research for forensic purposes, which have been used to help identify victims as well as perpetrators of terror and to exonerate wrongfully convicted criminals as part of the Innocence Project.[12] These same forensic resources have also been used for the purpose of racial profiling and incarceration.[13] Meanwhile, samples of various infectious agents like anthrax maintained by the US military or the Centers for Disease Control have been destroyed lest they fall into the wrong hands while samples of influenza and small pox are stored in the name of biosecurity preparedness.[14] Freezing has enabled the collection of still other potential sources of data that outpace the ability to put them to use. While we wait for their futures to arrive, practical challenges presented by the need to continue caring for these biodynamic archives have displaced and mutated the very problems they were initially created to solve.

Experimental systems that rely on human subjects, or parts of their bodies, can produce remarkable and unexpected insights. Those systems can also be confounded or halted by unanticipated ethical dilemmas. These dilemmas may be experienced as a disruption or even an impediment to the production of technical knowledge. However, they do not exist outside of or beyond science. They are not ancillary to it and they should not be treated as such. In any effort to redirect life in time, frictions are inevitable and there is always latent potential that may surprise or disrupt the routines of research. These moments are opportunities to make different and perhaps even better kinds of knowledge.

The research for this book began with an interest in recovering the invisible history of a mode of scientific work that had structured contemporary biomedical research. As I investigated the forces that shaped its frozen-tissue-based infrastructure, I began to ask a related question of the extent to which my own historical knowledge-production practices depended upon forms of extraction, preservation, and parasitism similar to those that had led me to the frozen archive.[15] In French, the word *parasite* has an additional meaning not present in English: interruption. To be a parasite is not only to dine at a table prepared by another but also to interrupt the meal. Philosopher Michel Serres's acknowledgment of the interruption points to a space of disruption in an otherwise self-regulating system.[16] He identifies this interruption as the source of novelty. It is also a possibility for reconfiguring a system of relations that can directly address the politicized nature of knowledge making that relies on a past that is imagined as constituted of stable, self-evident or frozen objects.

One important insight for researchers of all kinds—be they historians or scientists—is that when humans become our subjects, they too must be given a place at the table.[17] Even if that place is refused or remains otherwise unoccupied, having set it holds open a space to consider what cannot, or in some instances, should not be known. Moreover, as indigenous studies scholar Kim TallBear has recently asked, what new kinds of things could scientists and historians of science learn if we valued indigenous forms of knowledge as much as we value efforts to make knowledge about their bodies?[18] Other indigenous thinkers, such as Linda Tuhiwai Smith, have articulated critiques of research—of all kinds, not just biomedical forms—that point to its imbrication with colonialism but which also suggest new ways forward. "The word itself, 'research,'" she has argued, "is probably one of the dirtiest words in the indigenous world's vocabulary." Smith expressed emotions shared by many indigenous people: "It angers us when practices linked to the last century [the twentieth century] and the centuries before that, are still employed to deny the validity of indigenous peoples' claim to existence, to land and territories, to the right of self-determination, to the survival of our languages and forms of cultural knowledge, to our natural resources and systems for living within our environments."[19] Smith's position, a starting point for work in "decolonizing methodologies," calls into question who stands to benefit from archives of any kind and whether knowledge derived from them can, on its own, ever adequately address historical harms that continue to unfold.[20]

An unquestioned value of social and colonial history has been the ability to recover forms of harm associated with biomedical research that have been suppressed and to use that knowledge to bring new meaning to contemporary studies of health inequalities. While these harms have sometimes been deliberate, they often are invisible to those who pursued that research with their own sense of virtue. Yet, as federally funded institutions have attempted to address concerns about the appropriate use of research materials of all kinds, there have been suggestions that historians' access to public health archives be subject to IRB approval. Much like biomedical researchers before them, some historians were shocked and concerned by what they interpreted as a threat to their intellectual freedom.

Such suggestions were part of broad proposed changes to the US Department of Health and Human Services' Common Rule regulating human subjects. The most significant change would have required that, in the future, scientists obtain what is known as "broad" or "generic" consent from donors for research on any biospecimen they use to conduct research, even ones that have been made anonymous. In other words, researchers could reuse materials without first going back to donors. The fact that any person who seeks medical care may at some point find themselves asked to give permission for their tissues to be used makes it imperative that *all* patients—indigenous and otherwise—come to appreciate the impossibility of being able to predict the "as yet unknown" and the ways that the uncertainties of the future undermine the protections afforded by informed consent.[21]

There is value for both scientists and historians in contributing to examining the colonial politics of the archive as well as determining how it gets made in the future by considering the shared aspects of their knowledge-production practices. Doing so can make legible how those practices both constrain and enable the truths they each seek, creating space for other forms of expertise in the research enterprise. By asking, "What are the problems for which preservation provides solutions?" scientists and historians alike might begin to seek a wider range of answers. They might even come up with a more tractable, actionable set of questions. This does not necessarily mean abandoning the archive, or the collection, but it does mean being attentive to new relationships and forms of obligation that emerge through time. It means being able to let go of old ideas and materials in order to create space for new ones.

While life forms and their traces can be frozen, time, place, and most importantly, relationships, cannot. This insight is also an invitation to seek and be receptive to other, less frigid and more fluid ways of being in the world, fathoming new kinds of relationships with machines and each other, and embracing finitude. It is possible and necessary to create approaches to historical and scientific knowledge that, even as they reach into the past and strain toward the future, remain warm and alive to the circumstances of the present.

Acknowledgments

In the winter of 2005, I sent my DNA to National Geographic's Genographic Project. Several months later, I got an email instructing me how to access my results. With my mother, father, and sister looking over my shoulder, I learned my haplogroup. Making sense of what this information meant and how it had been possible to create it left me with questions I had not previously realized I needed to answer. Indeed, scratching at them ultimately drew me to blood and the history of efforts to negotiate its preservation. So it seems appropriate to begin the effort to honor the many relationships that have made this book possible by thanking the biological kin who were there from the start: Laura, Theodore, and Tina Radin.

My time as a student in the Department of History and Sociology of Science at the University of Pennsylvania gave me the training to answer these questions. Conversations with each and every one of the members of that department in the years between 2005 and 2012 made me a better scholar and a better person. Particular thanks are due to Susan Lindee, a generous mentor and, now, dear friend. This book would not exist without her wisdom, love, and steadfast support. Nor would it exist without John Tresch, who first encouraged me to think about what I could see by looking at the freezer, and Adriana Petryna, who insisted I think about what practices of freezing made hard to see. Ruth Schwartz Cowan taught me how the refrigerator got its hum, and Mark Adams helped me learn how to stop worrying and learn to love science fiction. It saddens me that Henrika Kuklick won't be able to read this book, but her ghost keeps me company.

Warwick Anderson's scholarship and mentorship have also made this book possible and meaningful. The high standards to which he holds my work have made me a stronger scholar. Since we met in 2009, Emma Kowal has been a true partner in thought and in friendship. Jenny Reardon hailed me with her work on the Human Genome Diversity Project and enthusiastically encouraged me to investigate its historical antecedents. Ann Kakaliouras's reflexivity and intellectual generosity is inspiring. I cherish my collaborations with each of them, from which I have learned so much.

At a very difficult time in his life, anthropologist Jonathan Friedlaender agreed to share his life story with me. He and Françoise Friedlaender introduced me to Ken and Judy Kidd, who welcomed me into their human genetics lab long before I joined them at Yale. I am grateful to them and the many scholars who spoke with me about the contemporary practice of archiving frozen blood. Special thanks are due to Ralph Garruto, Michael Little, Andrew Merriwether, Koji Lum, and their students. I am also grateful for the generosity of archivists at the National Academy of Sciences, the American Philosophical Society, the Melanesian Archive at the University of California at San Diego, Scripps Institute for Oceanographic Research, Yale's Sterling Memorial Library, and Harvard's Peabody Museum. Paul Nisson, Executive Director of the Biomedical Research Institute, Inc. (BRI), allowed me access to unpublished historical materials regarding the organization's role in the history of cryobiology and biobanking.

At various stages in my career a number of more experienced scholars made significant gestures of support. They may not remember, but I do. Thank you, Janet Browne, Melinda Cooper, Angela Creager, Soraya de Chadarevian, Nathaniel Comfort, Sheila Jasanoff, Andy Lakoff, Hannah Landecker, Frédéric Keck, Nikolai Krementsov, Staffan Müller-Wille, Lynn Nyhart, Bronwyn Parry, Ricardo Ventura Santos, Bruno Strasser, Kaushik Sunder Rajan, Kim TallBear, and Cathy Waldby. I am grateful to Amy Hinterberger for the language of "vital legacies." Michelle Murphy, Stefan Helmreich, David Jones, Ned Blackhawk, Bill Summers, Susan Lindee, and Audra Wolfe each read versions of the manuscript in its entirety and provided invaluable advice. Dan Bouk, Chie Sakakibara, Jenny Bangham, and Naomi Rogers lent their expertise to specific chapters. They and anonymous reviewers for University of Chicago Press helped me to make it better. Any and all shortcomings are, of course, my own.

The process of writing this book was made so much more pleasant by being able to spend several stints at the Max Planck Institute for the History of Science. For that privilege I thank Lorraine Daston, David Sepkoski, and Veronika Lipphardt. The Philadelphia Area Center for the History of

Science (now the Consortium for the History of Science, Technology, and Medicine) sponsored my research and writing in 2010–11. This book was published with assistance from the Frederick W. Hilles Publication Fund of Yale. Early versions of arguments presented in chapters 1 and 3 were published as Joanna Radin, "Latent Life: Concepts and Practices of Human Tissue Preservation in the International Biological Program," *Social Studies of Science*, 43, no. 4 (2013): 483–508. Portions of chapter 2 were published as Joanna Radin, "Unfolding Epidemiological Stories: How the WHO Made Frozen Blood into a Flexible Resource for the Future," *Studies in History and Philosophy of Biological and Biomedical Sciences* 47 (2014): 62–73.

I have been fortunate to complete the manuscript as a member of the Section for the History of Medicine and Program for the History of Science and Medicine at Yale. There, I have benefited from the collegiality of Paola Bertucci, Henry Cowles, Dan Kevles, Chitra Ramalingam, Bill Rankin, Naomi Rogers, Bill Summers, and John Harley Warner. Melissa Grafe has been unfailingly resourceful in locating esoteric archival resources. Ramona Moore, Dismayra Martinez, and Barbara McKay's doors were always open. Laura Waycott, Jonny Bunning, Jenna Healey, Tess Lanzarotta, Ann Sarnak, and Angel Rodriguez all contributed to the invisible labor of producing this book. Ava Kofman edited a draft of the entire manuscript and, in the process, became a friend.

Many other teachers, family, friends, and colleagues have inspired me to keep going when I wanted to stop: the Abramses, Rene Almeling, the Altunas, Robbie Aronowitz, Babak Ashrafi, Emily Banach, Jenny Bangham, Susanne Bauer, Ed Browne, Etienne Benson, Joshua Berson, the Birminghams, Ann-Emanuelle Birn, Sarah Blacker, Anna Bonnell-Freiden, Dan Bouk, Jen Brown, Luis Campos, Peter Collopy, Meggie Crnic, Deanna Day, Rosanna Dent, Erica Dwyer, Eli Gerson, Lesley and Ronny Grant cared for me during research in Binghamton, Jeremy Greene, Leslie Harkema, William Helfand, Matthew Hersch, Boris Jardine, Andi Johnson, Jon Kahn, Amy Kapczinski, Sarah Kaplan, Martine Lappe, Rebecca Lemov, Bruce Lewenstein, Ada Link, Katie Lofton, Scott Mackie, Katie McAuliffe, Erika Milam, Ali Miller, Projit Mukharji, Isaac Nakhimovsky, Thomas Near, Tamar Novick, Jason Oakes, Hans Pols, João Rangel de Almeida, Chitra Ramalingam, Ruth Rand, Sarah Richardson, David and Jenny Rittberg, Jody Roberts, Cynthia Rudin, Carrie Sewell, Eric Sargis, Robin Scheffler, Perrin Selcer, David Sepkoski, Mark Stern, Hallam Stevens, Karen-Sue Taussig, Dora Vargha, Kaitlyn Stack-Whitney, Sandra Widmer, Steve Wiseman, Rebecca Woods, Jason Zuzga, and my sounding board: Henry Cowles, Helen Curry, Lukas Rieppel, Alistair Sponsel, and especially Laura Stark.

Karen Darling believed in the potential of this book and her team made it manifest.

To my chosen kin: Kristoffer Whitney keeps my glass half full, Noreen Khawaja is good for the soul, Nadia Berenstein shows me the light, Jessica Helfand is magical, and Ariel Schwartz is unconditional. Matthew Grant deserves everything. I love you all.

Notes

PREFACE

1. Hannah Landecker, *Culturing Life: How Cells Became Technologies* (Cambridge, MA: Harvard University Press, 2007).
2. Rebecca Skloot, *The Immortal Life of Henrietta Lacks* (New York: Random House, 2010).
3. Avery Gordon, *Ghostly Matters: Haunting and the Sociological Imagination*, New University of Minnesota Press ed. (Minneapolis: University of Minnesota Press, 2008).

INTRODUCTION

1. Jaqcues Pepin, *The Origin of AIDS* (New York: Cambridge University Press, 2011); Edward Hooper, *The River: A Journey to the Source of HIV and AIDS* (Boston: Little, Brown, 1999).
2. Zhu Tuofo et al., "An African HIV-1 Sequence from 1959 and Implications for the Origin of the Epidemic," *Nature* 391, no. 6667 (1998): 594–97.
3. This history locates the emergence of a phenomenon known as "biomedicalization" in the Cold War. Biomedicalization refers to new "meso-level" infrastructural changes supported by technoscientific innovations that transform the relationship between the clinic and the lab as well as the lived experience of care. Adele Clarke et al., *Biomedicalization: Technoscience, Health, and Illness in the U.S.* (Durham, NC: Duke University Press, 2010).
4. For complementary treatments of the biobank as a sociocultural phenomenon see Catherine Waldby and R. Mitchell, *Tissue Economies: Blood, Organs, and Cell Lines in Late Capitalism* (Durham, NC: Duke University Press, 2006); Bronwyn Parry, *Trading the Genome: Investigating the Commodification of Bio-*

Information (New York: Columbia University Press, 2004); Kara W. Swanson, *Banking on the Body: The Market in Blood, Milk, and Sperm in Modern America* (Cambridge, MA: Harvard University Press, 2014); Klaus Hoeyer, *Exchanging Human Bodily Materials: Rethinking Bodies and Markets* (Dordrecht: Springer, 2013); Bruno Strasser, "Laboratories, Museums, and the Comparative Perspective: Alan A. Boyden's Quest for Objectivity in Serological Taxonomy, 1924–1962," *Historical Studies in the Natural Sciences* 40, no. 2 (2010): 149–82.

5. This refrain is the subject of chapter 3.
6. Donna Jeanne Haraway, "Remodeling the Human Way of Life: Sherwood Washburn and the New Physical Anthropology," in *Bones, Bodies, Behavior: Essays on Biological Anthropology*, ed. George W. Stocking, 206–60. Madison: University of Wisconsin Press, 1988.
7. A 1954 article in *Scientific American* announced, "The red fluid of the human body is a rich mixtures of cells and molecules. New methods of fractionating and preserving these many constituents increase their usefulness to medicine." Douglas M. Surgenor, "Blood," *Scientific American* 190, no. 2 (1954): 54.
8. These ideas were first articulated in dialogue with Georges Charbonnier, *Conversations with Claude Lévi-Strauss,* trans. John Weightman and Doreen Weightman (London: Jonathan Cape, 1969). See also Claude Lévi-Strauss, *The Savage Mind* (Chicago: University of Chicago Press, 1968).
9. This metaphorical language of hot and cold also appears in media theorist Marshall McLuhan's work. Marshall McLuhan, *Understanding Media: The Extensions of Man* (New York: McGraw-Hill, 1964). Despite the variation in degrees offered by thermal language it nonetheless contributed to the production of new kinds of binaries during the Cold War.
10. Claude Lévi-Strauss, *Tristes Tropiques* (New York: Penguin, 1992). Throughout this book I use the term "primitive" when my actors do. I may put it in scare quotes or modify it with words like "supposedly," or "so-called" to make clear the ways in which this label did work for human biologists. Examining the forms of violence associated with this mode of classifying humans is a central aspect of the inquiry into the phenomenon of life on ice.
11. Edward Dudley and M. E. Novak, *The Wild Man Within: An Image in Western Thought from the Renaissance to Romanticism* (Pittsburgh: University of Pittsburgh Press, 1972); George W. Stocking, *Victorian Anthropology* (New York: Free Press, 1987); Shepard Krech, *The Ecological Indian: Myth and History* (New York: W. W. Norton, 2000); Sadiah Qureshi, *Peoples on Parade: Exhibitions, Empire, and Anthropology in Nineteenth Century Britain* (Chicago: University of Chicago Press, 2012).
12. Cold storage is similarly backgrounded in accounts of the vast industry focused on preserving gametes for implantation and organs for transplantation. Kara W. Swanson, *Banking on the Body: The Market in Blood, Milk, and Sperm in Modern America* (Cambridge, MA: Harvard University Press, 2014); Susan E. Lederer, *Flesh and Blood: Organ Transplantation and Blood Transfusion in Twentieth-Century America* (New York: Oxford University Press, 2008); Lori B. Andrews

and Dorothy Nelkin, *Body Bazaar: The Market for Human Tissue in the Biotechnology Age* (New York: Crown, 2001); Klaus Hoeyer, *Exchanging Human Bodily Materials: Rethinking Bodies and Markets* (Dordrecht: Springer, 2013); Lesley A. Sharp, *Bodies, Commodities, and Biotechnologies: Death, Mourning, and Scientific Desire in the Realm of Human Organ Transfer* (New York: Columbia University Press, 2007); Rene Almeling, *Sex Cells: The Medical Market for Sperm and Eggs* (Berkeley: University of California Press, 2011). Margaret M. Lock and Judith Farquhar, *Beyond the Body Proper: Reading the Anthropology of Material Life*, Body, Commodity, Text (Durham, NC: Duke University Press, 2007); Kieran Healy, *Last Best Gifts: Altruism and the Market for Human Blood and Organs* (Chicago: University of Chicago Press, 2007); Douglas P. Starr, *Blood: An Epic History of Medicine and Commerce* (New York: Alfred A. Knopf, 1998).

13. Susan Leigh Star, "The Ethnography of Infrastructure," *American Behavioral Scientist* 43 (1999): 379. A recent "infrastructural turn" in the history of science suggests this is changing: Angela N. H. Creager and Hannah Landecker, "Technical Matters: Method, Knowledge and Infrastructure in Twentieth-Century Life Science," *Nature Methods* 6, no. 10 (2009); William J. Rankin, "Infrastructure and the International Governance of Economic Development, 1950–1965," in *Internationalization of Infrastructures*, ed. Jean-François Auger, Jan Jaap Bouma, and Rolf Kunneke (Delft: Delft University of Technology 2009), 61–75; Andrew Lakoff and S. J. Collier, "Infrastructure and Event: The Political Technology of Preparedness," in *Political Matter: Technoscience, Democracy, and Public Life* ed. B. Braun, S. J. Whatmore, and I. Stengers, (Minneapolis: University of Minnesota Press, 2010), 243–66; Ashley Carse, "Nature as Infrastructure: Making and Managing the Panama Canal Watershed," *Social Studies of Science* 42, no. 4 (2012): 539–63; Michelle Murphy, "Chemical Infrastructures of the St. Clair River," in *Toxicants, Health and Regulation since 1945,* ed. Soraya Boudia and Nathalie Jas (London: Pickering and Chatto, 2013).
14. Hannah Landecker, "Living Differently in Biological Time: Plasticity, Temporality, and Cellular Biotechnologies," *Culture Machine* 7 (2005); Bronwyn Parry, "Technologies of Immortality: The Brain on Ice," *Studies in History and Philosophy of Biological and Biomedical Sciences* 35, no. 2 (2004): 391–413.
15. Hannah Landecker, *Culturing Life: How Cells Became Technologies* (Cambridge, MA: Harvard University Press, 2007), 227.
16. Crosbie Smith and M. Norton Wise, *Energy and Empire: A Biographical Study of Lord Kelvin* (Cambridge: Cambridge University Press, 1989).
17. B. J. Luyet, "Some Basic Considerations on the Preservation of Biological Materials at Low Temperature," in *Long-Term Preservation of Red Blood Cells*, ed. Mary T. Sproul (Washington, DC: National Academies, 1965), 3–17.
18. B. J. Luyet and M. P. Gehenio, *Life and Death at Low Temperatures* (Normandy, MO: Biodynamica, 1940); Stephane Tirard, *Histoire de la vie latente: Des animauz ressuscitants du XVIIIe siècle aux embryons congeles du XXe siècle* (Paris: Vuibert, 2010).

19. On Freud's theories of latency, see Frank J. Sulloway, *Freud, Biologist of the Mind: Beyond the Psychoanalytic Legend* (Cambridge, MA: Harvard University Press, 1992); Crosbie Smith, *The Science of Energy: A Cultural History of Energy Physics in Victorian Britain* (Chicago: University of Chicago Press, 1998). Karl Marx, *Capital: A Critique of Political Economy*, vol. 2, *The Process of Circulation of Capital*, ed. Friedrich Engels, trans. Ernest Untermann (Chicago: Charles H. Kerr, 1909); Edward Said, *Orientalism* (New York: Vintage, 1979). Theories of latency in infectious disease abound, though they remain ripe for historical analysis. Among those perspectives on infection most relevant for this study are those of mid-twentieth-century disease ecologists F. Macfarlane Burnet and D. O. White, *Natural History of Infectious Disease* (Cambridge: Cambridge University Press, 1972).
20. Stefan Helmreich, "What Was Life? Answers from Three Limit Biologies," *Critical Inquiry* 37, no. 4 (2011), cited in Emma Kowal and Joanna Radin, "Indigenous Biospecimens and the Cryopolitics of Frozen Life," *Journal of Sociology* 51 (2015): 63–80.
21. Here, I am advancing an argument, grounded in attention to practice, that undermines the idea of science and religion as "non-overlapping magisteria" as propounded in Stephen Jay Gould, *Rock of Ages: Science and Religion in the Fullness of Life* (New York: Ballatine, 1999).
22. Sheila Jasanoff and Sang-Hyun Kim, *Dreamscapes of Modernity: Sociotehnical Imaginaries and the Fabrication of Power* (Chicago: University of Chicago Press, 2015).
23. William Cronon, *Nature's Metropolis: Chicago and the Great West* (New York: W. W. Norton, 1991). This is also an example of the role of time in the coproduction of social and technical orders. Sheila Jasanoff, *States of Knowledge: The Co-production of Science and Social Order* (New York: Routledge, 2004).
24. It comprised a kind of imperial ruin, which makes it possible to reconstruct biological and social features of late colonialism. Ann Laura Stoler, ed. *Imperial Debris: On Ruins and Ruination* (Durham, NC: Duke University Press, 2013).
25. Ulrich Beck, *Risk Society: Towards a New Modernity* (London: Sage, 1992); Mary Douglas, *Risk and Blame: Essays in Cultural Theory* (London: Routledge, 1992); Joseph Masco, "Atomic Health, or How the Bomb Altered American Notions of Death," in *Against Health: How Health Became the New Morality*, ed. Jonathan M. Metzl and Anna Kirkland (New York: New York University Press, 2010); Adriana Petryna, "What Is a Horizon? Navigating Thresholds in Climage Change Uncertainty," *Modes of Uncertainty: Anthropological Cases* (Chicago: University of Chicago Press, 2015), 147–64.
26. Jacob Gruber, "Ethnographic Salvage and the Shaping of Anthropology," *American Anthropologist* 72, no. 6 (1970): 1289–99; Rebecca Lemov, *Database of Dreams: The Lost Quest to Catalog Humanity* (New Haven, CT: Yale University Press, 2015).
27. Vincanne Adams, Michelle Murphy, and Adele Clarke, "Anticipation: Technoscience, Life, Affect, Temporality," *Subjectivity* 28 (2009): 246–65.

28. Hans-Jörg Rheinberger, *Toward a History of Epistemic Things: Synthesizing Proteins in the Test Tube*, Writing Science (Stanford, CA: Stanford University Press, 1997).
29. Geoffrey C. Bowker, *Memory Practices in the Sciences* (Cambridge, MA: MIT Press, 2005), 18.
30. M. Susan Lindee, "Voices of the Dead: James Neel's Amerindian Studies," in *Lost Paradises and the Ethics of Research and Publication*, ed. Francisco M. Salzano and A. Magdalena Hurtado (New York: Oxford University Press, 2004), 27–48.
31. K. J. Collins and J. S. Weiner, *Human Adaptability: A History and Compendium of Research in the International Biological Programme* (London: Taylor and Francis, 1977), 3.
32. Donna Jeanne Haraway, "Remodeling the Human Way of Life: Sherwood Washburn and the New Physical Anthropology," in *Bones, Bodies, Behavior: Essays on Biological Anthropology*, ed. George W. Stocking (Madison: University of Wisconsin Press, 1988), 206–60.
33. See for instance Karen K. Steinberg et al., "Use of Stored Tissue Samples for Genetic Research in Epidemiologic Studies," in *Stored Tissue Samples: Ethical, Legal and Public Implications*, ed. Robert F. Weir (Iowa City: University of Iowa Press, 1998); David Brown, "Asking Old Human Tissue to Answer New Scientific Questions," *Washington Post*, 16 April 2012; Elisa Eiseman and Rand Corporation, *Case Studies of Existing Human Tissue Repositories: "Best Practices" for a Biospecimen Resource for the Genomic and Proteomic Era* (Santa Monica, CA: RAND, 2003); Yvonne G. De Souza and John S. Greenspan, "Biobanking Past, Present and Future: Responsibilities and Benefits." *AIDS* 27, no. 3 (2013): 303–12.
34. Reinhart Koselleck, *Futures Past: On the Semantics of Historical Time*, trans. Keith Tribe (New York: Columbia University Press, 2004); John Zammito, "Koselleck's Philosophy of Historical Time(s) and the Practice of History," *History and Theory* 43, no. 1 (2004): 124–35, quote on 127.
35. Zammito, "Koselleck's Philosophy of Historical Time(s) and the Practice of History," 127, emphasis added.
36. Emma Kowal, Joanna Radin, and Jenny Reardon, "Indigenous Body Parts, Mutating Temporalities, and the Half-Lives of Postcolonial Technoscience," *Social Studies of Science* 43, no. 4 (2013): 465–83.
37. Ian Hacking, *Historical Ontology* (Cambridge, MA: Harvard University Press, 2002).
38. Annemarie Mol, *The Body Multiple: Ontology in Medical Practice* (Durham, NC: Duke University Press, 2002).
39. For example, Caroline Walker Bynum, *Wonderful Blood: Theology and Practice in Late Medieval Northern Germany and Beyond* (Philadelphia: University of Pennsylvania Press, 2007); Pierre Camporesi, *Juice of Life: The Symbolic and Magic Significance of Blood* (New York: Continuum, 1995).
40. Jules David Law, *The Social Life of Fluids: Blood, Milk, and Water in the Victorian Novel* (Ithaca, NY: Cornell University Press, 2010).
41. Spencie Love, *One Blood: The Death and Resurrection of Charles R. Drew* (Chapel Hill: University of North Carolina Press, 1996); Keith Wailoo, *Drawing Blood:*

Technology and Disease Identity in Twentieth-Century America (Baltimore: Johns Hopkins University Press, 1997).

42. J. Kēhaulani Kauanui, *Hawaiian Blood: Colonialism and the Politics of Sovereignty and Indigeneity* (Durham, NC: Duke University Press, 2008); Circe Dawn Sturm, *Blood Politics: Race, Culture, and Identity in the Cherokee Nation of Oklahoma* (Berkeley: University of California Press, 2002).
43. Kim TallBear, *Native American DNA: Tribal Belonging and the False Promise of Genomic Science* (Minneapolis: University of Minnesota Press, 2013).
44. Alondra Nelson, *The Social Life of DNA: Race, Reparations, and Reconciliation after the Genome* (New York: Beacon, 2015); Catherine Anne Bliss, *Race Decoded: The Genomic Fight for Social Justice* (Palo Alto, CA: Stanford University Press, 2012); Troy Duster, "Molecular Reinscription of Race: Unanticipated Issues in Biotechnology and Forensic Science," *Patterns of Prejudice* 40, nos. 4–5 (2006): 427–41; Joan H. Fujimura et al., "Clines without Classes: How to Make Sense of Human Variation," *Sociological Theory* 32, no. 3 (2014): 208–27; Nadia Abu El-Haj, *The Genealogical Science: The Search for Jewish Origins and the Politics of Epistemology* (Chicago: University of Chicago Press, 2012).
45. Laura Stark and Nancy Campbell, "Stowaways in the History of Science: The Case of Simian Virus 40 and Research on Federal Prisoners at the U.S. National Institutes of Health, 1960," *Studies in History and Philosophy of Biological and Biomedical Sciences* 48 (2014): 218–30.
46. Warwick Anderson, *The Collectors of Lost Souls: Turning Kuru Scientists into Whitemen* (Baltimore: Johns Hopkins University Press, 2008).
47. Warwick Anderson, "Objectivity and Its Discontents," *Social Studies of Science* 43 (2013): 557–76.
48. Jennifer Couzin-Frankel, "The Legacy Plan," *Science* 329 (2010): 135–37.
49. Emma Kowal, "Orphan DNA: Indigenous Samples, Ethical Biovalue and Postcolonial Science," *Social Studies of Science* 43, no. 5 (2013): 577–97.
50. Kowal and Radin, "Indigenous Biospecimens and the Cryopolitics of Frozen Life."
51. Deborah Bolnick et al., "The Science and Business of Genetic Ancestry Testing," *Science* 318, no. 5849 (2007): 399–400.
52. Lance Liotta, Mauro Ferrari, and Emanuel Petricoin, "Critical Proteomics: Written in Blood," *Nature* 425, no. 6961 (2003): 905.
53. Nasser Zakariya, "Is History Still a Fraud?" *Historical Studies in the Natural Sciences* 43, no. 5 (2013): 631–41. For arguments for ultralong durée history see Jo Guldi and David Armitage, *The History Manifesto* (Cambridge: Cambridge University Press, 2014).
54. Hans Ulrich Gumbrecht, *After 1945: Latency as Origin of the Present* (Stanford, CA: Stanford University Press, 2013). A related approach, which considers how bodily objects become cultural resources, is provided by Marianne Sommer, *History Within: The Science, Culture, and Politics of Bones, Organisms, and Molecules* (Chicago: University of Chicago Press, 2016).

55. Mel Chen, *Animacies: Biopolitics, Racial Mattering, and Queer Affect* (Durham, NC: Duke University Press, 2012).
56. Kim TallBear, "Beyond the Life / Not Life Binary: A Feminist-Indigenous Reading of Cryopreservation, Interspecies Thinking and the New Materialisms," in *Cryopolitics: Frozen Life in a Melting World*, ed. Joanna Radin and Emma Kowal (Cambridge, MA: MIT University Press, forthcoming).
57. Nikolas S. Rose, *The Politics of Life Itself: Biomedicine, Power, and Subjectivity in the Twenty-First Century* (Princeton, NJ: Princeton University Press, 2007).
58. John Tresch, *The Romantic Machine: Utopian Science and Technology after Napoleon* (Chicago: University of Chicago Press, 2012); Robert Mitchell, *Experimental Life: Vitalism in Romantic Science and Literature* (Baltimore: Johns Hopkins University Press, 2013). In the framework of twenty-first century medicine, a similar point is made in explicitly thermal terms: Jeannette Pols and I. Moser, "Cold Technologies versus Warm Care? On Affective and Social Relations with and through Care Technologies," *European Journal of Disability Research* 3, no. 2 (2009): 159–78.

CHAPTER 1

1. Biomedical Research Institute website, www.afbr-bri.com/biomedical-research-institute/, accessed 1 August 2015.
2. Ibid.
3. Ibid.
4. Bronwyn Parry, *Trading the Genome: Investigating the Commodification of Bio-Information* (New York: Columbia University Press, 2004).
5. David Keilin, "The Leeuwenhoek Lecture: The Problem of Anabiosis or Latent Life: History and Current Concept," *Proceedings of the Royal Society of London, Series B, Biological Sciences* 150, no. 939 (1959): 150.
6. Ibid.
7. Abou Ali Farman Farmaian, "Secular Immortal" (PhD diss., New York University, 2012); Sophia Roosth, "Life, Not Itself: Inanimacy and the Limits of Biology," *Grey Room* 57 (2014): 56–81; Stefan Helmreich, *Sounding the Limits of Life: Essays in the Anthropology of Biology and Beyond* (Princeton, NJ: Princeton University Press, 2016); Stephen Cave, *Immortality: The Quest to Live Forever and How It Drives Civilization* (New York: Crown, 2012).
8. D. Fraser Harris, "Latent Life, or, Apparent Death," *Scientific Monthly* 14, no. 5 (1922): 429–40.
9. In addition to latent life and, his own term "cryptobiosis," Keilin explained that the state of life poised at the precipice of death had also been known variously as "anabiosis," "viable lifelessness," "suspended animation," and "viability."
10. For other examples see Luis Campos, *Radium and the Secret of Life* (Chicago: University of Chicago Press, 2015); Elizabeth Grosz, *The Nick of Time: Politics, Evolution, and the Untimely* (Sydney: Allen and Unwin, 2004).

11. Donna Jeanne Haraway, *Modest_Witness@Second_Millennium. FemaleMan©_Meets_OncoMouse*TM*: Feminism and Technoscience* (New York: Routledge, 1997); Bruno Latour, *Science in Action: How to Follow Scientists and Engineers through Society* (Milton Keynes, Philadelphia: Open University Press, 1987).
12. Caroline Walker Bynum, *Wonderful Blood: Theology and Practice in Late Medieval Northern Germany and Beyond* (Philadelphia: University of Pennsylvania Press, 2007).
13. This insight extends recent arguments on bounded rationality into the realm of practice. Henry M. Cowles , et al., "Introduction,," *Isis* 106, no. 3 (2015): 621–22.
14. A. S. Parkes, "Some Biological Effects of Low Temperatures," *Advanced Science* 58 (September 1958): 1–8.
15. Basile Luyet, "Some Basic Considerations on the Preservation of Biological Materials at Low Temperatures," in *Long-Term Preservation of Red Blood Cells*, ed. M. T. Sproul (Washington, D.C., National Academy of Sciences-National Research Council, 1964), 3–17, quote on 3. The search for absolute zero and its technoscientific consequences is discussed in Hasok Chang, *Inventing Temperature: Measurement and Scientific Progress* (Oxford; New York: Oxford University Press, 2004). For a more popular account see Tom Shachtman, *Absolute Zero and the Conquest of Cold* (Boston: Houghton Mifflin, 1999).
16. B. J. Luyet, "Human Encounters with Cold, from Early Primitive Reactions to Modern Experimental Modes of Approach," *Cryobiology* 51 (1964): 4–10, quote on 8.
17. Beatrice Pellegrini Saparelli, Thomas Antoniette, and Jacques Dubochet, *Basile Luyet: Un vie pour la science, 1897–1974* (Sion, Switzerland: Editions de Musees Cantonaux du Valais, Sion, 1997).
18. Luyet to Harrison, 7 March 1931, MS 263, box 17, folder 1252, series I, Ross Granville Harrison Papers, Manuscripts and Archives, Yale University; B. J. Luyet, "The Effects of Ultra-Violet, X-, and Cathode Rays on the Spores of Mucoraceae," *Radiology* 18 (1932): 1019–22.
19. Alexis Carrel and Charles Claude Guthrie, "Results of the Biterminal Transplantation of Veins," *American Journal of the Medical Sciences* 132 (1906), 415–22.
20. Alexis Carrel, "Latent Life of Arteries," *Journal of Experimental Medicine* 12 (1910): 460–86, quote on 460.
21. Ibid., 460. On the contemporaneous biological roots of Freud's ideas about sexual latency, see Frank J Sulloway, *Freud, Biologist of the Mind: Beyond the Psychoanalytic Legend* (Cambridge, MA: Harvard University Press, 1992).
22. Carrel, "Latent Life of Arteries," 460.
23. Ibid., 462.
24. Stephane Tirard, *Histoire de la vie latente: Des animauz ressuscitants du XVIIIe siècle aux embryons congeles du XXe Siècle* (Paris: Vuibert, 2010).
25. Keilin, "The Leeuwenhoek Lecture."
26. Claude Bernard, *An Introduction to the Study of Experimental Medicine* (1927; repr., New York: Dover, 1957).
27. Keilin, "The Leeuwenhoek Lecture," 166.

28. Frederic L. Holmes, *Claude Bernard and Animal Chemistry: The Emergence of a Scientist* (Cambridge, MA: Harvard University Press, 1985).
29. See also the work of Carrel's colleagues Rockefeller Peyton Rous and J. R. Turner, "The Preservation of Living Red Blood Cells in Vitro. I. Methods of Preservation," *Journal of Experimental Medicine* 23, no. 2 (1916): 219–37.
30. Philip J. Pauly, *Controlling Life: Jacques Loeb and the Engineering Ideal in Biology* (New York: Oxford University Press, 1987); Keith Rodney Benson, Jane Maienschein, and Ronald Rainger, *The Expansion of American Biology* (New Brunswick, NJ: Rutgers University Press, 1991); Hannah Landecker, *Culturing Life: How Cells Became Technologies* (Cambridge, MA: Harvard University Press, 2007).
31. Martin Heidegger, *The Question Concerning Technology, and Other Essays* (New York: Garland, 1977).
32. L. K. Lozina-Lozinskii, *Studies in Cryobiology: Adaptation and Resistance of Organisms and Cells to Low Temperature,* trans. P. Harry (New York: Wiley, 1974), 33–34.
33. Nikolai Krementsov, *Revolutionary Experiments: The Quest for Immortality in Bolshevik Science and Fiction* (New York: Oxford University Press, 2014).
34. Ibid.
35. Carrel, "Latent Life of Arteries," 485.
36. Jonathan Rees, *Refrigeration Nation: A History of Ice, Appliances, and Enterprise* (Baltimore: Johns Hopkins University Press, 2013).
37. Ibid., 141.
38. For example, Graeme Gooday, "Placing or Replacing the Laboratory in the History of Science?" *Isis* 99, no. 4 (2008): 783–95.
39. Roger Thevenot, *A History of Refrigeration throughout the World*, trans. J. C. Fidler (Paris: International Institute of Refrigeration, 1979), 68.
40. Eric Wilson, *The Spiritual History of Ice: Romanticism, Science, and the Imagination* (New York: Palgrave Macmillan, 2003).
41. Crosbie Smith, *The Science of Energy: A Cultural History of Energy Physics in Victorian Britain* (Chicago: University of Chicago Press, 1998); Bruce Clarke, *Energy Forms: Allegory and Science in the Era of Classical Thermodynamics* (Ann Arbor: University of Michigan Press, 2001).
42. Stephen G. Brush, *The Temperature of History: Phases of Science and Culture in the Nineteenth Century* (New York: B. Franklin, 1978); Anson Rabinbach, *The Human Motor: Energy, Fatigue, and the Origins of Modernity* (New York: Basic Books, 1990); Crosbie Smith and M. Norton Wise, *Energy and Empire: A Biographical Study of Lord Kelvin* (Cambridge: Cambridge University Press, 1989).
43. M. Norton Wise and Crosbie Smith, "Work and Waste: Political Economy and Natural Philosophy in Nineteenth Century Britain," *History of Science* 27 (1989): 263–301.
44. Hans-Liudger Dienel, "Carl Linde and His Relationship with Georges Claude: The Cooperation between Two Independent Inventors in Cryogenics and Its Side Effects," in *History of Artificial Cold, Scientific, Technological and Cultural Issues*, ed. Kostas Gavroglu (Dordrecht: Springer, 2014), 171–88.

45. Today Linde is a global leader in the production of commercial refrigerants, primarily in the form of liquid gases. Hans-Liudger Dienel, *Linde: History of a Technology Corporation, 1879–2004* (New York: Palgrave Macmillan, 2004).
46. Mikael Hård, *Machines Are Frozen Spirit: The Scientification of Refrigeration and Brewing in the 19th Century; A Weberian Interpretation* (Boulder, CO: Westview, 1994).
47. Heidegger, *The Question Concerning Technology*.
48. For instance, Max Weber. *From Max Weber: Essays in Sociology*, ed. H. H. Gerth and C. W. Mills (New York: Oxford University Press, 1948), 32.
49. Hård, *Machines Are Frozen Spirit*, 22.
50. Hans-Jörg Rheinberger, *An Epistemology of the Concrete: Twentieth-Century Histories of Life* (Durham, NC: Duke University Press, 2010).
51. R. G. Scurlock, "A Matter of Degrees: A Brief History of Cryogenics," *Cryogenics* 30 (June 1990): 483–500.
52. Keilin, "The Leeuwenhoek Lecture," 182.
53. Thevenot, *A History of Refrigeration*, 154.
54. One of the only historical discussions of dry ice and its appeal is Jenny Leigh Smith, "Empire of Ice Cream: How Life Became Sweeter in the Postwar Soviet Union," in *Food Chains: From Farmyard to Shopping Cart*, ed. Warren Belasco and Roger Horowitz (Philadelphia: University of Pennsylvania Press, 2008).
55. Thevenot, *A History of Refrigeration*.
56. Mark Kurlansky, *Birdseye: The Adventures of a Curious Man* (New York: Anchor, 2012); Gabriella Petrick, "The Arbiters of Taste: Producers, Consumers and the Industrialization of Taste in America, 1900–1960" (PhD diss., University of Delaware, 2007).
57. Edwin William Williams, *Frozen Foods: A Biography of an Industry* (1968; Boston: Cahners 1970).
58. Susanne Freidburg, *Fresh: A Perishable History* (Cambridge, MA: Belknap Press of Harvard University Press, 2009).
59. Oscar Edward Anderson, *Refrigeration in America: A History of a New Technology and Its Impact* (Princeton, NJ: published for the University of Cincinnati by Princeton University Press, 1953).
60. Important contributions to the growing literature on refrigeration and consumer markets for food include Freidburg, *Fresh: A Perishable History*; Gabriella Petrick, "'Like Ribbons of Green and Gold': Industrializing Lettuce and the Quest for Quality in the Salinas Valley, 1920–1965," *Agricultural History* 80, no. 3 (2006): 269–95; Shane Hamilton, "Cold Capitalism: The Political Ecology of Frozen Concentrated Orange Juice," *Agricultural History* 77, no. 4 (2003): 557–81; Warren James Belasco and Roger Horowitz, *Food Chains: From Farmyard to Shopping Cart* (Philadelphia: University of Pennsylvania Press, 2009).
61. An extension of the momentum of large-scale socio-technical systems is described by Thomas P. Hughes, *Networks of Power: Electrification in Western Society, 1880–1930* (Baltimore: Johns Hopkins University Press, 1983).

62. Nicola Twilley, "The Coldscape," *Cabinet* 47 (2012): 78.
63. Byrd to George Mason, 25 July 1939, box RG 56.1, folder 7150, Papers of Admiral Richard E. Byrd, Ohio State University Archives.
64. Alexis Carrel, "The Preservation of Tissues and Its Applications in Surgery," *Journal of the American Medical Association* 59 (1912), 525; Hannah Landecker, "Building a New Type of Body in Which to Grow a Cell: Tissue Culture at the Rockefeller Institute, 1910–1914," in *Creating a Tradition of Biomedical Research: Contributions to the History of the Rockefeller University,* ed. Darwin Stapelton (New York: Rockefeller University Press, 2004), 151–74.
65. R. Chambers and H. P. Hale, "The Formation of Ice in Protoplasm," *Proceedings of the Royal Society* 110 (1932): 336–52. The choice of the term "shock" seems propitious given that chilled blood first facilitated the treatment of soldiers suffering from shock in World War I. William Schneider, "Blood Transfusion between the Wars," *Journal of the History of Medicine and Allied Sciences* 58, no. 2 (2003): 187–224.
66. John G. Baust and John M. Baust, eds., *Advances in Biopreservation* (Boca Raton, FL: CRC, 2007), 41.
67. B. J. Luyet and M. C. Gibbs, "On the Mechanism of Congelation and of Death in the Rapid Freezing of Epidermal Plant Cells," *Biodynamica* 1 (1937): 1–18; Barry J. Fuller, Nick Lane, and Erica E Benson, eds., *Life in the Frozen State* (Boca Raton, FL: CRC, 2004).
68. W. E. Garner, "Gustav Tammann, 1861–1938," *Journal of the Chemical Society* (1952), 1961–73.
69. B. J. Luyet and E. Hodapp, "Revival of Frog Spermatazoa Vitrified in Liquid Air," *Proceedings of the Society for Experimental Biology* 39 (1938): 433–44.
70. B. J. Luyet and M. P. Gehenio, *Life and Death at Low Temperatures* (Normandy, MO: Biodynamica, 1940).
71. Douglas P. Starr, *Blood: An Epic History of Medicine and Commerce* (New York: Alfred A. Knopf, 1998).
72. Kara W. Swanson, *Banking on the Body: The Market in Blood, Milk, and Sperm in Modern America* (Cambridge, MA: Harvard University Press, 2014).
73. Susan E. Lederer, *Flesh and Blood: Organ Transplantation and Blood Transfusion in Twentieth-Century America* (New York: Oxford University Press, 2008).
74. Basile Luyet, "Working Hypothesis on the Nature of Life," *Biodynamica* 1, no. 1 (1934): 1.
75. Ibid, 2.
76. John Tresch, "Technological World-Pictures: Cosmic Things and Cosmograms," *Isis* 98, no. 1 (2007): 84–99.
77. This quotation comes from Basile Luyet as editor in the front matter of the compendium of the first bound edition of the journal *Biodynamica,* which includes reprints of studies conducted between 1934 and 1948. *Biodynamica* 1, no. 1 (1934).
78. Basile Luyet, "Working Hypothesis on the Nature of Life," *Biodynamica* 1, no. 1 (1934): 1

79. Luyet and Gehenio, *Life and Death at Low Temperatures*.
80. Saparelli, Antoniette, and Dubochet, *Basile Luyet*.
81. Thomas F. Gieryn, *Cultural Boundaries of Science: Credibility on the Line* (Chicago: University of Chicago Press, 1999).
82. Saparelli, Antoniette, and Dubochet, *Basile Luyet*.
83. Ellen D. Langill, *Sub-Zero at Fifty: A History of the Sub-Zero Company, Incorporated, 1945–1995* (Madison, WI: Sub-Zero Freezer, 1995); Thevenot, *A History of Refrigeration*. See also the efforts to use plasma to create the blood banks of the American Red Cross. Spencie Love, *One Blood: The Death and Resurrection of Charles R. Drew* (Chapel Hill: University of North Carolina Press, 1996).
84. The bovine preceded the human in a range of activities connected with biological preservation. Kara W. Swanson, "Human Milk as Technology and Technologies of Human Milk: Medical Imaginings in the Early Twentieth-Century United States," *Women's Studies Quarterly* 37, nos. 1–2 (2009): 20–37.
85. Kostas Gavroglu, *History of Artificial Cold, Scientific, Technological and Cultural Issues* (Dordrecht: Springer, 2014).
86. "Ominous Refrigeration Need," *Refrigeration Engineering* 56 (December 1948): 495.
87. Mary T. Sproul, ed., *Long-Term Preservation of Red Blood Cells* (Washington, DC: National Academy of Sciences–National Research Council, 1964), iii.
88. M. Susan Lindee, "The Repatriation of Atomic Bomb Victim Body Parts to Japan: Natural Objects and Diplomacy," *Osiris* 13 (1998): 388.
89. Swanson, *Banking on the Body*.
90. Douglas M. Surgenor, "Blood," *Scientific American* 190, no. 2 (1954): 54–64.
91. Angela N. Creager, "Producing Molecular Therapeutics from Human Blood: Edwin Cohn's Wartime Enterprise," in *Molecularizing Biology and Medicine: New Practices and Alliances, 1910s–1970s*, ed. Soraya de Chadarevian and Harmke Kamminga (Amsterdam: OPA, 1998), 99–128.
92. Paul J. Schmidt, "Basile J. Luyet and the Beginnings of Transfusion Cryobiology," *Transfusion Medicine Reviews* 20, no. 3 (2006): 243–44.
93. C. Polge, A. Smith, and A. Parkes, "Revival of Spermatozoa after Vitrification and Dehydration at Low Temperatures," *Nature* 164 (1949): 666. Smith confirmed the finding, publishing the results in A. Smith, "Prevention of Hemolysis during Freezing and Thawing of Red Blood Cells," *Lancet* 2 (1950): 910–11. This "seminal" event is described both in Bronwyn Parry, "Technologies of Immortality: The Brain on Ice," *Studies in History and Philosophy of Biological and Biomedical Sciences* 35, no. 2 (2004): 391–413; and Landecker, *Culturing Life*.
94. Fuller, Lane, and Benson, *Life in the Frozen State*.
95. Jean Rostand, "Glycerine et resistance du sperme aux basses temperatures," *Comptes Rendus de l'Académie des Sciences* 222 (1946): 1524–25.
96. Audrey Ursula Smith, *Biological Effects of Freezing and Supercooling* (Baltimore: Williams and Wilkins, 1961).
97. V. Richards et al., "Initial Clinical Experiences with Liquid Nitrogen Preserved Blood, Employing PVP as a Protective Additive," *American Journal of*

Surgery 108, no. 2 (1964): 313–22. For a summary of contemporaneous studies of research on freezing and thawing blood, see Arthur R. Turner, *Frozen Blood: A Review of the Literature, 1949–1968* (New York: Gordon and Breach, 1970).

98. James Lovelock, "The Mechanism of Protective Action of Glycerol against Haemolysis by Freezing and Thawing," *Biochemica et Biophysica Acta II* 11 (1953): 28–36.
99. C. Polge and James Lovelock, "The Preservation of Bull Semen at –79c," *Veterinary Record* 64 (1952): 396.
100. Explained in C. Polge, "Sir Alan Sterling Parkes, 10 September 1900–17 July 1990: Elected FRS 1933," *Biographical Memoirs of Fellows of the Royal Society* 52 (2006): 263–83.
101. James Lovelock, "The Physical Instability of Human Red Blood Cells," *Biochemical Journal* 60 (1955): 692–96.
102. Harold Meryman in the foreword to Fuller, Lane, and Benson, *Life in the Frozen State*, ii.
103. D. Michael Strong, "The US Navy Tissue Bank: 50 Years on the Cutting Edge," *Cell and Tissue Banking* 1, no. 1 (2004): 9–16.
104. Walter W. Holland, "Karel Raska—the Development of Modern Epidemiology. The Role of the IEA," *Central European Journal of Public Health* 18, no. 1 (2010): 57–60.
105. D. Keilin and Y. L. Wang, "Stability of Haemoglobin and of Certain Endoerythrocytic Enzymes in Vitro,"*Biochemical Journal* 41 (1947): 491–500; Keilin, "The Leeuwenhoek Lecture," 182.
106. Keilin, "The Leeuwenhoek Lecture," 186.
107. Ibid., 187.
108. Nicolas Rasmussen, *Picture Control: The Electron Microscope and the Transformation of Biology in America, 1940–1960* (Stanford, CA: Stanford University Press, 1997).
109. Harold T. Meryman, "Basile J. Luyet: In Memoriam," *Cryobiology* 12, no. 4 (1975): 285–92.
110. Harold T. Meryman, "Mechanics of Freezing in Living Cells and Tissues," *Science* 124, no. 3321 (1956): 515–21, quote on 515.
111. Ibid., 521.
112. R. G. Scurlock, *History and Origins of Cryogenics* (Oxford: Clarendon Press; Oxford University Press, 1992), 497–98. See also Gavroglu, *History of Artificial Cold*.
113. All cited in Turner, *Frozen Blood: A Review of the Literature*.
114. James Lovelock and M. W. H. Bishop, "Prevention of Freezing Damage to Living Cells by Dimethyl Sulfoxide," *Nature* 183 (1959): 1394.
115. Turner, *Frozen Blood: A Review of the Literature*.
116. James Lovelock, *Homage to Gaia: The Life of an Independent Scientist* (Oxford: Oxford University Press, 2000), 109. Gaia was formulated in conjunction with research Lovelock was doing for NASA on methods for detecting life on Mars. James Lovelock, "A Physical Basis for Life Detection Experiments,"

Nature 207, no. 7 (1965): 568–70. The hypothesis was first popularized in Lovelock, *Gaia: A New Look at Life on Earth* (New York: Oxford University Press, 1979).

117. Bruno Latour has recently considered Lovelock's Gaia hypothesis as a form of natural religion, in ways that resonate with Luyet's vexed efforts to examine the relationship between time, temperature, and milieu as a Catholic scientist. Bruno Latour, *Face à Gaia: Huit conférences sur le nouveau régime climatique* (Paris: La Découverte, 2015).
118. Hannah Landecker, "On Beginning and Ending with Apoptosis: Cell Death and Biomedicine," in *Remaking Life and Death: Toward an Anthropology of the Biosciences,* ed. Sarah Franklin and Margaret M. Lock (Santa Fe, NM: School of American Research Press, 2003), 23–59.
119. "Embryo Is Frozen to Make Life," *Argus,* 14 February 1953; Thevenot, *A History of Refrigeration,* 295.
120. Luyet himself was dismissive of such "lurid" possibilities. "Deep-Freeze," *Time,* 28 April 1952, 70.
121. Cited in Lucy Kavaler, *Cold against Disease* (New York: John Day, 1971), 33.
122. Angus McLaren, *Reproduction by Design: Sex, Robots, Trees and Test-Tube Babies in Interwar Britain* (Chicago: University of Chicago Press, 2012).
123. See, for example, David Plotz, *The Genius Factory: The Curious History of the Nobel Prize Sperm Bank* (New York: Random House, 2005).
124. Hannah Arendt, *The Human Condition,* 2nd ed. (Chicago: University of Chicago Press, 1998), 2–3. On Arendt and Cold War science, see Waseem Yaqoob, "The Archimedean Point: Science and Technology in the Thought of Hannah Arendt," *Journal of European Studies* 44, no. 3 (2014): 199–224.
125. A. S. Parkes, *Off-Beat Biologist* (Cambridge: Galton Foundation, 1985): 344.
126. Hudson Hoagland and Gregory Pincus, "Revival of Mammalian Sperm after Immersion in Liquid Nitrogen," *Journal of General Physiology* 25, no. 3 (1942): 337–44.
127. Parkes, *Off-Beat Biologist,* 141.
128. Three calves, "Prima, Secundus, and Tertius," were created in this way. American Foundation for Biological Research (AFBR) Annual Report for 1975. Permission to access and quote historical materials associated with the Biomedical Research Institute, Inc. (BRI) and its antecedent organizations, including AFBR, were provided by the current BRI executive director, Paul Nisson. These materials, which are unprocessed, are maintained at BRI's Bethesda, MD, headquarters. Hereafter, this collection will be referred to as BRI.
129. ABS (American Breeders Service) Global website, www.absglobal.com/abs-history2, accessed 18 August 2011. See Adele Clarke, *Disciplining Reproduction: Modernity, American Life Sciences, and "the Problems of Sex"* (Berkeley: University of California Press, 1998); Clarke, "Research Materials and Reproductive Science in the United States, 1910–1940," in *Physiology in the American Context, 1850–1940,* ed. Gerald Geison (Bethesda, MD: American Physiological Society, 1987), 323–50.

130. Barbara Kimmelman, "The American Breeders' Association: Genetics and Eugenics in an Agricultural Context. 1903–13," *Social Studies of Science* 13, no. 2 (1983): 163–204.
131. "Refrigeration Makes Artificial Insemination Possible," *Refrigerating Engineering* 58 (May 1950).
132. ABS Global website, http://www.absglobal.com/abs-history, accessed 18 August 2011.
133. Parkes, *Off-Beat Biologist*, 150.
134. Ibid., 151.
135. Saparelli, Antoniette, and Dubochet, *Basile Luyet*, 31.
136. Meryman, "Basile J. Luyet: In Memoriam." Using freeze-dried sperm cells, reconstituted later with water, Meryman successfully impregnated a cow. He named the calf "Dessica," because she was born of sperm that had previously been desiccated. T. Rees Shapiro, "His Blood-Freezing Method Continues to Save Lives," *Washington Post*, 21 January 2010.
137. Meryman, "Basile J. Luyet: In Memoriam," 289.
138. James Leef and H. Meryman, "The American Foundation for Biological Research (AFBR): A Brief History," 22 February 2002. Internal document of the AFBR, BRI.
139. Ruth Schwartz Cowan, "How the Refrigerator Got Its Hum," in *The Social Shaping of Technology: How the Refrigerator Got Its Hum*, ed. Donald A. MacKenzie and Judy Wajcman (Philadelphia: Open University Press, 1985), 202–18.
140. ABS Global website, www.absglobal.com/abs-history2, accessed 18 August 2011.
141. Following World War II, liquid nitrogen use surpassed that of oxygen at a spectacular scale, growing from 507 million cubic feet in 1947 to 16 billion cubic feet in 1960. Scurlock, *History and Origins of Cryogenics*, 226.
142. See, for example, "Linde Cryobiology News: Report No. 2 from Linde Company, Division of Union Carbide," which was, in fact, one in a series of full-page advertisements published in the journal *Science* 139, no. 3549 (1963): 3.
143. "Stopping the Biological Clock," *AIBS Bulletin* 12, no. 5 (1962): 112.
144. Daniel Lee Kleinman, *Impure Cultures: University Biology and the World of Commerce* (Madison: University of Wisconsin Press, 2003).
145. Cited in Keilin, "The Leeuwenhoek Lecture," 164.
146. Robert C. W. Ettinger, *The Prospect of Immortality* (Garden City, NY: Doubleday, 1964).
147. Ibid., 88. On cryonics, see also Parry, "Technologies of Immortality"; Rodney Doyle, "Disciplined by the Future: The Promising Bodies of Cryonics," *Science as Culture* 6, no. 4 (1997): 582–616; Tiffany Romain, "Extreme Life Extension: Investing in Cryonics for the Long, Long Term," *Medical Anthropology* 29, no. 2 (2010): 194–215; Abou Farman, "Speculative Matter: Secular Bodies, Minds and Person," *Cultural Anthropology* 28, no. 4 (2013): 737–59; Jonny Bunning, "The Freezer Program: Value after Life," in *Cryopolitics: Frozen Life in a Melting World*, ed. Joanna Radin and Emma Kowal (Cambridge, MA: MIT University Press, forthcoming).

148. Mike Darwin, "Cold War: The Conflict between Cryonicists and Cryobiologists," ALCOR Life Extension Foundation website, http://www.alcor.org/Library/html/coldwar.html, accessed 11 November 2015.
149. Schmidt, "Basile J. Luyet and the Beginnings of Transfusion Cryobiology."
150. Malinin worked with Carrel and Charles Lindbergh on a perfusion pump, which allowed organs to be kept alive for study outside the body. He later wrote a biography of Carrel. Theodore I. Malinin, *Surgery and Life: The Extraordinary Career of Alexis Carrel* (New York: Harcourt Brace Jovanovich, Inc, 1979).
151. In the twenty-first century *Cryobiology* continues to be published under the title *Cell Preservation Technology*.
152. Other articles in this first issue covered various aspects of scientific practice, including "Glycerol Preservation of Red Blood Cells," "The Preservation of Bone Marrow for Clinical Use," "Historical Development of Cell and Tissue Culture Freezing," "Freeze Preservation of Cultured Animal Cells."
153. Cited in Kavaler, *Cold against Disease*, 28.
154. Harold Cook, "Time's Bodies," in *Merchants and Marvels: Commerce, Science and Art in Early Modern Europe*, ed. by Pamela Smith and Paula Findlen (New York: Routledge, 2002), 223–47, quote on 225.
155. Thevenot, *A History of Refrigeration*, 127
156. John Shannon, "The Role of Liquid Nitrogen and Refrigeration at the American Type Culture Collection" in *Roundtable Conference on the Cryogenic Preservation of Cell Cultures*, ed. A. P. Rinfret and B. LaSalle (Washington, DC: National Academy of Sciences, 1975), 1–8.
157. John O'Donnell, *Coriell: The Coriell Institute for Medical Research and a Half Century of Science* (Canton, MA: Science History, 2002).
158. Ibid.
159. Lewis L. Coriell, Arthur E. Greene, and Ruth K. Silver, "Historical Development of Cell and Tissue Culture Freezing," *Cryobiology* 1, no. 1 (1964): 72–79.
160. Ibid.
161. Leef and Meryman, "The American Foundation for Biological Research." BRI.
162. Ibid.
163. Ibid.
164. Michel Foucault, *The Archaeology of Knowledge* (London: Tavistock, 1972).
165. Carolyn Steedman, *Dust: The Archive and Cultural History* (New Brunswick, NJ: Rutgers University Press, 2002), 3.
166. Warwick Anderson, "The Frozen Archive, or Defrosting Derrida," *Journal of Cultural Economy* 8 (2015), 379–87.
167. Steedman, *Dust: The Archive and Cultural History*, 79.
168. Ibid; Jacques Derrida, *Archive Fever: A Freudian Impression* (Chicago: University of Chicago Press, 1996); Ann Laura Stoler, *Along the Archival Grain: Epistemic Anxieties and Colonial Common Sense* (Princeton, NJ: Princeton University Press, 2009); Michael Lynch, "Archives in Formation: Privileged Spaces, Popular Archives and Paper Trails," *History of the Human Sciences* 12, no. 2

(1999): 65–87; Warwick Anderson, "The Case of the Archive," *Critical Inquiry* 39, no. 3 (2013): 532–47; Anderson, "The Frozen Archive."

169. Vernon Perry, "A Message from the Director," in J. A. Panuska (president of the board of directors), "Status Report," American Foundation for Biological Research, 1975, BRI.
170. J. A. Panuska (president of the board of directors), "Status Report," American Foundation for Biological Research, 1975, BRI.
171. Ibid, 10.
172. "Biotechnology and the Cryosciences: An Absence of National Awareness," 6 April 1983, Corporate Records, BRI.
173. John G. Baust, vice president of the Society for Cryobiology, to Thomas Finnigan, Union Carbide, 6 April 1986, Corporate Records, BRI.
174. Meryman, "Basile J. Luyet: In Memoriam," 290.
175. Saparelli, Antoniette, and Dubochet, *Basile Luyet*, 64. My translation.
176. Victor Turner, "Variations on a Theme of Liminality," in *Secular Ritual*, ed. Sally Falk Moore and Barbara Myerhoff (Amsterdam: Van Gorcum, Assen, 1977), 36–52. See also Cori Hayden, "Suspended Animation: A Brine Shrimp Essay," in *Remaking Life and Death: Toward an Anthropology of the Biosciences*, ed. Sarah Franklin and Margaret M. Lock (Santa Fe, NM: School of American Research Press, 2003), 193–225.
177. Victor Turner, "Betwixt and Between: The Liminal Period in Rites of Passage," in *Betwixt and Between: Patterns of Masculine and Feminine Initiation,* ed. Louise Carus Madhi, Steven Foster, and Meredith Little (Peru, IL: Open Court, 1987), 6–7.
178. Susan Merrill Squier, *Liminal Lives: Imagining the Human at the Frontiers of Biomedicine* (Durham, NC: Duke University Press, 2004), 4.
179. Joanna Radin and Emma Kowal, "The Politics of Low Temperature," in *Cryopolitics: Frozen Life in a Melting World*, ed. Joanna Radin and Emma Kowal (Cambridge, MA: MIT University Press, forthcoming); Emma Kowal and Joanna Radin, "Indigenous Biospecimens and the Cryopolitics of Frozen Life," *Journal of Sociology* 51, no. 1 (2015): 63–80.
180. Grosz, *The Nick of Time*.

CHAPTER 2

1. The 1959 WHO report remarked explicitly on the role of poliomyelitis surveys in setting important precedent. World Health Organization, *Immunological and Haematological Surveys*, Technical Report Series, no. 181 (Geneva: World Health Organization, 1959), 10.
2. Ibid., 4.
3. Arguments in this chapter are adapted from Joanna Radin, "Unfolding Epidemiological Stories: How the WHO Made Frozen Blood into a Flexible Resource for the Future," *Studies in History and Philosophy of Biological and Biomedical Sciences* 47 (2014): 62–73.

4. A. C. Fleck and F. A. Ianni, "Epidemiology and Anthropology: Some Suggested Affinities in Theory and Method," *Human Organization* 16, no. 4 (1958): 38–40; Andrew J. Mendelsohn, "From Eradication to Equilibrium: How Epidemics Became Complex after World War I," in *Greater Than the Parts: Holism in Biomedicine, 1920–1950*, ed. Christopher Lawrence and George Weisz (New York: Oxford University Press, 1998), 303–34; Reuel A. Stallones, *Environment, Ecology, and Epidemiology*, Fourth PAHO/WHO Lecture on the Biomedical Sciences (Washington, DC: PAHO, 1971); Susanne Bauer, "Mining Data, Gathering Variables and Recombining Information: The Flexible Architecture of Epidemiological Studies," *Studies in History and Philosophy of Biological and Biomedical Sciences* 39 (2008): 415–28.
5. World Health Organization, *Immunological and Haematological Surveys*, 12.
6. Arturo Escobar, *Encountering Development: The Making and Unmaking of the Third World* (Princeton, NJ: Princeton University Press, 1995); Odd Arne Westad, *The Global Cold War: Third World Interventions and the Making of Our Times* (Cambridge: Cambridge University Press, 2005).
7. John Krige, *American Hegemony and the Postwar Reconstruction of Science in Europe* (Cambridge, MA: MIT Press, 2006).
8. James C. Scott, *Seeing Like a State: How Certain Schemes to Improve the Human Condition Have Failed* (New Haven, CT: Yale University Press, 1998).
9. As would be the case with later efforts to prospect the genomes of the entire national populations, as discussed in Michael Fortun, *Promising Genomics: Iceland and DeCODE Genetics in a World of Speculation* (Berkeley: University of California Press, 2008).
10. Giacomo Marramao, *Kairos: Towards an Ontology of "Due Time"* (Aurora, CO: Davies Group, 2007).
11. John R. Paul, "Clinical Epidemiology," *Journal of Clinical Investigation* 17 (1938): 539–41.
12. This early work in New Haven and abroad is described in Paul's *History of Poliomyelitis* (New Haven, CT: Yale University Press, 1971).
13. John R. Paul, *Clinical Epidemiology* (Chicago: University of Chicago Press, 1958). Paul planned to build on a tuberculosis vaccination program that had been conducted at Point Barrow and the vicinity in 1948.
14. Paul to Shelesnyak, 29 April 1949, MS 1333, box 13, folder 5, series II, John Rodman Paul Papers, Yale University Library Manuscripts and Archives (hereafter Paul Papers). See also Paul, *History of Poliomyelitis*.
15. Allen Brandt, *No Magic Bullet: A Social History of Venereal Disease in the United States since 1880*, expanded ed. (New York: Oxford University Press, 1987).
16. John Farley, *To Cast Out Disease: A History of the International Health Division of the Rockefeller Foundation (1913–1951)* (New York: Oxford University Press, 2004); Marcos Cueto, *Missionaries of Science: The Rockefeller Foundation and Latin America* (Bloomington: Indiana University Press, 1994).
17. Jenny Bangham, "Blood Groups and the Rise of Human Genetics in Mid-Twentieth Century Britain" (PhD diss., Cambridge University, 2013).

18. Paul to Shelesnyak, 29 April 1949, MS 1333, box 13, folder 5, series II, Paul Papers.
19. J. R. Paul, MD, "Preliminary Report of Expedition to Point Barrow, Alaska for the Purpose of Carrying out a Serological Survey Among Alaskan Eskimos," 12 August to 11 September 1949, box 13, folder 5, series II, Paul Papers.
20. For a critical assessment of epidemiological interest in virgin soil epidemics during this time, see David S. Jones, "Virgin Soils Revisited," *William and Mary Quarterly* 60, no. 4 (2003): 703–42.
21. John R. Paul, "Considerations with Regard to Epidemiological Investigations at the Arctic Research Laboratory, ONR," submitted to the Arctic Office of Naval Research, n.d. , box 13, folder 2, series II, Paul Papers.
22. Erwin H. Ackerknecht, MD, DIPL, ETH, Professor of Medical History, "The Eskimo," Ciba Foundation Symposium, vol. 10, no. 1, July–August 1948, by University of Wisconsin, Madison, box 13, folder 4, Alaska Serological Project, 1948, series II, Paul Papers.
23. Helen Tilley, *Africa as a Living Laboratory: Empire, Development, and the Problem of Scientific Knowledge, 1870–1950* (Chicago: University of Chicago Press, 2011).
24. M. C. Shelesnyak, "The History of the Arctic Research Laboratory," reprinted from *Arctic* 1, no. 2 (Autumn 1948), box 13, folder 4, Alaska Serological Project, 1948, series II, Paul Papers.
25. Chandra Mukerji, *A Fragile Power: Scientists and the State* (Princeton, NJ: Princeton University Press, 1989).
26. Melinda Cooper and Catherine Waldby, *Clinical Labor: Tissue Donors and Research Subjects in the Global Bioeconomy* (Durham, NC: Duke University Press, 2014).
27. Jenifer Van Vleck, *Empire of the Air: Aviation and the American Ascendancy* (Cambridge, MA: Harvard University Press, 2013).
28. Erwin H. Ackerknecht, MD, DIPL, ETH, Professor of Medical History, "The Eskimo," Ciba Foundation Symposium, vol. 10, no. 1, July–August 1948, University of Wisconsin, Madison, box 13, folder 4, Alaska Serological Project, 1948, series II, Paul Papers.
29. "Eskimos Resist Old Way of Life," *New York Times*, 21 November 1954, box 13, folder 1, Alaska Poliomyelitis Epidemic (1951–1961), series II, Paul Papers.
30. Paul to Shelesnyak, 29 April 1949, box 13, folder 5, series II, Paul Papers.
31. See also Naomi Rogers, "Polio Can Be Cured: Science and Health Propaganda in the United States from Polio Polly to Jonas Salk," in *Silent Victories: The History and Practice of Public Health in Twentieth Century America,* ed. John Ward and Christopher Warren (New York: Oxford University Press, 2007), 81–101.
32. Paul to Shelesnyak, 1 July 1949, "Check list of information on research proposals," box 13, folder 5, Alaska Serological Project, 1949, series II, Paul Papers.
33. Paul later communicated that a nurse at Point Barrow informed him that the local hospital and all its records had been destroyed by a fire nearly a decade prior to his field trip. Paul to Dr. J. H. Stickler, Division of Communicable and Preventable Disease Control, Alaska Department of Health in Juneau, 7 October 1949, box 13, folder 6, series II, Paul Papers.

34. An example of the "mechanical objectivity" that dominated elite knowledge-making during the era. Lorraine Daston and Peter Louis Galison, *Objectivity* (New York: Zone, 2007).
35. Among these were "antibodies to streptococcal disease, nonspecific antibodies as measured in the form of 'gamma globulin fractions' as determined by the Kunkel method, and a variety of other antibodies." Paul to Shelesnyak, 29 April 1949, box 13, folder 5, Alaska Serological Project, 1949, series II, Paul Papers.
36. Paul to Shelesnyak, 1 July 1949, "Check list of information on research proposals," box 13, folder 5, Alaska Serological Project, 1949, series II, Paul Papers.
37. Paul to G. E. MacGinitie, 21 July 1949, box 13, folder 5, Alaska Serological Project, 1949, series II, Paul Papers.
38. Ibid.
39. G. E. MacGinitie, scientific director of the Naval Arctic Research Laboratory, to Paul, 11 July 1949, box 13, folder 5, Alaska Serological Project, 1949, series II, Paul Papers.
40. Ibid.
41. Edith Egowa Tegoseak (and Evan Egowa), oral history, recorded 24 March 1966, Tanana-Yukon Historical Society Tapes, transcription made 18 February 2014, by Varpu Lotvonen, http://oralhistory.library.uaf.edu/97/97-66-27_SIDE_A_T01.pdf.
42. Ron Mancil, "Discoverer Seeks Support for Naming Dinosaur Site after Inupiat Grandparents," Alaska Science Outreach website, 28 October 2004, http://www.alaskascienceoutreach.com/index.php/main_pages/catchitem/discoverer_seeks_support_for_naming_dinosaur_site_after_inupiat_grandparent/, accessed 5 January 2016.
43. He later presented this work at the American Association of Physical Anthropology annual meeting. "Proceedings of the Eighteenth Annual Meeting of the AAPA," *American Journal of Physical Anthropology* 7, no. 2 (June 1949): 271–300.
44. J. R. Paul, MD, "Preliminary Report of Expedition to Point Barrow, Alaska for the Purpose of Carrying out a Serological Survey among Alaskan Eskimos," 12 August to 11 September 1949, p.10, box 13, folder 5, series II, Paul Papers.
45. Ibid.
46. Ibid.
47. Ibid.
48. Handwritten note from Edith Tegoseak to Paul, 30 August 1949, box 13, folder 5, Alaska Serological Project, 1949, series II, Paul Papers.
49. J. R. Paul, MD, "Preliminary Report of Expedition to Point Barrow, Alaska for the Purpose of Carrying out a Serological Survey among Alaskan Eskimos," 12 August to 11 September 1949, p.10, box 13, folder 5, series II, Paul Papers.
50. Jack C. Haldeman, senior surgeon, medical officer in charge, to Paul, 8 September 1949, box 13, folder 6, series II, Paul Papers.
51. M. C. Shelesnyak, "Problems of the Arctic, Joint Seminar of the Arctic Institute of North America and the Isaiah Bowman School of Geography,

The Johns Hopkins University," Rogers House, 25 May 1950, "The Arctic as a Strategic Scientific Area," 4, box 13, folder 7, series II, Paul Papers.

52. Liza Piper, "Chesterfield Inlet, 1949, and the Ecology of Epidemic Polio," *Environmental History* 101, no. 1 (2015): 1–28.
53. Paul to C. E. Van Rooyan, Department of Bacteriology, School of Hygiene, U. of Toronto, 12 January 1950, box 13, folder 7, series II, Paul Papers.
54. Gabrielle Hecht, *Being Nuclear: Africans and the Global Uranium Trade* (Cambridge, MA: MIT Press, 2012); Joseph Masco, *Nuclear Borderlands: The Manhattan Project in Post–Cold War New Mexico* (Princeton, NJ: Princeton University Press, 2006); Hecht, "On the Fallacies of Cold War Nostalgia," in *Entangled Geographies: Empire and Technopolitics in the Global Cold War,* ed. Gabrielle Hecht (Cambridge, MA: MIT Press, 2011), 75–94.
55. A. J. Rhodes, Eina M. Clark, Alice Goodfellow, and W. L. Donohue, "An Outbreak of Poliomyelitis in Canadian Eskimos in Wintertime," Technical Methods Reprint from the *Canadian Journal of Public Health,* October 1949, 418–19, box 13, folder 6, series II, Paul Papers.
56. Paul to C. E. Van Rooyan, Dept. of Bacteriology, School of Hygiene, University of Toronto, 12 January 1950, box 13, folder 7, series II, Paul Papers.
57. Ibid.
58. Paul to C. E. Van Rooyan, January 1950, box 13, folder 7, series II, Paul Papers.
59. John R. Paul, J. T. Riordan, and Lisabeth M. Kraft, "Serological Epidemiology: Antibody Patterns in North Alaskan Eskimos," *Journal of Immunology* 66 (1951): 695–713.
60. Ibid., 696.
61. Ibid., 711.
62. Javed Siddiqi, *World Health and World Politics: The World Health Organization and the UN System* (Columbia: University of South Carolina Press, 1995); Marcos Cueto, *Cold War, Deadly Fevers: Malaria Eradication in Mexico, 1955–1975* (Baltimore: Johns Hopkins University Press, 2007); Leo B. Slater, *War and Disease: Biomedical Research on Malaria in the Twentieth Century* (New Brunswick, NJ: Rutgers University Press, 2009); Nancy Stepan, *Eradication: Ridding the World of Diseases Forever?* (Ithaca, NY: Cornell University Press, 2011).
63. Susan Leigh Star and Karen Ruhleder, "Steps Towards an Ecology of Infrastructure: Design and Access for Large Information Spaces," *Information Systems Research* 7, no. 1 (1996): 111–34; Bauer, "Mining Data, Gathering Variables and Recombining Information"; Hannah Landecker and Angela N. H. Creager, "Technical Matters: Method, Knowledge and Infrastructure in Twentieth-Century Life Science," *Nature Methods* 6, no. 10 (2009): 701–5.
64. Alison Bashford, "Global Biopolitics and the History of World Health," *History of the Human Sciences* 19, no. 1 (2006): 67–88; Bashford, "Population, Geopolitics and International Organizations in the Mid-Twentieth Century," *Journal of World History* 19 (2008): 327–47; Valeska Huber, "The Unification of the Globe by Disease? The International Sanitary Conferences of 1851–1894," *Historical Journal* 49, no. 2 (2006): 453–76; Sunil S. Amrith, *Decolonizing*

International Health: India and Southeast Asia, 1930–65 (New York: Palgrave Macmillan, 2006).

65. Bashford, "Global Biopolitics and the History of World Health," 83.
66. Iris Borowy, "International Health Work: The Beginnings," *Michael* 8 (2011): 210–21; Borowy, *Coming to Terms with World Heatlh: The League of Nations Health Organization, 1921–1946* (New York: Peter Lang, 2009).
67. Viviane Quirke and Jean-Paul Gaudillière, "The Era of Biomedicine: Science, Medicine, and Public Health in Britain and France after the Second World War," *Medical History* 52, no. 4 (2008): 441–52; Adele Clarke et al., *Biomedicalization: Technoscience, Health, and Illness in the U.S.* (Durham, NC: Duke University Press, 2010); Jean-Paul Gaudillière, "From Propaganda to Scientific Marketing: Schering, Cortisone, and the Construction of Drug Markets," *History and Technology* 29, no. 2 (2013): 188–209.
68. Michael Bresalier, "Uses of a Pandemic: Forging the Identities of Influenza and Virus Research in Interwar Britain," *Social History of Medicine* 25, no. 2 (2011): 400–24; Christoph Gradmann, "Sensitive Matters: The World Heath Organisation and Antibiotic Resistance Testing, 1945–1975," *Social History of Medicine* 26, no. 3 (2013): 555–74; Gradmann and Jonathan Simon, *Evaluating and Standardizing Therapeutic Agents, 1890–1950* (New York: Palgrave Macmillan, 2010).
69. Axel Huntelmann, "Seriality and Standardizartion in the Production of '606,'" *History of Science* 48, no. 161 (2010): 435–60; Adrian F. Bristow, Trevor Barrowcliffe, and Derek R. Bangham, "Standardization of Biological Medicines: The First Hundred Years, 1900–2000," *Notes and Records of the Royal Society of London* 60, no. 3 (2006): 271–89; Steve Sturdy, "Reflections: Molecularization, Standardization and the History of Science," in *Molecularizing Biology and Medicine: New Practices and Alliances, 1910s–1970s,* ed. Soraya de Chadarevian and Harmke Kamminga (Amsterdam: OPA, 1998), 254–71.
70. J. H. Burn, D. J. Finney, and L. G. Goodwin, *Biological Standardization*, 2nd ed. (London: Oxford University Press, 1950).
71. William Schneider, "The History of Research on Blood Group Genetics: Initial Discovery and Diffusion," *History and Philosophy of the Life Sciences* 18 (1996): 277–03.
72. Peter Keating, "Holistic Bacteriology: Ludwick Hirszfeld's Doctorine of Serogenesis between the Two World Wars," in *Greater Than the Parts: Holism in Biomedicine 1920–1950,* ed. Christopher Lawrence and George Weisz (New York: Oxford University Press, 1998), 283–302.
73. Bangham, "Blood Groups and the Rise of Human Genetics." For accounts of the history of racial serology see Jonathan Marks, "The Legacy of Serological Studies in American Physical Anthropology," in *History and Philosophy of the Life Sciences*, 18 (1996): 345–62; Marks, "The Origins of Anthropological Genetics," *Current Anthropology,* 53, no. S5 (2012): S161–S172; Rachel Silverman, "The Blood Group 'Fad' in Post-War Racial Anthropology," in *Kroeber Anthropological Society Papers*, ed. Jonathan Marks (Berkeley: University of California Press, 2000), 11–27.

74. These first two groups are known as Smithies's haptoglobin types and Gm. Rune Grubb, "Interactions between Immunology and Genetics—Blood Group Systems as Important Early Models and Tools," in *Immunology, 1930–1980: Essays on the History of Immunology*, ed. Pauline H Mazumdar (Toronto: Wall and Thompson, 1989), 131–42; Schneider, "The History of Research on Blood Group Genetics."
75. Keating, "Holistic Bacteriology," 296.
76. W. Chas Cockburn, "The International Contribution to the Standardization of Biological Substances. I. Biological Standards and the League of Nations 1921–1946," *Biologicals* 19 (1991): 161–69; Derek R. Bangham, *History of Biological Standardization: Characterization and Measurement of Complex Molecules Important in Clinical and Research Medicine; Contributions from the UK, 1900–1995* (Bristol: published with the assistance of the Society for Endocrinology, 2000); Bristow, Barrowcliffe, and Bangham, "Standardization of Biological Medicines"; Pauline H. Mazumdar, "'In the Silence of the Laboratory': The League of Nations Standardizes Syphilis Tests," *Social History of Medicine* 16, no. 3 (2003): 437–58. In terms of model organisms, see especially Robert E. Kohler, *Lords of the Fly: Drosophila Genetics and the Experimental Life* (Chicago: University of Chicago Press, 1994); Angela N. H. Creager, *The Life of a Virus: Tobacco Mosaic Virus as an Experimental Model, 1930–1965* (Chicago: University of Chicago Press, 2002); Karen A. Rader, *Making Mice: Standardizing Animals for American Biomedical Research, 1900–1955* (Princeton, NJ: Princeton University Press, 2004); R. G. W. Kirk, "Wanted—Standard Guinea Pigs: Standardization and the Experimental Animal Market in Britain ca. 1919–1947," *Studies in History and Philosophy of Science, Part C* 39 (2008): 280–91.
77. Mazumdar, "In the Silence of the Laboratory," 437.
78. Paul N. Edwards, *The Closed World: Computers and the Politics of Discourse in Cold War America* (Cambridge, MA: MIT Press, 1996).
79. S. Declich and A. O. Carter, "Public Health Surveillance: Historical Origins, Methods and Evaluation," *Bulletin of the World Health Organization* 72, no. 2 (1994): 285–304.
80. Mark Pendergrast, *Inside the Outbreaks: The Elite Medical Detectives of the Epidemic Surveillance Service* (New York: Mariner, 2010).
81. Alexander D. Langmuir, "The Surveillance of Communicable Diseases of National Importance," *New England Journal of Medicine* 268 (1963): 182–83.
82. Walter W. Holland, "Karel Raska—the Development of Modern Epidemiology. The Role of the IEA," *Central European Journal of Public Health* 18, no. 1 (2010): 57.
83. Sanjoy Bhattacharya, "WHO-Led or WHO-Managed? Re-Assessing the Smallpox Eradication Program in India, 1960–1980," in *Medicine at the Border: Disease, Globalization and Security, 1850 to the Present*, ed. Alison Bashford (New York: Palgrave Macmillan, 2006), 60–75.
84. James H. Steele, "Karel Raska, 1909–1987, a Tribute," *Journal of Infectious Diseases* 158, no. 5 (1988): 915–16.

85. Karel Raska, “Epidemiologic Surveillance in the Control of Infectious Diseases,” *Reviews of Infectious Diseases* 5, no. 5 (1983): 1113.
86. This is in contrast to the kinds of experimental epidemiology that were innovated during the interwar period. There, acts of standardization were imposed upon strains of infection as well as approaches to inoculating experimental organisms like lab mice. Olga Amsterdamska, “Standardizing Epidemics: Infection, Inheritance, and Environment in Prewar Experimental Epidemiology,” in *Heredity and Infection: The History of Disease Transmission*, ed. Jean-Paul Gaudillière and Ilana Lowy (London: Routledge, 2001), 135–79.
87. For Chisholm’s vision for the WHO see Fraser Brockington, *World Health* (Baltimore: Penguin Books, 1958); John Farley, *Brock Chisholm, the World Health Organization, and the Cold War* (Vancouver: University of British Columbia Press, 2009).
88. World Health Organization, *Multipurpose Serological Surveys and Serum Reference Banks*, WHO Technical Report Series, no. 454 (Geneva: World Health Organization, 1970), 9.
89. This period of reconfiguration and intensification of international organizations has been termed the “New Internationalism.” Diplomatic historian Akira Iriye writes, “At the very moment when humankind was becoming frightened by the spectacle of nuclear war, serious and at least partially successful attempts were being made to save lives through international cooperation.” Akira Iriye, *Global Community: The Role of International Organizations in the Making of the Contemporary World,* (Berkeley: University of California Press, 2002), 75.
90. Amrith, *Decolonizing International Health.*
91. Alfred S. Evans, “Serological Surveys: The Role of the WHO Reference Serum Banks,” *World Health Organization Chronicle* 21 (1968): 190.
92. Amy L. S. Staples, “Constructing International Identity: The World Bank, Food and Agriculture Organization, and World Health Organization, 1945–1965” (PhD diss., Ohio State University, 1998).
93. Phillips to Horstmann, 9 December 1958, box 8, folder 154, series I, Paul Papers.
94. World Health Organization, *Immunological and Haematological Surveys.*
95. Ibid., 6.
96. Herbert C. Dessauer and Mark S. Hafner, eds., *Collections of Frozen Tissues: Value, Management, Field and Laboratory Procedures, and Directory of Existing Collections* (Lawrence, KS: Association of Systematics Collections, 1984). I explore the logic of “planned hindsight” in the context of nonhuman specimens in Joanna Radin, “Planned Hindsight: Vital Valuations of Frozen Tissue at the Zoo and the Natural History Museum,” *Journal of Cultural Economy* 8, no. 3 (2015).
97. Reinhart Koselleck, *Futures Past: On the Semantics of Historical Time*, trans. Keith Tribe (New York: Columbia University Press, 2004).
98. World Health Organization, *Immunological and Haematological Surveys*, 4.

99. Gina Kolata, *Flu: The Story of the Great Influenza Pandemic of 1918 and the Search for the Virus That Caused It* (New York: Farrar, Straus and Giroux, 1999).
100. Jeffery K. Taubenberger, Johan V. Hultin, and David M. Morens, "Discovery and Characterization of the 1918 Pandemic Influenza in Historical Context," *Antiviral Therapy* 12, no. 4, part B (2007): 581.
101. Edwards, *The Closed World*.
102. World Health Organization, *Immunological and Haematological Surveys*, 28.
103. For example, in Mourant's early hemoglobin work with Lehmann, described in Soraya de Chadarevian, "Following Molecules: Hemoglobin Between the Clinic and the Laboratory" in *Molecularizing Biology and Medicine: New Practices and Alliances, 1910s–1970s*, ed. Soraya de Chadarevian and H. Kamminga (Amsterdam, OPA, 1998), 171–201. Mourant drafted an early version of preservation protocol, drawing on his experiences handling materials in the field and in the lab. This protocol was adapted for WHO Technical Reports.
104. William J. Elser, Ruth A. Thomas, and Gustav I. Steffen, "The Desiccation of Sera and Other Biological Products (Including Microorganisms) in the Frozen State with the Preservation of the Original Qualities of Products So Treated," *Journal of Immunology* 28, no. 6 (1935): 433–73.
105. Earl W. Flosdorf and Stuart Mudd, "Biologics Now Preserved by Drying from the Frozen State," *Refrigerating Engineering* 36 no. 6 (December 1938): 379.
106. This technical note would be cited in a World Health Organization report that dealt with collection and storage of serum for global epidemiological surveillance. World Health Organization Technical Report Series, no. 454, "Multipurpose Serologic Surveys and Serum Reference Banks" (1970): 1–95.
107. T. Guthe, "Freezing and Transport of Sera in Liquid Nitrogen at –150c to –196c," *Bulletin of the World Health Organization* 33, no. 6 (1965): 864.
108. Ibid.
109. Ibid.
110. Ibid., 865.
111. Ibid., 867.
112. Ibid.
113. John R. Paul, "The Story to Be Learned from Blood Samples: Its Value to the Epidemiologist," *Journal of the American Medical Association* 175, no. 7 (1961): 148.
114. In official WHO documents they are referred to as serum reference banks, but the Yale group referred to theirs without apparent comment as a reference serum bank.
115. Raska, "Epidemiological Surveillance." The histories of these other banks remain elusive.
116. Local universality involves "being in several locales at the same time, yet always being the product of contingent negotiations and pre-existing institutional and material locations." Stefan Timmermans and Marc Berg Timmermans, "Standardization in Action: Local Universality through Medical Protocols," *Social Studies of Science* 27, no. 2 (1997): 273–305.

117. There is an international history of biomedical computing, yet to be told, that emerges from this enterprise. For the best treatments of the subject in the American context see Joseph November, *Biomedical Computing: Digitizing Life in the United States* (Baltimore: Johns Hopkins University Press, 2012); Hallam Stevens, *Life Out of Sequence: A Data-Driven History of Bioinformation* (Chicago: University of Chicago Press, 2013).
118. John R. Paul, “Annual Report 1964–1965 of the WHO Reference Serum Bank, Yale University,” box 233, folder 11, series I, Paul Papers.
119. Monica Casper and Adele Clarke, “Making the Pap Smear into the ‘Right Tool’ for the Job: Cervical Cancer Screening in the USA, Circa 1940–95,” *Social Studies of Science* 28, no. 2 (1998): 269.
120. Especially as a move was made toward automation, as described in Evans, “Serological Surveys.”
121. John R. Paul, MD, director, “Provisional Outline Prepared for the Use of the Staff of the WHO Reference Serum Bank at Yale University,” “Ideas for Blood Study,” p. 2, MS coll. 96, box 90, folder 6, series II, James V. Neel Papers, American Philosophical Society (hereafter James V. Neel Papers).
122. Paul, “Annual Report.”
123. Paul, “Provisional Outline.”
124. F. Macfarlane Burnet, “Men or Molecules: A Tilt at Molecular Biology,” *Lancet* 1 (1966): 37–39.
125. Warwick Anderson and Ian R. MacKay, “Fashioning the Immunological Self: The Biological Individuality of F. Macfarlane Burnet,” *Journal of the History of Biology* 47, no. 1 (2013): 147–75; Warwick Anderson, “Natural Histories of Infectious Disease: Ecological Vision in Twentieth-Century Biomedical Science,” *Osiris* 19 (2004): 39–64.
126. Burnet, “Men or Molecules,” 38–39.
127. In correspondence, Paul made clear a vision indebted to British epidemiologist Jerry Morris, known for his promotion of an approach focused on social inequalities. Dorothy Porter, “Calculating Health and Social Change: An Essay on Jerry Morris and Late-Modernist Epidemiology,” *International Journal of Epidemiology* 36 (2007): 1180–84.
128. There has been a surprising lack of historical attention to disease ecology, particularly from historians of medicine. Notable exceptions include Warwick Anderson, “Natural Histories of Infectious Disease: Ecological Vision in Twentieth-Century Biomedical Science,” *Osiris* 19 (2004): 39–64; and Linda Nash, *Inescapable Ecologies: A History of Environment, Disease, and Knowledge* (Berkeley: University of California Press, 2007). For an account that attempts to cultivate interests in the subject among environmental historians see Gregg Mitman, “In Search of Health: Landscape and Disease in American Environmental History,” *Environmental History* 20, no. 2 (2005): 184–210.
129. John R. Paul, *Clinical Epidemiology* (Chicago: University of Chicago Press, 1958), 152.

130. Henry Fountain, "Alfred S. Evans, 78, Expert on Origins of Mononucleosis," *New York Times*, 25 January 1996.
131. Alfred S. Evans, "Epidemiology and the Public Health Laboratory," *American Journal of Public Health* 57, no. 6 (1967): 1041–52.
132. Ibid., 1051.
133. M. Terris and S. Blatt, "Differences in Serum Cholesterol in Young White and Negro Adults," *American Journal of Public Health* 54, no. 12 (1964): 1996–2008.
134. Paul, "Annual Report."
135. Ibid.
136. Ibid.
137. World Health Organization, *Multipurpose Serological Surveys and Serum Reference Banks*, 10.
138. "Seroepidemiology," *Journal of Parasitology*, section 2, part 3: Supplement: Second International Congress of Parasitology, Technical Reviews 56, no. 4 (1970): 8.
139. "World Health Organization, *Multipurpose Serological Surveys and Serum Reference Banks*, 27.
140. Similar to radioisotopes. Angela N. H. Creager, "Radioisotopes as Political Instruments, 1946–1953," *Dynamis* 29 (2009): 219–39.
141. "World Health Organization, *Multipurpose Serological Surveys and Serum Reference Banks*, 18.
142. Hans-Jörg Rheinberger, *Toward a History of Epistemic Things: Synthesizing Proteins in the Test Tube* (Stanford, CA: Stanford University Press, 1997), 28.
143. In keeping with Soraya de Chadarevian, "Following Molecules: Hemoglobin between the Clinic and the Laboratory," in *Molecularizing Biology and Medicine: New Practices and Alliances, 1910s–1970s*, ed. Soraya de Chadarevian and Harmke Kamminga (Amsterdam: Harwood Academic, 1998), 171–201; Bruno Strasser, "The Experimenter's Museum: Genbank, Natural History, and the Moral Economies of Biomedicine," *Isis* 102, no. 1 (2011): 60–96.
144. World Health Organization, *Multipurpose Serological Surveys and Serum Reference Banks*, 24.
145. Ibid., 25.
146. Linda Nash, *Inescapable Ecologies: A History of Environment, Disease, and Knowledge* (Berkeley: University of California Press, 2006), 30.
147. Ibid.
148. This section based on analysis of materials in Paul, "Annual Report."
149. Paul, "Annual Report."
150. Baruch S. Blumberg, *Proceedings of the Conference on Genetic Polymorphisms and Geographic Variations in Disease* (New York: Grune and Stratton, 1961).
151. Baruch S. Blumberg, *Hepatitis B: The Hunt for a Killer Virus* (Princeton, NJ: Princeton University Press, 2002).
152. Warwick Anderson, "The Possession of Kuru: Medical Science and Biocolonial Exchange," *Comparative Studies in Society and History* 42 (2000): 713–44.

153. M. Susan Lindee, *Suffering Made Real: American Science and the Survivors at Hiroshima* (Chicago: University of Chicago Press, 1994); Lindee, "Voices of the Dead: James Neel's Amerindian Studies," in *Lost Paradises and the Ethics of Research and Publication,* ed. Francisco M. Salzano and A. Magdalena Hurtado (New York: Oxford University Press, 2004), 27–48.
154. Francis Black, "Serological Epidemiology in Measles," *Yale Journal of Biology and Medicine* 32, no. 1 (September 1959): 44–50.
155. World Health Organization, *Immunological and Haematological Surveys*, 9–10.
156. Karl Marx made famous the term "primitive accumulation" to describe the land and resource grab that set the initial conditions for industrial capitalism. To Marx, primitive accumulation "plays the same role in political economy as original sin does in theology," and "So-called primitive accumulation, therefore, is nothing else than the historical process of divorcing the producers from the means of production. It appears as 'primitive' because it forms the pre-history of capital, and the mode of production corresponding to capital." Karl Marx, *Capital, A Critique of Political Economy* vol. 1 (New York: Penguin, 1976): 873–75. On the role of primitive accumulation and its recurrent properties see Catherine Waldby and Melinda Cooper, "The Biopolitics of Reproduction," *Australian Feminist Studies* 23 no. 55 (2008): 57–73; Massimiliano Tomba, *Marx's Temporalities*, trans. Peter D. Thomas and Sara R. Farris (Chicago: Haymarket Books, 2013).
157. Anthony M. Payne, "Serum Surveys," *Milbank Memorial Fund Quarterly* 43, no. 2 (1965): 345–50.
158. Walter Sullivan, "Dying Language Being Recorded," *New York Times*, 26 October 1961, box 13, folder 1, Alaska Poliomyelitis Epidemic (1951–1961), series II, Paul Papers.
159. Paul to R. D. Hamilton, undated, box 13, folder 6, series II, Paul Papers.
160. John R. Paul, "Considerations with Regard to Epidemiological Investigations at the Arctic Research Laboratory, ONR," submitted to the Arctic Office of Naval Research, n.d., p. 7, box 13, folder 2, series II, Paul Papers.
161. Paul to Shelesnyak, 24 March 1949, box 13, folder 5, series II, Paul Papers.

CHAPTER 3

1. Talk given at symposium of Nuclear Proteins, Viruses and Atypical Growth at Yale, box 158, folder: Talks, 1962–1963, series III, James V. Neel Papers.
2. M. Susan Lindee, "Scaling Up: Human Genetics as a Cold War Network," *Studies in the History and Philosophy of Biological and Biomedical Sciences* 47 (2014): 185–90.
3. Barry Commoner, *Science and Survival* (New York: Viking, 1963), 25.
4. Ibid., 28.
5. In the 1990s, sociologist Ulrich Beck would diagnose this era as having birthed "the risk society"—a mode of modernity in which the unanticipated forces of industrialization stood to harm those who had contributed to their

production. Ulrich Beck, *Risk Society: Towards a New Modernity* (London: Sage, 1992).

6. Commoner, *Science and Survival*, 28.
7. Beck, *Risk Society*.
8. Commoner, *Science and Survival*, 25
9. James V. Neel, *Physician to the Gene Pool: Genetic Lessons and Other Stories* (New York: J. Wiley, 1994), 118–19.
10. Such concerns were also articulated by theologians, including Reinhold Niebuhr, *The Nature and Destiny of Man* (New York: Charles Scribner's Sons, 1964); Ewert H. Cousins, *Hope and the Future of Man* (Philadelphia: Fortress, 1972); Benedict XVI, *Faith and the Future* (Chicago: Franciscan Herald Press, 1971).
11. This was one among many "evolutionary futures" that animated mid-twentieth-century biology. Erika Milam, "The Ascent of Man and the Politics of Humanity's Evolutionary Future," paper given at the 2015 History of Science Society Annual Meeting in San Francisco, CA. See also Harrison Brown, *The Challenge of Man's Future: An Inquiry Concerning the Condition of Man during the Years That Lie Ahead* (1954; New York: Viking, 1963); Commoner, *Science and Survival*; Garrett Hardin, *Nature and Man's Fate* (New York: New American Library, 1959); Peter Medawar, *The Future of Man*, BBC Reith Lectures (New York: Basic Books, 1959); W. G. Smillie, *Public Health: Its Promise for the Future* (New York: Macmillan, 1955); Gordon Wolstenholme, ed., *Man and His Future* (Boston: Little, Brown, 1963); William L. Thomas, *Man's Role in Changing the Face of the Earth* (Chicago: University of Chicago Press, 1956).
12. "The Proposed International Biological Program, An Evaluation by an Ad Hoc Committee of the National Academy of Sciences," 14, folder: USNC/IBP: Ad hoc, Report: Evaluation of Proposed IBP, series 5, National Academy of Sciences, International Biological Program Papers, (hereafter NAS).
13. Ibid.
14. Ibid.
15. The seven sections were Terrestrial Conservation, Terrestrial Productivity, Marine Productivity, Freshwater Productivity, Use & Management of Resources, Production Processes, and Human Adaptability. Toby A. Appel, *Shaping Biology: The National Science Foundation and American Biological Research, 1945–1975* (Baltimore: Johns Hopkins University Press, 2000); James H. Capshew and Karen A. Rader, "Big Science: Price to the Present," *Osiris* 7 (1992): 2–25; Frank Greenaway, *Science International: A History of the International Council of Scientific Unions* (Cambridge: Cambridge University Press, 1996).
16. Joel Hagen, *An Entangled Bank: The Origins of Ecosystem Ecology* (New Brunswick, NJ: Rutgers University Press, 1992). Sharon Kingsland, *The Evolution of American Ecology, 1890–2000* (Baltimore: Johns Hopkins University Press, 2005).
17. E. B. Worthington, *The Evolution of IBP*, vol. 1 (Cambridge: Cambridge University Press, 1975), 56.

18. Chunglin Kwa, "Representations of Nature Mediating between Ecology and Science Policy: The Case of the International Biological Programme," *Social Studies of Science* 17, no. 3 (1987): 413–42; Elena Aronova, Karen S. Baker, and Naomi Oreskes, "Big Science and Big Data in Biology: From the International Geophysical Year to the International Biological Program to the Long Term Ecological Research (LTER) Network, 1957–Present," *Historical Studies in the Natural Sciences* 40, no. 2 (2010): 183–224; David C. Coleman, *Big Ecology: The Emergence of Ecosystem Science* (Berkeley: University of California Press, 2010).
19. K. J. Collins and J. S. Weiner, *Human Adaptability: A History and Compendium of Research in the International Biological Programme* (London: Taylor and Francis, 1977), 14.
20. Ibid., 3.
21. An early important conceptual encounter was a 1950 meeting at Cold Spring Harbor Labs on the "Origin and Evolution of Man," described in Jenny Reardon, *Race to the Finish: Identity and Governance in an Age of Genomics* (Princeton, NJ: Princeton University Press, 2005).
22. Lindor Brown, preface to P. T. Baker and J. S. Weiner, eds., *The Biology of Human Adaptability* (Oxford: Clarendon, 1966), v.
23. James V. Neel and Francisco M. Salzano, "A Prospectus for Genetic Studies on the American Indians," in Baker and Weiner, *The Biology of Human Adaptability*.
24. Ibid., 249.
25. See also Stephen G. Brush, *Choosing Selection: The Revival of Natural Selection in American Evolutionary Biology, 1930–1970* (Philadelphia: American Philosophical Society, 2009), 72.
26. Neel, *Physician to the Gene Pool*, 160.
27. Damon, "Proposal for Use of R/V Alpha Helix," box 2, folder: Travel Corr. and Grant Appl., 1971–1972, Solomon Islands Expedition (W. W. Howells files), Peabody Museum of Archaeology and Ethnology at Harvard University, Archival Paper Collections (hereafter HPM).
28. James V. Neel, "The Study of Natural Selection in Primitive and Civilized Human Populations," *Human Biology* 3 (1958): 59.
29. Donna Jeanne Haraway, "Remodeling the Human Way of Life: Sherwood Washburn and the New Physical Anthropology," in *Bones, Bodies, Behavior: Essays on Biological Anthropology*, ed. George W. Stocking (Madison: University of Wisconsin Press, 1988), 206–60.
30. *Cold Spring Harbor Symposia on Quantitative Biology*, vol. 15, *Origin and Evolution of Man*, ed. Katherine Brehme Warren (Cold Spring Harbor, NY: Biological Laboratory, 1950); Michael Little and Kenneth R. Kennedy, eds., *Histories of American Physical Anthropology in the Twentieth Century* (New York: Lexington Books, 2010); Vassiliki Betty Smocovitis, "Humanizing Evolution: Anthropology, the Evolutionary Synthesis and the Prehistory of Biological Anthropology, 1927–1962," *Current Anthropology* 53, S5 (2012): S108–S125.
31. Haraway, "Remodeling the Human Way of Life," 242.

32. Neel and Salzano, "A Prospectus for Genetic Studies on the American Indians," 253.
33. Collins and Weiner, *Human Adaptability*, 4.
34. Ibid.
35. In the tradition of "atoms for peace," described in Jacob Darwin Hamblin, "Exorcising Ghosts in the Age of Automation: United Nations Experts and Atoms for Peace," *Technology and Culture* 47, no. 4 (2006): 734–56; Richard Hewlett and Jack M. Holl, *Atoms for Peace and War, 1953–1961: Eisenhower and the Atomic Energy Commission* (Berkeley: University of California Press, 1989); Angela N. H. Creager, *Life Atomic: A History of Radioisotopes in Science and Medicine* (Chicago: University of Chicago Press, 2013).
36. In this regard they were engaged in "big science" of the sort more commonly associated with postwar physics. For example, Derek de Solla Price, *Little Science, Big Science* (New York: Columbia University Press, 1963); Peter Louis Galison and Bruce William Hevly, *Big Science: The Growth of Large-Scale Research* (Stanford, CA: Stanford University Press, 1992); Alvin Weinberg, *Reflections on Big Science* (Cambridge, MA: MIT Press, 1967); Robert Seidel, "A Home for Big Science: The Atomic Energy Commission's Laboratory System," *Historical Studies in Physical and Biological Sciences* 16, no. 1 (1985): 135–75; Daniel J. Kevles, "Big Science and Big Politics in the United States: Reflections on the Death of the SSC and the Life of the Human Genome Project," *Historical Studies in the Physical and Biological Sciences* 2 (1997): 269–97.
37. Roger D. Reid, "Biology—a Weapon for Peace* or Why We Need an 'IBY,'" Naval Research Reviews (Office of Naval Research, Department of the Navy, 1961), folder: IBP Publicity, 1961–1967, series 5, NAS.
38. Aldous Huxley, *Brave New World* (New York: Doubleday, 1932), quoted in Reid, "Biology—a Weapon for Peace," unpaginated.
39. See, for example Brown, *The Challenge of Man's Future*; Hardin, *Nature and Man's Fate.*
40. Reid, "Biology—a Weapon for Peace."
41. Worthington, *The Evolution of IBP*, 1.
42. For an in-depth analysis of IGY, particularly its polar dimensions, see Dian Olson Belanger, *Deep Freeze: The United States, the International Geophysical Year, and the Origins of Antarctica's Age of Science* (Boulder: University Press of Colorado, 2007).
43. Jacob Darwin Hamblin, *Arming Mother Nature: The Birth of Catastrophic Environmentalism* (New York: Oxford University Press, 2013), 94.
44. Also, Barton Worthington, who had previously played a role in colonial ecological interventions in Africa. Helen Tilley, *Africa as a Living Laboratory: Empire, Development, and the Problem of Scientific Knowledge, 1870–1950* (Chicago: University of Chicago Press, 2011).
45. Conrad Waddington, *The Scientific Attitude* (1941; London: Penguin 1948), vii.
46. Mary Midgley, *Science as Salvation: A Modern Myth and Its Meaning* (New York: Routledge, 1992). On the subject of early twentieth-century biological

humanism, see Marianne Sommer, "Biology as a Technology of Social Justice in Interwar Britain: Arguments from Evolutionary History, Heredity, and Human Diversity," *Science, Technology, and Human Values* 39, no. 4 (2014): 561–86.

47. In his classic essay on Merton's norms of science, David Hollinger notes that "although Waddington was distressed with the state of existing democratic societies and called for their reconstruction along the lines of the values more fully embodied in science, he identified himself as a 'democrat' and insisted that it was by democratic standards that the political philosophy of Marxism was to be judged." David A. Hollinger, *Science, Jews, and Secular Culture: Studies in Mid-Twentieth-Century American Intellectual History* (Princeton, NJ: Princeton University Press, 1996), 88.
48. Haraway, "Remodeling the Human Way of Life"; Perrin Selcer, "The View from Everywhere: Disciplining Diversity in Post–World War II International Social Science," *Journal of the History of the Behavioral Sciences* 45, no. 4 (2009): 309–29; Selcer, "Beyond the Cephalic Index," *Current Anthropology* 53, no. S5 (2012): S173–84.
49. Worthington, *The Evolution of IBP*, 1.
50. G. Ledyard Stebbins, "International Horizons in the Life Sciences," *AIBS Bulletin* 12, no. 6 (1962): 13. See also Stebbins, "International Biological Program," *Science* 137, no. 3532 (1962): 768–70.
51. Stebbins, "International Biological Program," 768.
52. Collins and Weiner, *Human Adaptability*.
53. Stebbins, "International Horizons in the Life Sciences," 16.
54. Worthington, *The Evolution of IBP*, 3.
55. D. F. Roberts and G. A. Harrison, eds., *Natural Selection in Human Populations* (New York: Pergamon Press, 1959), 59
56. R. MacArthur and E. O. Wilson, *The Theory of Island Biogeography* (Princeton, NJ: Princeton University Press, 1967); Janet Browne, *The Secular Ark: Studies in the History of Biogeography* (New Haven, CT: Yale University Press, 1983).
57. O. H. Frankel and E. Bennett, eds., *Genetic Resources in Plants—Their Exploration and Conservation*, IBP Handbook 11 (Philadelphia: F. A. Davis, 1970), 476.
58. Ibid., 470.
59. Ibid.
60. Collins and Weiner, *Human Adaptability*, 3.
61. Odd Arne Westad, *The Global Cold War: Third World Interventions and the Making of Our Times* (Cambridge: Cambridge University Press, 2005).
62. W. F. Blair, *Big Biology: The US/IBP* (Stroudsburg, PA: Dowden, Hutchinson, and Ross, 1977), 71. Blair did emphasize that cooperation did not just mean going to collect samples and then leaving, though this is often what happened.
63. Worthington, *The Evolution of IBP*, 53–54.
64. See also Sebastián Gil-Riaño, "Historicizing Anti-Racism: UNESCO's Campaigns against Race Prejudice in the 1950s" (PhD diss., University of Toronto, 2014).
65. Haraway, "Remodeling the Human Way of Life."

66. Brush, *Choosing Selection*, 99. See also the discussion of Neel's work in Daniel J. Kevles, *In the Name of Eugenics: Genetics and the Uses of Human Heredity* (Berkeley: University of California Press, 1986). The tension between relief of human suffering and human improvement in medical genetics is described in Nathaniel Comfort, *The Science of Human Perfection: How Genes Became the Heart of American Medicine* (New Haven, CT: Yale University Press, 2012).
67. George W. Stocking, *Race, Culture, and Evolution: Essays in the History of Anthropology* (New York: Free Press, 1968).
68. Peter Medawar, foreword to *Human Biology: An Introduction to Human Evolution, Variation, and Growth*, by G. A. Harrison et al., v–vi (New York: Oxford University Press, 1964).
69. Peter Medawar, "Problems of Adaptation," *New Biology* 11 (1951): 10–26.
70. Harrison to Fejos, 13 July 1964, "The Biology of Populations of Anthropological Importance" (June 29–July 12, 1964), no. 23, folder: Administrative Papers, Wenner-Gren Foundation for Anthropological Research (hereafter WGF).
71. Gabriel W. Lasker, "Human Biological Adaptability," *Science* 166, no. 3912 (1964): 1484. Twenty-first-century human biologists credit the published proceedings of the Burg Wartenstein meeting, entitled *Human Adaptability*, along with Lasker's article, with laying the foundations for the modern study of human variability: Stanley Ulijaszek and Rebecca Huss-Ashmore, *Human Adaptability: Past, Present, and Future* (Oxford: Oxford University Press, 1997).
72. Michael Little, "Human Population Biology in the Second Half of the Twentieth Century," *Current Anthropology* 53, no. S5 (2012): S126–S138.
73. Brush, *Choosing Selection*, 99.
74. Roberts and Harrison, *Natural Selection in Human Populations*.
75. E. B. Ford, "Polymorphism," *Biological Reviews* 20 (1945): 73–88.
76. Alice M. Brues, "Selection and Polymorphism in the A-B-O Blood Groups," *American Journal of Physical Anthropology* 12, no. 4 (1954): 559–97.
77. Theodosius Dobzhansky, "Evolution at Work," *Science* 127 (1958), 1093, cited in Brush, *Choosing Selection*, 99. See also Dobzhansky, *Evolution, Genetics and Man* (New York: John Wiley and Sons, 1955).
78. Philip Sheppard, "Natural Selection and Some Polymorphic Characters in Man," in *Natural Selection in Human Populations*, ed. D. F. Roberts and G. A. Harrison, 45–46 (New York: Pergamon, 1959) .
79. Keith Wailoo and Stephen Gregory Pemberton, *The Troubled Dream of Genetic Medicine: Ethnicity and Innovation in Tay-Sachs, Cystic Fibrosis, and Sickle Cell Disease* (Baltimore: Johns Hopkins University Press, 2006).
80. G. Montalenti, "Comment on Haldane, JBS; Disease and Evolution," *Rice Science* 19, suppl. (1949): 333–34; J. B. S. Haldane, "The Rate of Mutation of Human Genes," *Proceedings of the International Congress on Genetic Heredity* 35, suppl. (1949): 267–73.
81. G. Montalenti, *Infectious Diseases as Selective Agents: A Symposium* (Edinburgh: Oliver and Boyd, 1965).

82. Melbourne Tapper, *Sickle Cell Anemia and the Politics of Race* (Philadelphia: University of Pennsylvania Press, 1999); Keith Wailoo, *Dying in the City of the Blues: Sickle Cell Anemia and the Politics of Race and Health* (Chapel Hill: University of North Carolina Press, 2001); Wailoo, *Drawing Blood: Technology and Disease Identity in Twentieth-Century America* (Baltimore: Johns Hopkins University Press, 1997).
83. James V. Neel, "The Inheritance of Sickle-Cell Anemia," *Science* 110, no. 2846 (1949), 64–66.
84. Harvey A. Itano and James V. Neel, "A New Inherited Abnormality of Human Hemoglobin," *Proceedings of the National Academy of Sciences* 36 (1950): 613–17.
85. L. Pauling et al., "Sickle-Cell Anemia: A Molecular Disease," *Science* 110 (1949): 543–47; Bruno Strasser, "Linus Pauling's 'Molecular Diseases': Between History and Memory," *American Journal of Medical Genetics* 115 (2002): 83–93; Simon D. Feldman and Alfred I. Tauber, "Sickle Cell Anemia: Reexamining the First 'Molecular Disease,'" *Bulletin of the History of Medicine* 71 (1997): 623–50.
86. Strasser, "Linus Pauling's 'Molecular Diseases.'"
87. A. C. Allison et al., "Blood Groups in Some East African Tribes," *Journal of the Royal Anthropological Institute of Great Britain and Ireland* 82, no. 1 (1952): 55–61.
88. Jenny Bangham. "Blood Groups and the Rise of Human Genetics in Mid-Twentieth Century Britain" (PhD diss., Cambridge University, 2013).
89. A. C. Allison, "Protection Afforded by Sickle-Cell Trait against Subtertian Malarial Infection," *British Medical Journal* 1 (1954): 290–94; Allison, "The Distribution of the Sickle-Cell Trait in East Africa and Elsewhere, and Its Apparent Relationship to the Incidence of Subtertian Malaria," *Transactions of the Royal Society for Tropical Medicine and Hygiene* 48 (1954): 312–18; Allison, "Notes on Sickle-Cell Polymorphism," *Annals of Human Genetics* 19 (1954): 39–57. The finding meshed with arguments for the superior fitness of the heterozygote described by agricultural geneticist I. M. Lerner, *Genetic Homeostasis* (Berkeley: University of California Press, 1954).
90. J. M. Vandepitte et al., "Evidence Concerning the Inadequacy of Mutation as an Explanation of the Frequency of the Sickle Cell Gene in the Belgian Congo" *Blood* 10, no. 4 (1955): 341–50.
91. F. B. Livingstone, "Anthropological Implications of Sickle Cell Gene Distribution in West Africa," *American Anthropologist* 60 (1958): 533–62.
92. The collection of the 1959 samples in question is described in Arno G. Motulsky et al., "Population Genetic Studies in the Congo," *American Journal of Human Genetics* 18, no. 6 (1966): 514–37. The discovery of antibodies for HIV-1 was initially published as A. J. Nahmias, et al., "Evidence for Human Infection with HTLV III/LAV-like virus in Central Africa," *Lancet* 327, no. 8492 (1986): 1279–80.
93. F. B. Livingstone et al., "The Distribution of Several Blood Group Genes in Liberia, the Ivory Coast and Upper Volta," *American Journal of Physical Anthropology* 18 (1960): 161–78.

94. James V. Neel, "The Study of Human Mutation Rates," *American Naturalist* 86, no. 828 (1952): 129.
95. Ibid., 142.
96. Ibid.
97. James V. Neel, "On Some Pitfalls in Developing an Adequate Genetic Hypothesis," *American Journal of Human Genetics* 7, no. 1 (1955): 1.
98. Ibid.
99. Ibid., 10.
100. Claude Lévi-Strauss, "Race et culture," *Revue internationale des sciences sociales* 23, no. 4 (1971): 647–66.
101. Ibid., 618–19, cited in Staffan Müller-Wille, "Claude Lévi-Strauss on Race, History and Genetics," *BioSocieties* 5, no. 3 (2010): 330–47.
102. Müller-Wille, "Claude Lévi-Strauss on Race, History and Genetics."
103. Georges Canguilhem, *The Normal and the Pathological* (New York: Zone Books, 1989), 264.
104. Ibid.
105. Ibid.
106. Ibid., 243.
107. Neel, *Physician to the Gene Pool*, 96.
108. Ian Hacking, "Making People Up," in *The Science Studies Reader*, ed. Mario Biagioli (New York: Routledge, 1999), 161–71; Hacking, "Canguilhem amid the Cyborgs," *Economy and Society* 27 (1998): 202–16.
109. Canguilhem, *The Normal and the Pathological*, 239.
110. Duana Fullwiley, *The Enculturated Gene: Sickle Cell Health Politics and Biological Difference in West Africa* (Princeton, NJ: Princeton University Press, 2011).
111. John Beatty, "Scientific Collaboration, Internationalism, and Diplomacy: The Case of the Atomic Bomb Casualty Commission," *Journal of the History of Biology* 26, no. 2 (1993): 205–31; M. Susan Lindee, *Suffering Made Real: American Science and the Survivors at Hiroshima* (Chicago: University of Chicago Press, 1994).
112. He had also been influenced by H. J. Muller, "Our Load of Mutations," *American Journal of Human Genetics* 2 (1950): 111–76.
113. James V. Neel et al., "Studies on the Xavante Indians of the Brazilian Matto Grosso," *American Journal of Physical Anthropology* 16, no. 1 (1964): 52–140; Neel and Salzano, "A Prospectus for Genetic Studies on the American Indians."
114. Neel to Salzano, 28 December 1955, box 66, folder: Salzano, Francisco M. 1955–1962, series I, James V. Neel Papers.
115. Ibid.
116. Neel, "The Study of Natural Selection in Primitive and Civilized Human Populations."
117. Neel would also admit that, apart from his "elaborate rationalizations," his Amerindian studies presented a situation in which he could "test" himself and perhaps "ameliorate a vanishing world." Neel, *Physician to the Gene Pool*, 122.
118. Marcos Cueto, *Missionaries of Science: The Rockefeller Foundation and Latin America* (Bloomington: Indiana University Press, 1994); Anne-Emanuelle

Birn, *Marriage of Convenience: Rockefeller International Health and Revolutionary Mexico* (Rochester, NY: University of Rochester Press, 2006); Stephen Palmer, *Launching Global Health: The Caribbean Odyssey of the Rockefeller Foundation* (Ann Arbor: University of Michigan Press, 2010).

119. Salzano to Neel, 25 April 1959, box 66, folder: Salzano, Francisco M. 1955–1962, series I, James V. Neel Papers.
120. Neel to Salzano, 8 May 1959, box 66, folder: Salzano, Francisco M. 1955–1962, series I, James V. Neel Papers.
121. William Laughlin, ed., *Papers of the Physical Anthropology of the American Indian* (New York: Viking Fund, 1949), v.
122. Neel to Salzano, 8 May 1959, box 66, folder: Salzano, Francisco M. 1955–1962, series I, James V. Neel Papers; Sherwood Washburn, "The New Physical Anthropology," *Transactions of the New York Academy of Sciences* 13, no. 7 (1951): 298–304; Smocovitis, "Humanizing Evolution"; Haraway, "Remodeling the Human Way of Life."
123. F. B. Livingstone, *Abnormal Hemoglobins in Human Populations* (Chicago: Aldine, 1967), xi.
124. Neel to Salzano, 28 May 1959, Neel to Salzano, 23 November 1959, Salzano to Neel, 31 March 1960, box 66, folder: Salzano, Francisco M. 1955–1962, series I, James V. Neel Papers. On Motulsky and other human life scientists' subsequent contributions to pharmacogenomics see David S. Jones, "How Personalized Medicine Became Genetic, and Racial: Werner Kalow and the Formations of Pharmacogenetics," *Journal of the history of Medicine and Allied Sciences* (2011); Ricardo Ventura Santos, "Pharmacogenomics, Human Genetic Diversity and the Incorporation and Rejection of Color/Race in Brazil," *BioSocieties* 10 (2014): 48–69.
125. Claude Lévi-Strauss, *Tristes Tropiques* (New York: Penguin, 1992).
126. Neel to Salzano, 20 March 1962, box 66, folder: Salzano, Francisco M. 1955–1962, series I, James V. Neel Papers.
127. One of the earliest articulations of this thermodynamism was in Georges Charbonnier, *Conversations with Claude Lévi-Strauss*, trans. John Weightman and Doreen Weightman (London: Jonathan Cape, 1969). It was more widely disseminated in Claude Lévi-Strauss, *The Savage Mind* (Chicago: University of Chicago Press, 1968). Another famous thermodynamic metaphor of the period was Canadian media theorist Marshall McLuhan's theory of hot and cold media. Marshall McLuhan, *Understanding Media: The Extensions of Man* (New York: McGraw-Hill, 1964).
128. On the perspectives of Cold War modernization theorists as their efforts impacted indigenous peoples see Daniel Lerner, *The Passing of Traditional Society: Modernizing the Middle East* (Glencoe, IL: Free Press, 1958); Tom Brass. *Labor Regime Change in the Twenty-First Century: Unfreedom, Capitalism and Primitive Accumulation* (Leiden: Brill, 2011); Michael E. Latham. *The Right Kind of Revolution: Modernization, Development and U.S. Foreign Policy From the Cold War to the Present* (Ithaca, NY: Cornell University Press, 2011).

129. Bo Lindell and R. Lowry Dobson, *Ionizing Radiation and Health*, Public Health Papers 6 (Geneva: World Health Organization, 1961).
130. "Both WHO and PAHO would welcome the opportunity of collaborating with any research project that you propose. Perhaps you are already aware that the PAHO has a similar arrangement as WHO with the NIH to expand research in the Americas." Abraham Horwitz, director of PAHO, to Neel, 26 January 1962, box 66, folder: Salzano, Francisco M. 1955–1962, series I, James V. Neel Papers,
131. Neel to Salzano, 20 March 1962, box 66, folder: Salzano, Francisco M. 1955–1962, series I, James V. Neel Papers.
132. Ibid.
133. Horwitz to Neel, 26 January 1962, box 66, folder: Salzano, Francisco M. 1955–1962, series I, James V. Neel Papers
134. Neel to Salzano, 21 September 1962, box 66, folder: Salzano, Francisco M. 1955–1962, series I, James V. Neel Papers
135. Neel et al., "Studies on the Xavante Indians of the Brazilian Matto Grosso." For an account of the political circumstances in Brazil that would have shaped Neel and Salzano's encounter with the Xavante in the early 1960s, see Seth Garfield, *Indigenous Struggle at the Heart of Brazil: State Policy, Frontier Expansion, and the Xavante Indians, 1937–1988* (Durham, NC: Duke University Press, 1988). The statement about "physical specimens" comes from Neel, *Physician to the Gene Pool*, 129.
136. In this way, the genetic studies undertaken by Neel became linked to the "urgent anthropology" advocated by film archivists such as E. Richard Sorenson, the focus of Adrianna Link, "Salvaging a Record for Humankind: Urgent Anthropology at the Smithsonian Institution, 1964–1984" (PhD diss., Johns Hopkins University, 2016).
137. Chapters 1 and 2 of Rosanna Dent's University of Pennsylvania history and sociology of science dissertation in progress, currently titled "Research Subjectivities and the Political Economy of Science in Indigenous Brazil, 1958–2016," describe Neel and Salzano's encounters with Xavante communities in depth.
138. Neel, *Physician to the Gene Pool*, 121.
139. Ibid., 129.
140. For a related account, with particular attention to cytological research, see Soraya de Chadarevian, "Human Population Studies and the World Health Organization," *Dynamis* 35, no. 2 (2015): 359–88.
141. The Human Adaptability subgroup of IBP was formed in the spring of 1962. During this time period, the task was to formulate realistic proposals for fieldwork. Collins and Weiner, *Human Adaptability*, 4.
142. Representatively: D. Carleton Gajdusek and V. Zigas, "Kuru: Clinical, Pathological and Epidemiological Study of an Acute Progressive Degenerative Disease of the Central Nervous System among Natives of the Eastern Highlands of New Guinea," *American Journal of Medicine* 26, no. 3 (1959):

442–69; A. E. Mourant, "The Use of Blood Groups in Anthropology," *Journal of the Royal Anthropological Institute of Great Britain and Ireland* 77, no. 2 (1947): 139–44; James V. Neel and William Schull, *The Effect of Exposure to the Atomic Bombs on Pregnancy Termination in Hiroshima and Nagasaki* (Washington, DC: National Research Council, 1956).Other notable participants included Robert Kirk, Nigel Barnicot, and Francisco Salzano. William Laughlin (who would become head of the US National Committee for Human Adaptability in IBP) and Joseph Weiner (head of the International Committee for Human Adaptability in IBP) were in attendance as "consultants."

143. WHO, "Scientific Group on Research in Population Genetics of Primitive Groups," 27 November to 3 December 1962, Report to director general, folder: IBP, section HA 1963–1967, series 5, NAS. The version in the IBP archives at the US National Academy of Sciences is dated 18 March 1963, titled "Report to Director-General" and labeled "Restricted—For Internal Use." The final version of the report appeared in 1964 and is essentially the same as the 1963 document.
144. World Health Organization, *Research in Population Genetics of Primitive Groups*, World Health Organization Technical Report Series 279 (Geneva: World Health Organization, 1964), 5.
145. Ibid. For a detailed discussion of each of these points in the context of mid-twentieth-century human biology see Joanna Radin, "Latent Life: Concepts and Practices of Human Tissue Preservation in the International Biological Program," *Social Studies of Science* 43, no. 4 (2013): 483–508.
146. World Health Organization, *Research in Population Genetics of Primitive Groups*, 23.
147. Ibid., 26.
148. Neel, *Physician to the Gene Pool*, 130.
149. Neel, "The Study of Natural Selection in Primitive and Civilized Human Populations," 45.
150. Gajdusek to Neel, 6 February 1963, box 24, folder: D. Carleton Gajdusek, series I, James V. Neel Papers.
151. Collins and Weiner, *Human Adaptability*, 287.
152. Neel to Gajdusek, 23 April 1969, box 24, folder: D. Carleton Gajdusek, series I, James V. Neel Papers.
153. Critiques from as early as 1960 included Catherine H Berndt, "The Concept of Primitive," *Sociologus* 10, no. 1 (1960): 50–69; Stanley Diamond, "On the Origins of Modern Theoretical Anthropology," *American Anthropologist* 66 (1964): 127–29; R. H. Edwards, "Primitive," *Current Anthropology* 2 (1962): 396; Ashley Montagu, "The Fallacy of the Primitive," *Journal of the American Medical Association* 179, no. 12 (1962): 962–63; Francis L. K. Hsu, "Rethinking the Concept 'Primitive,'" *Current Anthropology* 179 (June 1964): 169–78.
154. Stanley Diamond, *In Search of the Primitive: A Critique of Civilization* (New Brunswick, NJ: Transaction Books, 1974), 297, 300.

155. Lasker to Campbell, 4 March 1964, folder: USNC/IBP: Ad hoc, Membership, Laughlin, WS, Survey of Biologists re Interest in IBP, 1964, series 1, NAS.
156. Motoo Kimura and James F. Crow, "The Number of Alleles That Can Be Maintained in a Finite Population," *Genetics* 49 (1964): 725–38; Motoo Kimura, *The Neutral Theory of Molecular Evolution* (New York: Cambridge University Press, 1983); William Provine, "The Neutral Theory of Molecular Evolution in Historical Perspective," in *Population Biology of Genes and Molecules*, ed. Naoyuki Takahata and James F. Crow (Tokyo: Baifukan, 1990).
157. Lasker to Campbell, 4 March 1964, folder USNC/IBP, Ad hoc, Membership, Laughlin, WS, Survey of Biologists re Interest in IBP, 1964, series 1, NAS.
158. Ibid.
159. Gabriel Ward Lasker and Michael A. Little, *Happenings and Hearsay: Experiences of a Biological Anthropologist* (Detroit: Savoyard Books, 1999).
160. Jonathan Marks, "The Human Genome Diversity Project: Good for If Not Good as Anthropology?" *Anthropology Newsletter*, April 1995; Marks, *Human Biodiversity: Genes, Race, and History* (New York: Alaine de Gruyter, 1995).
161. Lasker and Little, *Happenings and Hearsay*, 181.
162. Arturo Escobar, *Encountering Development: The Making and Unmaking of the Third World* (Princeton, NJ: Princeton University Press, 1995).
163. Alfred Sauvy, "Trois mondes, une planete," *L'Observateur* 118 (1952): 5.
164. Ronald Niezen, *The Origins of Indigenism: Human Rights and the Politics of Identity* (Berkeley: University of California Press, 2003).
165. Tom Brass, *Labor Regime Change in the Twenty-First Century: Unfreedom, Capitalism and Primitive Accumulation* (Leiden: Brill, 2011), 255.
166. See, for instance, Rosanna Dent's University of Pennsylvania dissertation in progress, a longitudinal study of the relationship between the Xavante, the first group Neel and Salzano studied, and academic research up through the present.
167. The implications of the importance of such bodies to science are in line with postcolonial science and technology studies arguments that the development of modern science and technologies have long depended upon the knowledge of native informants. David Hess has referred to this in terms of "epistemic primitive accumulation." David Hess, "Science in an Era of Globalization: Alternative Pathways," in *The Postcolonial Science and Technology Studies Reader*, ed. Sandra Harding (Durham, NC: Duke University Press, 2011). See also, Gabriela Soto Laveaga, *Jungle Laboratories: Mexican Peasants, National Projects, and the Making of the Pill* (Durham, NC: Duke University Press, 2009). In the realm of "biocapital," this process has recurred. See Melinda Cooper and Cathy Waldby *Clinical Labor: Tissue Donors and Research Subjects in the Global Bioeconomy* (Durham, NC: Duke University Press, 2014). On the phenomenon of biocapital more broadly see Kaushik Sunder Rajan, *Biocapital: The Constitution of Postgenomic Life* (Durham, NC: Duke University Press, 2006).
168. As described in Collins and Weiner, *Human Adaptability*, 151.

169. Johannes Fabian, *Time and the Other: How Anthropology Makes Its Object* (New York: Columbia University Press, 1983); Eric Wolf, *Europe and the People without History* (Berkeley: University of California Press, 1982).
170. World Health Organization, "Research on Human Population Genetics," World Health Organization Technical Report Series no. 387 (Geneva: WHO, 1968).
171. Around the same time, Sol Tax, the progressive editor of *Current Anthropology*, had begun a campaign to disentangle the study of indigenous peoples from the language of "primitiveness." He commissioned and published a series of articles meant to reckon with what he saw as an outmoded category. See, for example: Hsu, "Rethinking the Concept 'Primitive'"; Laura Thompson et al., "Steps toward a Unified Anthropology [and Comments and Reply]," *Current Anthropology* 8, nos. 1–2 (1967): 169–78.
172. Ricardo Ventura Santos, M. Susan Lindee, and Vanderlei Sabastiao de Souza, "Varieties of the Primitive: Human Biological Diversity Studies in Cold War Brazil (1962–1970)," *American Anthropologist* 116, no. 4 (2014): 723–35.
173. Ibid.
174. World Health Organization. "Research on Human Population Genetics."
175. Newton Morton, "Problems and Methods in the Genetics of Primitive Groups," *American Journal of Physical Anthropology* 28 (1968): 191–202.
176. Ibid., 192.
177. Ibid., 198.
178. Baruch Blumberg, for instance, who would win the Nobel Prize in 1976 for his ability to discern the viral etiology of hepatitis B—based on the analysis of thousands of blood samples from Pacific populations—was a proponent of this new mode of research. Baruch S. Blumberg, *Hepatitis B: The Hunt for a Killer Virus* (Princeton, NJ: Princeton University Press, 2002).
179. Morton, "Problems and Methods in the Genetics of Primitive Groups," 201.
180. Ibid., 196.
181. Ibid., 200.
182. The full text of this report was later published as "Research on Human Population Genetics Report of a WHO Scientific Group," in *Current Anthropology*, 11, no. 2 (April 1970): 225–33. The guidelines also informed a practical textbook called J. S. Weiner and John Adam Lourie, *Human Biology: A Guide to Field Methods* (Oxford: published for the International Biological Programme by Blackwell Scientific, 1969). Santos, Lindee, and de Souza, "Varieties of the Primitive."
183. "Research on Human Population Genetics," 5–6. David Jones has demonstrated the violence of the persistence of "virgin soil" narratives throughout the period in question. David S. Jones, "Virgin Soils Revisited," *William and Mary Quarterly*, series 3, 60, no. 4 (2003): 703–42. Virgin soil narratives, he argues, have been part of a larger project of legitimating colonial violence. David S. Jones *Rationalizing Epidemics: Meanings and Uses of American Indian Mortality since 1600* (Cambridge, MA: Harvard University Press, 2004).
184. Patrick Tierney, *Darkness in El Dorado: How Scientists and Journalists Devastated the Amazon* (New York: Norton, 2000); "Response to Allegations against

James V. Neel in *Darkness in El Dorado*, by Patrick Tierney," *American Journal of Human Genetics* 70, no. 1 (2002): 1–10; Robert Borofsky, *Yanomami: The Fierce Controversy and What We Might Learn from It*, California Series in Public Anthropology 12 (Berkeley: University of California Press, 2005).

185. James V. Neel, "The American Indian in the International Biological Program," paper presented at the Biomedical Challenges Presented by the American Indian, Pan American Sanitary Bureau, Regional Office of the WHO, 1968.
186. Ibid., 8.
187. Ibid., 9.
188. Ibid., 11.
189. Ibid.
190. M. Susan Lindee, "Voices of the Dead: James Neel's Amerindian Studies," in *Lost Paradises and the Ethics of Research and Publication*, ed. Francisco M. Salzano and A. Magdalena Hurtado (New York: Oxford University Press, 2004), 27–48.
191. Cited in Trudy Turner, "The Measles Epidemic of 1968," http://anthroniche.com/darkness_documents/0574.pdf.
192. Tierney, *Darkness in El Dorado*, 59. Cited also in Turner, "The Measles Epidemic of 1968."
193. Francisco M. Salzano and A. Magdalena Hurtado, eds., *Lost Paradises and the Ethics of Research and Publication* (New York: Oxford University Press, 2004).
194. James V. Neel, "Lessons from a 'Primitive' People: Do Recent Data Concerning South American Indians Have Relevance to Problems of Highly Civilized Communities?" *Science* 170, no. 3960 (1970): 819.
195. Review of first draft of Neel's "Lessons from a Primitive People" for *Science*, 1970, box 116, folder: Lessons from a "Primitive" People, series IIIa, James V. Neel Papers.
196. Ibid.
197. Peter Sachs Collopy, "Race Relationships: Collegiality and Demarcation in Physical Anthropology," *Journal of the History of the Behavioral Sciences* 51, no. 3 (2015): 237–60.
198. Ashley Motangu to Neel, 27 January 1971, box 116, folder: Lessons From a "Primitive" People, Series IIIa, James V. Neel Papers.
199. Ernest Becker, *The Denial of Death* (New York: Simon and Schuster, 1973).
200. Gruber, "Ethnographic Salvage and the Shaping of Anthropology," *American Anthropologist* 72, no. 6 (1970): 1290.
201. Adam Kuper, *The Reinvention of Primitive Society: Transformations of a Myth* (London: Routledge, 2005).
202. Gruber, "Ethnographic Salvage and the Shaping of Anthropology," 1290. For instance: Peter J. Bowler, "From 'Savage' to 'Primitive': Victorian Evolutionism and the Interpretation of Marginalized Peoples," *Antiquity* 66 (1992): 721–29; Kuper, *The Reinvention of Primitive Society*.
203. Gruber, "Ethnographic Salvage and the Shaping of Anthropology," 1289.

204. The papers contained in a 2014 special issue on human heredity all reckon with the rise of population thinking after World War II. The argument is summarized in Jenny Bangham and Soraya de Chadarevian, "Human Heredity after 1945: Moving Populations Centre Stage," *Studies in History and Philosophy of Science, Part C* 47 (September 2014).
205. Gruber, "Ethnographic Salvage and the Shaping of Anthropology," 1289.
206. Frank Spencer, ed., *A History of American Physical Anthropology, 1930–1980* (New York: Academic, 1982).
207. John C. Burke, "The Wild Man's Pedigree: Scientific Method and Racial Anthropology," in *The Wild Man Within: An Image in Western Thought from the Renaissance to Romanticism*, ed. Edward Dudley and Maximillian E. Novak (Pittsburgh: University of Pittsburgh Press, 1972), 277–78.
208. Thompson et al., "Steps toward a Unified Anthropology."
209. D. Carleton Gajdusek, "Urgent Opportunistic Observations: The Study of Changing, Transient and Disappearing Phenomena of Medical Interest in Disrupted Primitive Human Communities," in *Health and Disease in Tribal Societies*, compiled in Ciba Foundation Symposium 49 (Amsterdam: Excerpta Medica, 1977), 69–102.
210. "Man's Survival in a Changing World," no. 2, US Participation in the International Biological Program, personal collection of author.

CHAPTER 4

1. The black-white binary common in America was also registered in Melanesia (named for the dark pigment of its inhabitants); indeed scientists have long sought to draw connections between Melanesian islanders and Africans. Douglas Oliver, *Black Islanders: A Personal Perspective of Bougainville, 1937–1991* (Honolulu: University of Hawaii Press, 1991).
2. This and other films are available upon request from Scripps Institute of Oceanography Archives (hereafter SIO).
3. Jenny Bangham, "Blood Groups and the Rise of Human Genetics in Mid-Twentieth Century Britain" (PhD diss., Cambridge University, 2013).
4. Sandra Bamford, *Biology Unmoored: Melanesian Reflections on Life and Biotechnology* (Berkeley: University of California Press, 2007).
5. K. J. Collins and J. S. Weiner, *Human Adaptability: A History and Compendium of Research in the International Biological Programme* (London: Taylor and Francis, 1977).
6. Jonathan Rees, *Refrigeration Nation: A History of Ice, Appliances and Enterprise* (Baltimore: Johns Hopkins University Press, 2013).
7. Douglas P. Starr, *Blood: An Epic History of Medicine and Commerce* (New York: Alfred A. Knopf, 1998).
8. Brigadier General Douglas B. Kendrick, *Medical Department, U.S. Army, Blood Program in World War II* (Washington, DC: Office of the Surgeon General, 1989).
9. Starr, *Blood: An Epic History of Medicine and Commerce.*

10. Kendrick, *Blood Program in World War II.*
11. Knut Schmidt-Nielsen, *Per Scholander, 1905–1980: A Biographical Memoir* (Washington, DC: National Academy of Sciences, 1987); P. F. Scholander, *Enjoying a Life in Science: The Autobiography of P. F. Scholander* (Fairbanks: University of Alaska Press, 1990). On the IBP described by practitioners as "Big Biology" see W. F. Blair, *Big Biology: The US/IBP* (Stroudsburg, PA: Dowden, Hutchinson, and Ross, 1977).
12. "The Floating Biological Laboratory: Concept and Realization: An Assessment of the First Five Years of Operation of the Research Vessel ALPHA HELIX, 1966–1970 and Recommendations for Its Future Mission," report requested by the National Science Foundation, prepared by the National Advisory Board for the R/V *Alpha Helix*, May 1971, 27–28, SIO Reading Room.
13. Thereby solving the problem of coordinating displacements necessary to "raise the world" via the lab. Here, the *Alpha Helix* confronted the conceptualization held by human biologists that, to a certain extent, the lab *was* the world. Bruno Latour, "Give Me a Laboratory and I Will Raise the World," in *Science Observed: Perspectives on the Social Study of Science*, ed. Michael Mulkay and K. Knorr-Cetina (London: Sage, 1983), 141–70.
14. "The Floating Biological Laboratory," 33.
15. Richard Sorrenson, "Ship as a Scientific Instrument in the Eighteenth Century," *Osiris* 11 (1996): 221–36; Warwick Anderson, "Hybridity, Race, and Science: The Voyage of the Zaca, 1934–1935," *Isis* 103, no. 2 (2012): 229–53; Antony Adler, "The Ship as Laboratory: Making Space for Field Science at Sea," *Journal of the History of Biology* 47, no. 2 (2014): 333–62.
16. Thomas F. Gieryn, "Three Truth Spots," *Journal of the History of the Behavioral Sciences* 38, no. 2 (2002): 113–32.
17. Without a doubt, the *Alpha Helix* was an instrument for scientific imperialism to establish American networks across the globe. Nathan Reingold and Marc Rothenberg, *Scientific Colonialism: A Cross-Cultural Comparison* (Washington, DC: Smithsonian Institution Press, 1987). See also Sujit Sivasundaram, "Science," in *Pacific Histories: Ocean, Land, People,* ed. David Armitage and Alison Bashford (London: Pallgrave Macmillan, 2014), 237–62.
18. In terms of electrical power the ship had two 100 kW diesel electric generators that produced 440 V of sixty-cycle AC electricity, transformed to 115 V. A 10 kW generating plant supplied emergency power.
19. Ann Laura Stoler, *Along the Archival Grain: Epistemic Anxieties and Colonial Common Sense* (Princeton, NJ: Princeton University Press, 2009).
20. "The Floating Biological Laboratory."
21. The findings of the longitudinal project were published as Jonathan Scott Friedlaender, W. W. Howells, and John G. Rhoads, *The Solomon Islands Project: A Long-Term Study of Health, Human Biology, and Culture Change* (New York: Oxford University Press, 1987).
22. World Health Organization, *Research in Population Genetics of Primitive Groups*, WHO Technical Report Series 279 (Geneva: World Health Organiza-

tion, 1964). Damon participated in the 1968 working group that created "Research on Human Population Genetics: Report of a WHO Scientific Group," *Current Anthropology* 11, no. 2 (1970).

23. Department of Health, Education, and Welfare, Public Health Service. Application for Training Grant, "Biological Anthropology at Harvard," 10 August 1967, box 4, folder: Grant Applications, 1967–1975, HPM.
24. Damon to Howells, 24 July 1966, box 2, folder: Corr. & Study, 1966–1976, HPM.
25. Damon to Harvey Levy at Harvard's Kennedy Labs, 18 July 1972, box 2, folder, Levy, HPM.
26. Gajdusek to Damon, 25 March 1966, box 5, folder: Solomons Corresp Misc 1966, HPM.
27. Damon, "Proposal for Use of R/V Alpha Helix," box 2, folder: Travel Corr. and Grant Appl., 1971–1972, HPM.
28. Ibid.
29. Damon to William Nierenberg, 5 January 1973, box 1, folder 1.2, HPM.
30. Osborne to Damon, 17 January 1965, box 2, folder: Solomons Corresp. 1965, HPM.
31. Ibid.
32. Ibid.
33. Damon to John Peterson at American Breeders Service, 20 April 1966, box 5, folder: Solomons Corresp. Misc. 1966, HPM.
34. Rinfret to Damon, 28 April 1966, box 5, folder: Solomons Corresp Misc, 1966, HPM.
35. Oliver, *Black Islanders*.
36. W. W. Howells, "Anthropometry and Blood Types in Fiji and Solomon Islands: Based Upon Data of Dr. William L. Moss," *Anthropological Papers of the American Museum of Natural History* 33, no. 4 (1933).
37. For a comparative analysis of indigenous versus European conceptions of the Pacific, see Sivasundaram, "Science."
38. Damon, "Proposal for Use of R/V Alpha Helix."
39. Ibid.
40. Decades later, Pierre Maranda would find himself torn between past and present when the Christian descendants of the Lau elders entrusted him with secrets they did not want shared with Christians. Charles Montgomery, "The Octopus: Can the Myths of the Lau Survive Their Preservation," *Walrus*, May 2006, 52–59.
41. Damon to Pierre and Elli Maranda, 4 April 1968, box 2, folder: Corresp., Maranda, HPM.
42. A tendency that reached back at least to the nineteenth century. Niel Gunson, "British Missionaries and Their Contribution to Science in the Pacific Islands," in *Darwin's Laboratory: Evolutionary Theory and Natural History in the Pacific*, ed. Roy Macleod and P. F. Rehbock (Honolulu: University of Hawaii Press, 1994), 283–316.

43. Simon Schaffer et al., eds., *The Brokered World: Go-Betweens and Global Intelligence, 1770–1820* (Sagamore Beach, MA: Science History, 2009).
44. David L. Hilliard, "Colonialism and Christianity: The Melanesian Mission in the Solomon Islands," *Journal of Pacific History* 9 (1974): 93–116; A. R. Tippett, *Solomon Islands Christianity: A Study in Growth and Obstruction* (London: Lutterworth, 1967); Gunson, "British Missionaries."
45. L. L. Cavalli-Sforza, "Population Structure and Human Evolution," *Proceedings of the Royal Society of London, Series B, Biological Sciences* 164 (1965), described in Cavalli-Sforza, "Genes, Peoples and Languages," *Scientific American* 265, no. 5 (November 1991).
46. Marilyn Strathern, *Kinship, Law and the Unexpected: Relatives Are Always a Surprise* (New York: Cambridge University Press, 2005). See also Margaret Jolly, "Imagining Oceania: Indigenous and Foreign Representations of a Sea of Islands," *Contemporary Pacific* 19, no. 2 (2007): 508–45. The classic account of Euro-American biogenetic kinship is David Schneider, *American Kinship: A Cultural Account* (Englewood Cliffs, NJ: Prentice Hall, 1968).
47. Joshua Berson, "Linguistic Liberalism: Ethnography, Property, Northern Australia, and the Making of the Endangered Language, 1919–1992" (PhD diss., University of Pennsylvania, 2009).
48. Jonathan Scott Friedlaender and Joanna Radin, *From Anthropometry to Genomics: Reflections of a Pacific Fieldworker* (Bloomington, IN: IUniverse, 2009).
49. Keesing would publish a life history of one member of the Kwaio. Roger M. Keesing, *'Elota's Story: The Life and Times of a Solomon Islands Big Man* (1978; New York: Holt, Rinehart and Winston, 1983). The royalties from this book went to the Kwaio and 'Elota's family. Keesing would later coauthor a radical account of the Kwaio's involvement in the massacre of a colonial officer. Roger M. Keesing and Peter Corris, *Lightning Meets the West Wind: The Malaita Massacre* (New York: Oxford University Press, 1980).
50. Douglas Oliver to Keesing, 20 November 1963, Research Notes and Records—Correspondence in the Field, Photocopies, 1961–1980, box 33, folder 2, Roger Keesing Papers, Tuzin Archive for Melanesian Anthropology, University of California, San Diego (hereafter TAM).
51. Ibid.
52. Oliver to Eugene Ogan, 15 November 1964, Research Notes and Records—Correspondence in the Field, Photocopies, 1961–1980, box 33, folder 2, Roger Keesing Papers, TAM.
53. Ogan to Keesing, 13 May 1963, Research Notes and Records—Correspondence in the Field, Photocopies, 1961–1980, box 33, folder 2, Roger Keesing Papers, TAM.
54. Hal Ross to Nancy Lubin and Buffy Ellis, 12 April 1974, in unprocessed portion of Harvard Solomon Islands Project papers, HPM.
55. Ogan to Keesing, 15 May 1963, Research Notes and Records—Correspondence in the Field, Photocopies, 1961–1980, box 33, folder 2, Roger Keesing Papers, TAM.
56. Ogan to Keesing, 25 June 1963, Research Notes and Records—Correspondence in the Field, Photocopies, 1961–1980, box 33, folder 2, Roger Keesing Papers, TAM.

57. Ibid.
58. Henrika Kuklick has described the early importance of "energetic" conceptions of the relationship between a fieldworker and research subject. Malinowski and other social anthropologists in the British tradition drew on widely circulating ideas about the newly articulated laws of thermodynamics. Henrika Kuklick, "Personal Equations: Reflections on the History of Fieldwork, with Special Reference to Sociocultural Anthropology," *Isis* 102 (2011): 1–33.
59. Pierre Maranda to Damon, 24 February 1968, box 2, folder: Corresp., Maranda, HPM.
60. Keesing to Jonathan Friedlaender, 16 August 1977, unprocessed papers of Jonathan Friedlaender, APS.
61. Jonathan Scott Friedlaender, "Commentary: Changing Standards of Informed Consent: Raising the Bar," in *Biological Anthropology and Ethics: From Repatriation to Genetic Identity,* ed. Trudy Turner (Albany: State University of New York Press, 2005), 263–74.
62. Friedlaender and Radin, *From Anthropometry to Genomics.*
63. Keesing to Jonathan Friedlaender, 16 August 1977, unprocessed papers of Jonathan Friedlaender, APS.
64. Ibid.
65. These relations of power were not dissimilar to those that would emerge between contract research organizations and treatment of naïve patients living in resource-poor environments. Adriana Petryna, *When Experiments Travel: Clinical Trials and the Global Search for Human Subjects* (Princeton, NJ: Princeton University Press, 2009). See also J. Fairhead, M. Leach, and M. Small, "Where Techno-Science Meets Poverty: Medical Research and the Economy of Blood in the Gambia, West Africa," *Social Science and Medicine* 63, no. 4 (2006): 1109–20.
66. Presaging the relationships described in Johanna T. Crane, "Unequal 'Partners': AIDS, Academia and the Rise of Global Health," *Behemoth* 3, no. 3 (2010): 78–97.
67. Damon sent a clipping from the Massachusetts General Hospital Newsletter with the title "Of Ships & Skins & Rays, Cannibals & Cults" to Lot Page, which reported on the initial expedition. Across the top, Damon wrote "apologies for the hyperbole." Box 2, folder: SOLS publicity HPM.
68. Damon, "Proposal for Use of R/V Alpha Helix."
69. Ibid.
70. Ruth Oldenziel, "Islands: The United States as a Networked Empire," in *Entangled Geographies,* ed. Gabrielle Hecht (Cambridge, MA: MIT Press, 2011), 13–42.
71. Damon, "Proposal for Use of R/V Alpha Helix."
72. Ibid.
73. Ibid.
74. Dentition expert Howard Bailit to Jonathan Friedlaender, as described in Friedlaender and Radin, *From Anthropometry to Genomics.*

75. Damon, "Proposal for Use of R/V Alpha Helix."
76. Warwick Anderson, *The Collectors of Lost Souls: Turning Kuru Scientists into Whitemen* (Baltimore: Johns Hopkins University Press, 2008).
77. Warwick Anderson, "Objectivity and Its Discontents," *Social Studies of Science* 43 (2013): 557–76.
78. Nicholas Thomas, *Entangled Objects: Exchange, Material Culture, and Colonialism in the Pacific* (Cambridge, MA: Harvard University Press, 1991).
79. Damon to Gajdusek, 31 August 1972, in D. Carleton Gajdusek, *Journal of a Medical and Population Genetic Survey Expedition of the Research Vessel 'Alpha Helix' to the Banks and Torres Islands of the New Hebrides, Southern Islands of the British Solomon Islands Protectorate and Pingelap Atoll, Eastern Caroline Islands, 1972* (Bethesda: Study of Child Growth and Development and Disease Patterns in Primitive Cultures, National Institute of Neurological and Communicative Disorders and Stroke, 1985), 212. All subsequent references in this section cited as Gajdusek, *Journal*, are from this source.
80. Port Vila, Efate, New Hebrides, 13 September 1972, Gajdusek, *Journal*, 12.
81. Sorrenson, "Ship as a Scientific Instrument in the Eighteenth Century"; Neil Safier, "Global Knowledge on the Move: Itineraries, Amerindian Narratives, and Deep Histories of Science," *Isis* 101, no. 1 (2010): 133–45; Bruno Latour, *Science in Action: How to Follow Scientists and Engineers through Society* (Milton Keynes, Philadelphia: Open University Press, 1987).
82. Gajdusek to Eldon at Kuru Research Center, Papua New Guinea, 8 May 1971, Gajdusek, *Journal*, 186. Emphasis added. Anthropologist Sandra Widmer's analysis of these field notes lends support to arguments made in chapter 3 that Gajdusek deployed static ideas of culture to construct his subject populations as "primitive." Sandra Widmer, "Making Blood 'Melanesian': Fieldwork and Isolating Techniques in Genetic Epidemiology (1963–1976)," *Studies in History and Philosophy of Biological and Biomedical Sciences* 47 (2014): 118–29.
83. Warwick Anderson, "The Possession of Kuru: Medical Science and Biocolonial Exchange," *Comparative Studies in Society and History* 42 (2000): 713–44.
84. Gajdusek to William A. Nierenberg, 18 March 1971, Gajdusek, *Journal*, 169.
85. Damon's liquid nitrogen was not brought on the team's voyage.
86. Gajdusek to staff at NIH, 14 September 1972, Gajdusek, *Journal*, 218.
87. Port Vila, Efate, New Hebrides, 15 September 1972, Gajdusek, *Journal*, 13.
88. Merig Island, Levolvol village, 24 September 1972, Gajdusek, *Journal*, 40.
89. Merelava, St. Paul Lekwel, 21 September 1972, Gajdusek, *Journal*, 32.
90. Ibid.
91. Hu Island, 15 October 1972, Gajdusek, *Journal*, 76. The nature of Gajdusek's sexual relations with those he encountered in the field has been the subject of much interest. However, his intimacy with certain of those he encountered—however problematic—was connected to his ability to successfully form and sustain meaningful personal and intellectual relationships.

For a sensitive and perceptive treatment of the significance of this dynamic, see Anderson, *The Collectors of Lost Souls*.

92. Merig Island, Levolvol village, 24 September 1972, Gajdusek, *Journal*, 40–41.
93. Hu Island, 15 October 1972, Gajdusek, *Journal*, 78.
94. K. Knorr-Cetina, *Epistemic Cultures: How the Sciences Make Knowledge* (Cambridge, MA: Harvard University Press, 1999). See also Gieryn, "Three Truth Spots."
95. Marilyn Strathern, *Property, Substance, and Effect: Anthropological Essays on Persons and Things* (London: Athlone, 1999).
96. Graciosa Bay, Santa Cruz, Nende, 6 November 1972, Gajdusek, *Journal*, 121.
97. Ibid.
98. Gajdusek to Kirk, 15 November 1972, Gajdusek, *Journal*, 237.
99. Ponape, Caroline Islands, 21 November 1972, Gajdusek, *Journal*, 146.
100. Undated transcript of message from Kirk, ca. November 1972, Alpha Helix Journals 1972, unprocessed material of the D. (Daniel Carleton) Gajdusek Correspondence, 1934–1988, MS coll. 58, APS.
101. Neel submitted his proposal to NSF for the Amazon trip but indicated that he had also talked with Garey about "the possibility that we utilize the Alpha Helix for a much shorter time during the summer of 1976 off the coast of New Guinea." See Neel to C. Ladd Prosser, 3 July 1974, box 1, folder 28, SIO.
102. This was a marked change from research practices earlier in the century, as described in S. E. Lederer, *Subjected to Science: Human Experimentation in America before the Second World War* (Baltimore: Johns Hopkins University Press, 1995). See also Wolfgang U. Eckart. *Man, Medicine, and the State: The Human Body as an Object of Government Sponsored Medical Research in the 20th century* (Stuttgart: Steiner, 2006).
103. Laura Stark, *Behind Closed Doors: IRBs and the Making of Ethical Research* (Chicago: University of Chicago Press, 2011).
104. "Request for Human Use Authorization by the Committee to Review Grants for Clinical Research and Investigation Involving Human Beings," 1 March 1974, box 217, folder: National Science Foundation—International Biological Program—1973–1977, James V. Neel Papers.
105. Ibid., 3.
106. Ibid., 4.
107. The subject of refusal would come to be appreciated as a form of activism and resistance. Sherry B. Ortner, "Resistance and the Problem of Ethnographic Refusal," *Comparative Studies in Society and History* 37, no. 1 (1995): 173–93; Audra Simpson, "On Ethnographic Refusal: Indigeneity, 'Voice' and Colonial Citizenship," *Junctures: The Journal for Thematic Dialogue* 9 (December 2007): 67–80.
108. L. H. Woolsey, "The Leticia Dispute between Colombia and Peru," *American Journal of International Law* 29, no. 1 (1935): 94–99.
109. Neel to Jose Alberto de Mallo, 11 May 1976, box 1, folder 28, SIO.

110. Kendall W. Brown, *History of Mining in Latin America: From the Colonial Era to the Present* (Albequerque: University of New Mexico Press, 2012); Susanna B. Hecht, *Scramble for the Amazon and the "Lost Paradise" of Euclides da Cuhna* (Chicago: University of Chicago Press, 2013). On natural history as a form of resource extraction, see Hugh Raffles, *In Amazonia: A Natural History* (Princeton, NJ: Princeton University Press, 2002).
111. "Research Proposal and Budget," 14 March 1973, p. 5, box 1, folder 30, SIO.
112. Ibid., 2.
113. Neel to Walter Garey, 11 February 1976, box 1, folder 28, SIO.
114. Vera Alexander to Neel, 3 March 1976, box 1, folder 28, SIO.
115. Among the critiques leveraged at the IBP in its early days was that its participants were playing at "bio-politics," by which they meant using science to serve state power. E. B. Worthington, *The Evolution of IBP*, vol. 1 (Cambridge: Cambridge University Press, 1975), 53–54.
116. Garey to Neel, 25 May 1976, box 1, folder 28, SIO.
117. Survival International website, "FUNAI—National Indian Foundation (Brazil)," background briefing, http://www.survivalinternational.org/about/funai, accessed 12 January 2011. FUNAI has a complicated relationship to the peoples it claims to protect, critiqued in works like Jonathan W. Warren, *Racial Revolutions: Antiracism and Indian Resurgence in Brazil* (Durham, NC: Duke University Press, 2001); John Fred Peters, *Life Among the Yanomami* (New York: Broadview Press, 1998); John Hemming, *Die If You Must: Brazilian Indians in the Twentieth Century* (London: Pan Macmillan, 2003).
118. Survival International website, "FUNAI—National Indian Foundation (Brazil)."
119. Telecon from Neel to Garey, 8 June 1976, box 1, folder 24, Alpha Helix Amazon Expedition, 1976–7, Phase II, Neel/Forhan 5226, September 1975–June 1976, SIO.
120. One of the most sophisticated commentators on the unfolding legacy of violence of the Putumayo region—a few hundred miles from where Neel was set to launch his expedition—is anthropologist Michael Taussig. See Michael Taussig, *Shamanism, Colonialism, and the Wild Man: A Study in Terror and Healing* (Chicago: University of Chicago Press, 1986); Taussig, *The Devil and Commodity Fetishism in South America* (1983; Chapel Hill: University of North Carolina Press, 2010).
121. It is this idea of El Dorado that inspired a critique of Neel as an agent not only of extraction but of willful destruction. Patrick Tierney, *Darkness in El Dorado: How Scientists and Journalists Devastated the Amazon* (New York: W. W. Norton, 2000). While there are clearly disturbing parallels between the mining of mineral and bodily resources, I strongly disagree with Tierney's claims that human biologists like Neel had any intention of inducing disease in the populations they studied. For studies of this controversy and the evidence against Tierney's interpretation, see Francisco M. Salzano and A. Magdalena Hurtado, eds., *Lost Paradises and the Ethics of Research and Publication* (New York: Oxford University Press, 2004); Robert Borofsky, *Yanomami: The Fierce*

Controversy and What We Might Learn from It, California Series in Public Anthropology 12 (Berkeley: University of California Press, 2005).

122. S. Kirsch, "Lost Tribes: Indigenous People and the Social Imaginary," *Anthropological Quarterly* 70 no. 2 (1997): 58–67; B. M. Knauft, *From Primitive to Postcolonial in Melanesia and Anthropology* (Ann Arbor: University of Michigan Press, 1999). Anthropologist and activist David Maybury-Lewis, who collaborated with Neel on his early studies of the Xavante, argued that they had retreated from the onset of European colonialism in the nineteenth century. James V. Neel et al., "Studies on the Xavante Indians of the Brazilian Matto Grosso," *American Journal of Physical Anthropology* 16, no. 1 (1964): 52–140.
123. For example, Charles L. Briggs and Clara Mantini-Briggs, *Stories in the Time of Cholera: Racial Profiling during a Medical Nightmare* (Berkeley: University of California Press, 2004).
124. Telecon from Neel to Garey, 8 June 1976, box 1, folder 24, Alpha Helix Amazon Expedition, 1976–7, Phase II, Neel/Forhan 5226, September 1975–June 1976, SIO.
125. Neel to Salzano, 8 June 1976, box 1, folder 28, SIO.
126. Ibid.
127. Ibid.
128. Ibid.
129. The socio-political implications of hybridizing spiritual practices have been demonstrated in Jean Comaroff and John Comaroff, *Of Revelation and Revolution*, vol. 1, *Christianity, Colonialism, and Consciousness in South Africa* (Chicago: University of Chicago Press, 1991), and vol. 2, *The Dialectics of Modernity on a South African Frontier* (Chicago: University of Chicago Press, 1997).
130. Neel to Salzano, 8 June 1976, box 1, folder 28, SIO.
131. See the case of the Tasaday, who in 1971 were celebrated as a lost tribe, only to be revealed otherwise a decade later. John Nance, *The Gentle Tasaday: A Stone Age People in the Philippine Rain Forest* (New York: Harcourt Brace Jovanovich, 1975); Robin Hemley, *Invented Eden: The Elusive, Disputed History of the Tasaday* (New York: Farrar, Straus and Giroux, 2003).
132. Richard Sorrenson, "Ship as a Scientific Instrument in the Eighteenth Century."
133. Neel to Mary Johrde, 24 August 1976, box 1, folder 28, SIO.
134. Washburn, Sherwood. "The New Physical Anthropology." *Transactions of the New York Academy of Sciences* 13, no. 7 (1951): 298–304.
135. "Report on Phase II of the Alpha Helix Cruise of 1976–1977," p. 3, box 1, folder 26, SIO.
136. Ibid., 4.
137. Ibid. Emphasis added.
138. Ibid.
139. Ibid.
140. Pauline Turner Strong, "Fathoming the Primitive," *Ethnohistory* 33 (1986): 175–94.

141. *Health and Disease in Tribal Societies*, Ciba Foundation Symposium 49 (Amsterdam: Elsevier, 1977), 3.
142. Ibid., 121.
143. A highly relevant and important early critique is Mary Louise Pratt, *Imperial Eyes: Travel Writing and Transculturation* (New York: Routledge, 1992).
144. For an incisive summary of the controversy see Robert Borofsky, "Cook, Lono, Obeyesekere, and Sahlins," *Current Anthropology* 38, no. 2 (1997): 255–82.
145. Appel suggests that the *Alpha Helix* ultimately failed because biologists, more generally, were resistant to the administrative burden associated with national or regional facilities. Toby A. Appel, *Shaping Biology: The National Science Foundation and American Biological Research, 1945–1975* (Baltimore: Johns Hopkins University Press, 2000), 205.
146. Bamford, *Biology Unmoored*; Stefan Helmreich, "Trees and Seas of Information: Alien Kinship and the Biopolitics of Gene Transfer in Marine Biology and Biotechnology," *American Ethnologist* 30, no. 3 (2003): 340–58.

CHAPTER 5

1. Kenneth M. Weiss and Ranajit Chakraborty, "Genes, Populations, and Disease, 1930–1980: A Problem Oriented Review," in *A History of American Physical Anthropology*, ed. Frank Spencer (New York: Academic, 1982): 381.
2. Hans-Jörg Rheinberger, *Toward a History of Epistemic Things: Synthesizing Proteins in the Test Tube*, Writing Science (Stanford, CA: Stanford University Press, 1997).
3. The volume summarizing the findings of the multidimensional Harvard Solomon Islands Project, for instance, indicated that "[s]urprisingly, the most successful diagnostic material is dermatoglyphics. . . . We arrive at the unsatisfactory conclusion that most of the biological variation cannot be systematized." Jonathan Scott Friedlaender, W. W. Howells, and John G. Rhoads, *The Solomon Islands Project: A Long-Term Study of Health, Human Biology, and Culture Change* (New York: Oxford University Press, 1987), 12.
4. Francisco M. Salzano, ed., *The Role of Natural Selection in Human Evolution* (New York: Elsevier, 1975).
5. James V. Neel, "'Private' Genetic Variants and the Frequency of Mutation among South American Indians," *Proceedings of the National Academy of Sciences* 70 (1973): 3311–15.
6. Nicholas B. King, "The Scale Politics of Emerging Disease," *Osiris* 19 (2004): 62–76; Edward Hooper, *The River: A Journey to the Source of HIV and AIDS* (Boston: Little, Brown, 1999); Jaqcues Pépin, *The Origin of AIDS* (New York: Cambridge University Press, 2011).
7. Alfred S. Evans and Nancy E. Mueller, "The Past Is Prologue: Use of Serum Banks in Cancer Research," *Cancer Research* 57 (1992), 5560.
8. Neel to Salzano, 22 February 1991, box 66, folder: Salzano, Francisco M., series I, James V. Neel Papers.

9. Jenny Reardon, "The 'Persons' and 'Genomics' of Personal Genomics," *Personalized Medicine* 8, no. 1 (2011): 95–107; Michael Fortun, *Promising Genomics: Iceland and DeCODE Genetics in a World of Speculation* (Berkeley: University of California Press, 2008); Daniel J. Kevles and European Group on Ethics in Science and New Technologies to the European Commission, *A History of Patenting Life in the United States with Comparative Attention to Europe and Canada: A Report to the European Group on Ethics in Science and New Technologies* (Luxembourg: Office for Official Publications of the European Commission, 2002); Judith F. Minkove, "Safe Keeping," *Hopkins Medicine*, 1 October 2013, 20–23.
10. Ann Laura Stoler, ed., *Imperial Debris: On Ruins and Ruination* (Durham, NC: Duke University Press, 2013), 7.
11. James Clifford, *Returns: Becoming Indigenous in the Twenty-First Century* (Cambridge, MA: Harvard University Press, 2013); Ronald Niezen, *The Origins of Indigenism: Human Rights and the Politics of Identity* (Berkeley: University of California Press, 2003); Marisol de la Cadena and Orin Starn, *Indigenous Experience Today* (New York: Berg, 2007).
12. Cori Hayden, "A Biodiversity Sampler for the Millennium," in *Reproducing Reproduction: Kinship, Power and Technological Innovation*, ed. Sarah Franklin and Helena Ragone (Philadelphia: University of Pennsylvania Press, 1996), 173–206; Hayden, *When Nature Goes Public: The Making and Unmaking of Bioprospecting in Mexico*, In-Formation Series (Princeton, NJ: Princeton University Press, 2003).
13. Beth Conklin, "Shamans versus Pirates in the Amazonian Treasure Chest," *American Anthropologist* 104, no. 4 (2002): 1050–61. See also David Hess, "Science in an Era of Globalization: Alternative Pathways," in *The Postcolonial Science and Technology Studies Reader*, ed. Sandra Harding (Durham, NC: Duke University Press, 2011), 419–38.
14. Cressida Fforde, Jane Hubert, and Paul Turnbull, eds., *The Dead and Their Possessions: Repatriation in Principle, Policy and Practice* (London: Routledge, 2002). For differences between federally recognized and unrecognized Native American tribes, see Thomas Biolsi, "Imagined Geographies: Sovereignty, Indigenous Space, and American Indian Struggle," *American Ethnologist* 32, no. 2 (2005): 239–59; Mark Edwin Miller, *Forgotten Tribes: Unrecognized Indians and the Federal Acknowledgement Process* (Lincoln: University of Nebraska Press, 2004).
15. Rob Borofsky, "Returning Blood Samples to the Yanomami," Center for a Public Anthropology website, http://center-yanomami.publicanthropology.org, accessed 3 January 2015. A thoughtful perspective on the ethical issues associated disposition of the blood samples from Weiss's lab was published on 21 April 2010 by Anne Buchanan as "The Fierce Non-Controversy" at http://ecodevoevo.blogspot.com/2010/04/fierce-non-controversy.html, accessed 3 January 2015.
16. For theories of mutation in postcolonial science and technology studies, see Joseph Masco, "Mutant Ecologies: Radioactive Life in Post–Cold War New

Mexico," *Cultural Anthropology* 19, no. 4 (2004): 517–50; Emma Kowal, Joanna Radin, and Jenny Reardon, "Indigenous Body Parts, Mutating Temporalities, and the Half-Lives of Postcolonial Technoscience," *Social Studies of Science* 43, no. 4 (2013): 465–83.

17. On the distinction between labor and work see Arendt's critique of Marx, in which she theorizes reproductive activity and maintenance as forms of labor. For Marx, work is performed by those who build the world through engaging in the sphere of political action. Labor is done to sustain that world but does not produce new forms of action. Hannah Arendt, *The Human Condition*, 2nd ed. (Chicago: University of Chicago Press, 1998).
18. David Edgerton, *The Shock of the Old: Technology and Global History since 1900* (Oxford: Oxford University Press, 2007).
19. Marx, Karl. *Capital: A Critique of Political Economy*, vol. 2, *The Process of Circulation of Capital*, ed. Friedrich Engels, trans. Ernest Untermann (Chicago: Charles H. Kerr, 1909); William Cronon, *Nature's Metropolis: Chicago and the Great West* (New York: W. W. Norton, 1991).
20. On the future orientation and promissory value of biobanked materials, see Michael Fortun, "Mediated Speculations in the Genomics Futures Markets," *New Genetics and Society* 20, no. 2 (2001): 139–56; Fortun, *Promising Genomics*; Cathy Waldby, "Stem Cells, Tissue Cultures and the Production of Biovalue," *Health: An Interdisciplinary Journal for the Social Study of Health, Illness and Medicine* 6, no. 3 (2002): 305–23.
21. Donna Jeanne Haraway, "Situated Knowledges: The Science Question in Feminism and the Privilege of Partial Perspective," *Feminist Studies* 14 (1988): 575–99.
22. D. Andrew Merriwether, "Freezer Anthropology: New Uses for Old Blood," *Philosophical Transactions: Biological Sciences* 354, no. 1379 (1999): 121–29.
23. Paul Rabinow, *Making PCR: A Story of Biotechnology* (Chicago: University of Chicago Press, 1996), 169.
24. Merriwether, "Freezer Anthropology."
25. Rheinberger has made explicit the recursive dimensions of experimental systems. Hans-Jörg Rheinberger, *An Epistemology of the Concrete: Twentieth-Century Histories of Life* (Durham, NC: Duke University Press, 2010).
26. Merriwether, "Freezer Anthropology," 121.
27. Ibid.
28. Moses Schanfield, interview by author, 13 April 2009.
29. William Wills et al., "Hepatitis B Surface Antigen (Australia Antigen) in Mosquitoes Collected in Senegal, West Africa," *American Journal of Tropical Hygiene* 25 (1976): 186–90. I thank Marissa Mika for bringing this dimension of Blumberg's collection activities to my attention.
30. Oonagh Corrigan and Richard Tutton, *Genetic Databases: Socio-Ethical Issues in the Collection and Use of DNA* (London: Routledge, 2004); Geoffrey C. Bowker, *Memory Practices in the Sciences* (Cambridge, MA: MIT Press, 2005); Robin Bunton and Alan Petersen, *Genetic Governance: Health, Risk and Ethics in the Biotech Era* (New York: Routledge, 2005).

31. Merriwether, "Freezer Anthropology," 121.
32. Ibid.
33. Jacob S. Sherkow and Henry T. Greely, "The History of Patenting Genetic Material," *Annual Review of Genetics* 49 (November 2015): 161–82.
34. Rene Almeling, *Sex Cells: The Medical Market for Sperm and Eggs* (Berkeley: University of California Press, 2011).
35. Jonathan Scott Friedlaender, "Genes, People, and Property: Furor Erupts over Genetic Research on Indigenous Groups," *Cultural Survival Quarterly* 20, no. 2 (1996); M. M. Lock, "The Alienation of Body Tissue and the Biopolitics of Immortalized Cell Lines," *Body and Society* 7, nos. 2–3 (2001): 63–91; Priscilla Wald, "What's in a Cell? John Moore's Spleen and the Language of Bioslavery," *New Literary History* 36 (2005): 205–25.
36. Giving new coordinates for the species of biocapital mapped by Stefan Helmreich, "Species of Biocapital," *Science as Culture* 17, no. 4 (2008): 463–78; Alondra Nelson, *The Social Life of DNA: Race, Reparations, and Reconciliation after the Genome* (New York: Beacon, 2015).
37. For example, R. Santos, "Indigenous Peoples, Postcolonial Contexts, and Genomic Research in the Late Twentieth Century: A View from Amazonia (1960–2000)," *Critique of Anthropology* 22, no. 1 (2002): 81–104.
38. L. Luca Cavalli-Sforza et al., "Call for a Worldwide Survey of Human Genetic Diversity: A Vanishing Opportunity for the Human Genome Project," *Genomics* 11 (Summer 1991).
39. Leslie Roberts, "A Genetic Survey of Vanishing Peoples," *Science* 252 (1991): 1614–17.
40. Catherine Anne Bliss, *Race Decoded: The Genomic Fight for Social Justice* (Palo Alto, CA: Stanford University Press, 2012).
41. Jenny Reardon, *Race to the Finish: Identity and Governance in an Age of Genomics*, In-Formation Series (Princeton, NJ: Princeton University Press, 2005).
42. Jonathan Marks, "The Human Genome Divesity Project: Good for If Not Good as Anthropology?" *Anthropology Newsletter* 36 (April 1995): 72.
43. Margaret M. Lock, "Interrogating the Human Genome Diversity Project," *Social Science and Medicine* 39 (1994): 603–6.
44. Ibid.; Hayden, "A Biodiversity Sampler for the Millennium"; Donna Jeanne Haraway, *Modest_Witness@Second_Millennium. FemaleMan©_Meets_OncoMouse*TM*: Feminism and Technoscience* (New York: Routledge, 1997).
45. International organizations' role in fostering the indigenous-peoples' movement have at times also been based on nostalgic, Victorian-era ideas of the primitive. Adam Kuper, "The Return of the Native," *Current Anthropology* 44, no. 3 (2003).
46. Rural Advancement Fund International (RAFI), "Patents, Indigenous Peoples, and Human Genetic Diversity," *RAFI Communique*, May 1993. On the subject of bioprospecting, indigeneity, and activism see Hayden, "A Biodiversity Sampler for the Millennium"; Hayden *When Nature Goes Public*.

47. Their press releases can be accessed at the ETC group website, http://www.etcgroup.org/. The organization split and subsequently became known as the ETC Group, through which members have since policed innovations in synthetic biology, nanotechnology, and geoengineering. RAFI-USA continues to pursue food justice: http://rafiusa.org.
48. RAFI, "The Patenting of Human Genetic Material" in *RAFI Communiqué*, 1 February 1994: 1–12, http://www.etcgroup.org/content/patenting-human-genetic-material, accessed 10 June 2016.
49. Indigenous Peoples Council on Biocolonialism, "Indigenous People's Opposition to the HGDP," http://www.ipcb.org/resolutions/htmls/summary_indig_opp.html, accessed 29 September 2014.
50. Ibid. The 1968 report in question is "Biomedical Challenges Presented by the American Indian," paper presented at the PAHO Advisory Committee on Medical Research, Washington, DC, 1968.
51. Indigenous People's Council on Biocolonialism, "Indigenous People's Opposition to the HGDP."
52. Henry T. Greely, "Genes, Patents, and Indigenous Peoples: Biomedical Research and Indigenous Peoples' Rights," *Cultural Survival Quarterly* 20, no. 2 (1996).
53. Reardon, *Race to the Finish.*
54. F. Dukepoo, "The Trouble with the Human Genome Diversity Project," *Molecular Medicine Today* 4 (1998).
55. Reardon, *Race to the Finish.*
56. Robert Borofsky, *Yanomami: The Fierce Controversy and What We Might Learn from It*, California Series in Public Anthropology 12 (Berkeley: University of California Press, 2005); Michael M. J. Fischer, *Emergent Forms of Life and the Anthropological Voice* (Durham, NC: Duke University Press, 2003); Francisco M. Salzano and A. Magdalena Hurtado, eds., *Lost Paradises and the Ethics of Research and Publication* (New York: Oxford University Press, 2004).
57. This work dovetailed with other efforts to use the DNA of commensal species—from rats to pigs—as proxies for questions about human variation. See M. S. Allen and E. Matisoo-Smith, "Pacific 'Babes': Issues in the Origins and Dispersal of Pacific Pigs and the Potential of Mitochondrial DNA Analysis," *International Journal of Osteoarchaeology* 11, nos. 1–2 (2001): 4–13; E. Matisoo-Smith, "Animal Translocations, Genetic Variation, and the Human Settlement of the Pacific," in *Genes, Language, and Culture History in the Southwest Pacific*, ed. Jonathan Scott Friedlaender (Oxford: Oxford University Press, 2007), 157–70.
58. For a first-person account of his involvement in these controversies, see Jonathan Scott Friedlaender and Joanna Radin, *From Anthropometry to Genomics: Reflections of a Pacific Fieldworker* (Bloomington, IN: IUniverse, 2009).
59. Hannah Landecker and Angela N. H. Creager, "Technical Matters: Method, Knowledge and Infrastructure in Twentieth-Century Life Science," *Nature Methods* 6, no. 10 (2009): 701–5.

60. Hannah Landecker, *Culturing Life: How Cells Became Technologies* (Cambridge, MA: Harvard University Press, 2007).
61. "HGDP-CEPH Human Genome Diversity Cell Line Panel," Foundation Jean Dausset, Human Polymorphism Study Center, http://www.cephb.fr/en/hgdp_panel.php, accessed 10 June 2016.
62. Richard Yanagihara et al., "Isolation of HTLV-1 from Members of a Remote Tribe in Papua New Guinea," *New England Journal of Medicine* 323 (1990): 993–94; Yanagihara et al., "Human T Lymphotrophic Virus Type I Infection in Papua New Guinea: High Prevalence among the Hagahai Confirmed by Western Analysis," *Journal of Infectious Disease* 162 (1990): 649–54.
63. Neel to Salzano, 22 February 1991, box 67, folder: Salzano correspondence, James V. Neel Papers.
64. *John Moore, Plaintiff and Appellant, v. The Regents of California, et al., Defendants and Respondents*, 51 Cal 3d. 120, 210 *Cal. Rptr.* 146, 793 P.2d 479.
65. No perspectives of any member of the Hagahai were articulated in the accusations, only those of members associated with RAFI. Jenkins affirmed that the community was supportive of her research. See Jenkins correspondence with Friedlaender in his unprocessed papers at the American Philosophical Society.
66. G. Taubes, "Scientists Attacked for 'Patenting' Pacific Tribe," *Science* 270, no. 17 (November 1995): 1112.
67. For example, Margaret Lock, "The Alienation of Body Tissue and the Biopolitics of Immortalized Cell Lines"; Sandra Bamford, *Biology Unmoored: Melanesian Reflections on Life and Biotechnology* (Berkeley: University of California Press, 2007). The controversy also called attention to the issue of just how lost or isolated the Hagahai actually were. David Boyd, "A Tale of 'First Contact': The Hagahai of Papua New Guinea," *Research in Melanesia* 20 (1996): 103–40.
68. "Brazil: Indians' Genetic Material Sold on Internet," Cultural Survival website, https://www.culturalsurvival.org/news/brazil-indians-genetic-material-sold-internet, accessed 10 June 2016. Cell lines from these groups are also maintained by the Kidd lab, which makes its materials freely available.
69. A. Pottage, "The Inscription of Life in Law: Genes, Patents, and Bio-Politics," *Modern Law Review* 61 (1998).
70. David Maybury-Lewis, *Akwe-Shavante Society* (Oxford: Clarendon, 1967).
71. David Maybury-Lewis, *The Savage and the Innocent* (New York: Beacon, 1958).
72. Friedlaender, "Genes, People, and Property."
73. Ruth Liloqula, "Value of Life: Saving Genes versus Saving Indigenous Peoples," *Cultural Survival Quarterly* 20, no. 2 (1996): 42–45.
74. Aroha Te Pareake Mead, "Genealogy, Sacredness, and the Commodities Market," *Survival Quarterly* 20, no. 2 (1996): 46–51.
75. Emma Kowal and Joanna Radin, "Indigenous Biospecimens and the Cryopolitics of Frozen Life," *Journal of Sociology* 51, no. 1 (2015): 63–80.
76. Ryk Ward et al., "Extensive Mitochondrial Diversity within a Single Amerindian Tribe," *Proceedings of the National Academy of Science, USA* 88 (1991): 8720–24. Discussed in Alain Froment, "Anthropobiological Surveys in the

Field: Reflections on the Bioethics of Human Medical and DNA Surveys," in *Centralizing Fieldwork: Critical Perspectives from Primatology, Biological, and Social Anthropology*, ed. Jeremy MacClancy and Agustin Fuentes (New York: Berghahn, 2011): 186–99.

77. David Wiwchar, "Nuu-Chah-Nulth Blood Returns to West Coast," *Ha-Shilth-Sa* 31, no. 25 (2004): 1–4.
78. Patrick Tierney, *Darkness in El Dorado: How Scientists and Journalists Devastated the Amazon*, (New York: Norton, 2000).
79. Calls for the return of blood sparked debates among historians and anthropologists of science. See M. Susan Lindee, "Voices of the Dead: James Neel's Amerindian Studies," in *Lost Paradises and the Ethics of Research and Publication*, ed. Francisco M. Salzano and A. Magdalena Hurtado (New York: Oxford University Press, 2004), 27–48; Michael M. J. Fischer, "In the Science Zone: The Yanomami and the Fight for Representation," *Anthropology Today* 17, no. 4 (2001): 9–14.
80. Borofsky, *Yanomami: The Fierce Controversy*, 62.
81. Quoted in ibid., 64.
82. Quoted in ibid., 64.
83. Quoted in "Stolen Spirit: The Issue of Yanomami Blood," https://www.culturalsurvival.org/news/stolen-spirit-issue-yanomami-blood, accessed 14 June 2016.
84. Lindee, "Voices of the Dead."
85. Summarized in a report on the proceedings by Donald Brenneis, "On the El Dorado Taskforce Papers" *Anthropology News*, 43, no. 6 (2002): 8.
86. "Request for Human Use Authorization by the Committee to Review Grants for Clinical Research and Investigation Involving Human Beings," 1 March 1974, box 217, folder: National Science Foundation—International Biological Program—1973–1977, James V. Neel Papers.
87. David Glenn, "Blood Feud—A Controversy over South American DNA Samples Held in North America Ripples through Anthropology," Comissão Pró-Yanomami website, http://www.proyanomami.org.br/v0904/index.asp?pag=noticia&id=4300, accessed 14 June 2016.
88. Froment, "Anthropobiological Surveys in the Field," 196.
89. Jenny Reardon and Kim TallBear, "'Your DNA Is Our History': Genomics, Anthropology, and the Construction of Whiteness as Property," *Current Anthropology* 53, no. S5 (2012): 233–45; Kim TallBear, *Native American DNA: Tribal Belonging and the False Promise of Genomic Science* (Minneapolis: University of Minnesota Press, 2013); Donna Jeanne Haraway, "Remodeling the Human Way of Life: Sherwood Washburn and the New Physical Anthropology," in *Bones, Bodies, Behavior: Essays on Biological Anthropology*, ed. George W. Stocking (Madison: University of Wisconsin Press, 1988), 206–60.
90. Ann Kakaliouras, "An Anthropology of Repatriation: Contemporary Physical Anthropology and Native American Ontologies of Practice," *Current Anthropology* 53, no. S5 (2012): S213.

91. Warwick Anderson, "The Frozen Archive, or Defrosting Derrida," *Journal of Cultural Economy* 8 (2015): 379–87.
92. This section draws on material published as Joanna Radin and Emma Kowal, "Indigenous Blood and Ethical Regimes in the United States and Australia since the 1960s," *American Ethnologist* 42, no. 4 (2015): 749–65.
93. "Graduate Program in Biomedical Anthropology," Binghamton University website, http://biomedical.binghamton.edu/serum.htm, accessed 22 October 2008.
94. Ibid.
95. Klaus Hoeyer has argued that in the twenty-first century, biospecimens are increasingly coming to be regarded as rights-bearing extensions of persons. Klaus Hoeyer, *Exchanging Human Bodily Materials: Rethinking Bodies and Markets* (Dordrecht: Springer, 2013).
96. Rebecca Tsosie, "Cultural Challenges to Biotechnology: Native American Genetic Resources and the Concept of Cultural Harm," *Journal of Law, Medicine, and Ethics* 35, no. 3 (2007), 405.
97. For example, Debra Harry, "Acts of Self-Determination and Self-Defense: Indigenous Peoples' Responses to Biocolonialism," in *Rights and Liberties in the Biotech Age: Why We Need a Genetic Bill of Rights*, ed. Sheldon Krimsky and Peter Shorett (Lanham, MD: Rowan and Littlefield, 2005), 87–97; Sarah Franklin, "Ethical Biocapital: New Strategies of Cell Culture," in *Remaking Life and Death: Toward an Anthropology of the Biosciences,* ed. Sarah Franklin and Margaret M. Lock (Santa Fe, NM: School of American Research Press, 2003), 97–128.
98. Emma Kowal, Ashley Greenwood, and Rebekah E. McWhirter, "All in the Blood: A Review of Aboriginal Australians' Cultural Beliefs about Blood and Implications of Biospecimen Research," *Journal of Empirical Research on Human Research Ethics* 10, no. 4 (2015): 347–59.
99. Beth Povinelli, *Labor's Lot: The Power, History, and Culture of Aboriginal Action* (Chicago: University of Chicago Press, 1993).
100. H. McDonald, "East Kimberley Concepts of Health and Illness: A Contribution to Intercultural Health Programs in Northern Australia," *Australian Aboriginal Studies* 2 (2006): 86–97.
101. Kowal, Greenwood, and McWhirter, "All in the Blood," 355. See also Joan Cunningham and Terry Dunbar, "Consent for Long-Term Storage of Blood Samples by Indigenous Australian Research Participants: The Druid Study Experience," *Epidemiologic Perspectives and Innovations* 4, no. 7 (2007).
102. Schwarz's research on blood donation and transfusion among those who identify as Navajo suggests that concerns about contamination and sorcery from "enemy outsiders" impact who can give and receive blood. M. T. Schwarz, "Emplacement and Contamination: Mediation of Navajo Identity through Excorporated Blood," *Body and Society* 15, no. 2 (2009): 145–68.
103. Puneet Chawla Sahota, "Body Fragmentation: Native American Community Members' Views on Specimen Disposition in Biomedical/Genetics Research," *AJOB Empirical Bioethics* 5, no. 3 (2014): 19–30.

104. V. Y. Hiratsuka et al., "Alaska Native People's Perceptions, Understandings, and Expectations for Research Involving Biological Specimens," *International Journal for Circumpolar Health* 71 (2012). Similar findings apply to Indigenous Hawaiians. M. Tauali'i et al., "Native Hawaiian Views on Biobanking," *Journal of Cancer Education* 29, no. 3 (2014): 570–76.
105. Amy Harmon, "Where'd You Go with My DNA?" *New York Times*, 24 April 2010.
106. Adina Hoffman and Peter Cole, *Sacred Trash: The Lost and Found World of the Cairo Geniza* (New York: Shocken 2011).
107. Emma Kowal, "Orphan DNA: Indigenous Samples, Ethical Biovalue and Postcolonial Science," *Social Studies of Science* 43, no. 5 (2013).
108. Interview with Ralph Garruto by author, Binghamton, NY, 4 February 2010.
109. Radin and Kowal, "Indigenous Blood and Ethical Regimes."
110. I also obtained an IRB from University of Pennsylvania and from Binghamton to conduct research at the serum archive.
111. On the importance of "paper technologies" in the history of scientific practice see, Staffan Müller-Wille and Isabelle Charmantier. "Lists as Research Technologies," *Isis* 103, no. 4 (2012): 743–52; Volker Hess and J. Andrew Mendelsohn, "Case and Series: Medical Knowledge and Paper Technology, 1600–1900," *History of Science* 48, nos. 3–4 (2010): 287–314; Jenny Bangham, "Writing, Printing, Speaking: Rhesus Bloodgroup Genetics and Nomenclatures in the Mid-twentieth Century," *British Journal for the History of Science* 47, no. 2 (2014): 335–61.
112. For example, Adriana Petryna, "Ethical Variability: Drug Development and the Globalization of Clinical Trials," *American Ethnologist* 32, no. 2 (2005): 183–97; Laura Stark, *Behind Closed Doors: IRBs and the Making of Ethical Research* (Chicago: University of Chicago Press, 2011).
113. Deepa Reddy, "Good Gifts for the Common Good: Blood and Bioethics in the Market of Genetics Research," *Cultural Anthropology* 22, no. 3 (2007): 429–72.
114. Kowal, "Orphan DNA."
115. Laura Arbour and Doris Cook, "DNA on Loan: Issues to Consider When Carrying out Genetic Research with Aboriginal Families and Communities," *Community Genetics* 9 (2006): 153–60.
116. Rebecca L. Cann and J. Koji Lum, "Dispersal Ghosts in Oceania," *American Journal of Human Biology* 16 (2004): 440–51.
117. Randall Packard, *Making of a Tropical Disease: A Short History of Malaria* (Baltimore: Johns Hopkins University Press, 2007).
118. For example, C. W. Chan et al., "Flashback to the 1960s: Utility of Archived Sera to Explore the Origin and Evolution of *Plasmodium falciparum* Chloroquine Resistance in the Pacific," *Acta Tropica* 99 (2006): 15–22; K. Lum, "Contributions of Population Origins and Gene Flow to the Diversity of Neutral and Malaria Selected Autosomal Genetic Loci of Pacific Island Populations," in *Genes, Language, and Culture History in the Southwest Pacific*, ed. Jonathan Scott Friedlaender (New York: Oxford University Press, 2007), 219–30.

119. Angela N. H. Creager, "Adaptation or Selection? Old Issues and New Stakes in the Postwar Debates over Bacterial Drug Resistance," *Studies in History and Philosophy of Biological and Biomedical Sciences* 38, no. 1 (2007); Hannah Landecker, "Antibiotic Resistance and the Biology of History," *Body and Society*, published online before print, (2015), 1–34.
120. Chan et al., "Flashback to the 1960s."
121. Hannah Landecker, "Antibiotic Resistance and the Biology of History."
122. The long history of failed malaria eradication has been amply documented. See, for example, Packard, *Making of a Tropical Disease*; Leo B. Slater, *War and Disease: Biomedical Research on Malaria in the Twentieth Century* (New Brunswick, NJ: Rutgers University Press, 2009); Marcos Cueto, *Cold War, Deadly Fevers: Malaria Eradication in Mexico, 1955–1975* (Baltimore: Johns Hopkins University Press, 2007).
123. Much as nonhuman commensal species such as rats and pigs are being used as proxies for studies of human evolution and migration, so is malaria. For example, K. Tanabe et al., "*Plasmodium falciparum* Accompanied the Human Expansion out of Africa," *Current Biology* 20, no. 14 (2010): 1283–89.
124. This is happening in Australia, for example, in ways that are different than in the United States. Radin and Kowal, "Indigenous Blood and Ethical Regimes."
125. *The Genius and the Boys*, directed by Bosse Lindquist (SVT Sales, 2009).
126. John G. Baust, interview by author,, Binghamton, NY, 9 February 2010.
127. Ibid. I do not have evidence to confirm or deny this claim.
128. Melinda Cooper, *Life as Surplus: Biotechnology and Capitalism in the Neoliberal Era* (Seattle: University of Washington Press, 2008); Warwick Anderson, *The Collectors of Lost Souls: Turning Kuru Scientists into Whitemen* (Baltimore: Johns Hopkins University Press, 2008).
129. Jonathan Friedlander's own blood collections were adopted by the Marshfield Clinic, the site of a major personalized medicine initiative. "'One in a Million' Project Will Add 80,000 New Patients to Clinic Biobank," Marshfield Clinic Research Foundation website, 28 August 2015, http://www.marshfieldresearch.org/News/one-in-a-million-project-will-add-80-000-new-patients-to-clinic-biobank.
130. Kowal and Radin, "Indigenous Biospecimens and the Cryopolitics of Frozen Life."
131. Vanessa Hayes, "Indigenous Genomics," *Science* 332, no. 6030 (2011); Katrina Claw and Nánibaa Garrison, "Bringing Indigenous Researchers to the Forefront of Genomics," *SACNAS* (Winter 2015), http://sacnas.org/about/stories/sacnas-news/winter-2015-indigenous-genomics, accessed 10 June 2016. See also the Australian National Centre for Indigenous Genomics, http://ncig.anu.edu.au; Indigenous Genomics Alliance at University of Washington, http://depts.washington.edu/cghe/IGA.
132. Including Donna Jeanne Haraway, *When Species Meet* (Minneapolis: University of Minnesota Press, 2008); Sandra Harding, *Sciences from Below: Feminisms, Postcolonialities, and Modernities* (Durham, NC: Duke University

Press, 2008); Kim TallBear, "Beyond the Life / Not Life Binary: A Feminist-Indigenous Reading of Cryopreservation, Interspecies Thinking and the New Materialisms," in *Cryopolitics: Freezing Life in a Melting World,* ed. Joanna Radin and Emma Kowal (Cambridge, MA: MIT University Press, forthcoming); S. Eben Kirksey and Stefan Helmreich, "The Emergence of Multispecies Ethnography," *Cultural Anthropology* 25, no. 4 (2010): 545–76.

133. Bruno Latour, *We Have Never Been Modern* (Cambridge, MA: Harvard University Press, 1993); Donna Jeanne Haraway, *Simians, Cyborgs, and Women: The Reinvention of Nature* (New York: Routledge, 1991).
134. Rosamund Rhodes, Nada Gligorov, and Abraham Schwab, eds., *The Human Microbiome: Ethical, Legal, and Social Concerns* (Oxford: Oxford University Press, 2013), 1–2.
135. Ibid., 2.
136. E. B. Worthington, *The Evolution of IBP*, vol. 1 (Cambridge: Cambridge University Press, 1975).
137. Georges Canguilhem, "The Living Being and Its Environment (Milieu)," in *La connaissance de la vie* (1952; Paris: J. Vrin, 1980).
138. Peter J. Turnbaugh et al., "The Human Microbiome Project: Exploring the Microbial Part of Ourselves in a Changing World," *Nature* 449, no. 7164 (2007): 804.
139. Jose C. Clemente et al., "The Microbiome of Uncontacted Amerindians," *Science Advances* 1, no. 3 (2015), http://advances.sciencemag.org/content/1/3/e1500183.full, accessed 1 January 2016.
140. Quoted in Rose Eveleth, "Genetic Testing and Tribal Identity," *Atlantic*, 26 January 2015, http://www.theatlantic.com/technology/archive/2015/01/the-cultural-limitations-of-genetic-testing/384740/.

EPILOGUE

1. "Brazil: Blood Samples Returned to Yanomami after Nearly 50 Years," Survival website, 13 April 2015, http://www.survivalinternational.org/news/10727.
2. Robert Borofsky, "Returning Blood Samples to the Yanomami," Center for a Public Anthropology website, http://center-yanomami.publicanthropology.org.
3. Rick Kearns, "Yanomami of Brazil Honor Return of Stolen Blood," *Indian Country Today*, 10 April 2015, http://indiancountrytodaymedianetwork.com/2015/04/10/yanomami-brazil-honor-return-stolen-blood-159958, accessed 15 June 2015.
4. Quoted in "Yanomami Controversy," Asia-Pacific Economics Blog, 21 August 2014, http://apecsec.org/yanomami-controversy/.
5. Elisa Eiseman and Susanne Haga, *Handbook of Human Tissue Sources: A National Resource of Human Tissue Samples* (Santa Monica, CA: Rand, 1999).
6. Jim Vaught, *Testimony before the Subcommittee on Investigations and Oversight Committee on Science and Technology United States House of Representatives:*

Biorepository Policies and Practices of the National Cancer Institute, ed. Office of Biorepositories and Biospecimen Research (Bethesda: National Cancer Institute, National Institutes of Health, 2008), 5.

7. Ibid., 5.
8. An argument suggested by Jenny Reardon, "The 'Persons' and 'Genomics' of Personal Genomics," *Personalized Medicine* 8, no. 1 (2011). This line of reasoning is the subject of her forthcoming study of the landscape of personalized genomics, "The Postgenomic Condition."
9. Robert F. Weir and Robert S. Olick with Jeffrey C. Murray, *The Stored Tissue Issue: Biomedical Research, Ethics and Law in the Era of Genomic Medicine* (New York: Oxford University Press, 2004); Oonagh Corrigan and Richard Tutton, *Genetic Databases: Socio-Ethical Issues in the Collection and Use of DNA* (London: Routledge, 2004); Robert Mitchell and Catherine Waldby, "National Biobanks: Clinical Labor, Risk Production and the Creation of Biovalue," *Science, Technology and Human Values* 35, no. 3 (2010): 330–55.
10. Melissa Healy, "FDA Wins High-Profile Support in Consumer Genetics Kerfuffle," *Los Angeles Times*, 13 March 2014.
11. Jenny Reardon and Kim TallBear, "'Your DNA Is Our History': Genomics, Anthropology, and the Construction of Whiteness as Property," *Current Anthropology* 53, no. S5 (2012): 233–45.
12. See Nathan James, "DNA Testing in Criminal Justice: Background, Current Law, Grants, and Issues," Congressional Research Service, 6 December 2012, http://fas.org/sgp/crs/misc/R41800.pdf; Exonerations Worldwide, *The Innocence Project*, http://www.innocenceproject.org/exonerate/, accessed 10 June 2016.
13. Dorothy Roberts, *Fatal Invention: How Science, Politics, and Big Business Re-Create Race in the Twenty-First Century* (New York: New Press, 2011).
14. See US Food and Drug Administration, "Updates on Findings in the FDA Cold Storage Area on the NIH Campus," 16 July 2014, http://www.fda.gov/NewsEvents/Newsroom/PressAnnouncements/ucm405434.htm, accessed 21 October 2014; Maryn McKenna, "Update on the Found Vials: There Weren't 6; There Were 327 (Not All of Them Were Smallpox)," *Wired*, 16 July 2014, http://www.wired.com/2014/07/pox-four/, accessed 21 October 2014. On biosecurity preparedness see Andrew Lakoff, "The Generic Biothreat, or, How We Became Unprepared," *Cultural Anthropology* 23 no. 3 (2008): 399–428; Frédéric Keck, "Stockpiling as a Technique of Preparedness: Conserving the Past for an Unpredictable Future" in *Cryopolitics: Frozen Life in a Melting World*, ed. by Joanna Radin and Emma Kowal. (Cambridge, MA, MIT Press, forthcoming).
15. A perspective on the possibility for overcoming such relations in anthropological knowledge projects has been provided by a colleague of Garruto's at Binghamton, Douglas Holmes. Douglas R. Holmes and George E. Marcus, "Collaboration Today and the Re-Imagination of the Classic Scene of Fieldwork Encounter," *Collaborative Anthropologies* 1 (2008): 81–101. See also

George E. Marcus, ed., *Para-Sites: A Casebook against Cynical Reason* (Chicago: University of Chicago Press, 2000); Bruno Latour, "Why Has Critique Run out of Steam? From Matters of Fact to Matters of Concern," *Critical Inquiry* 30, no. 2 (2004): 225–48.

16. Michel Serres, *The Parasite*, trans. Lawrence R. Schehr (Minneapolis: University of Minnesota Press, 2007).
17. In the fall of 2015, this was the subject of an international workshop I co-organized at Yale with Ned Blackhawk, Ricardo Ventura Santos, and Veronika Lipphardt. It brought indigenous experts together with historians of science and medicine to discuss ways to expand our knowledge-making strategies.
18. Kim TallBear, "Beyond the Life / Not Life Binary: A Feminist-Indigenous Reading of Cryopreservation, Interspecies Thinking and the New Materialisms," in *Cryopolitics: Freezing Life in a Melting World*, ed. Joanna Radin and Emma Kowal (Cambridge, MA: MIT Press, forthcoming).
19. Linda Tuhiwai Smith, *Decolonizing Methodologies: Research and Indigenous Peoples*, 2nd ed. (London: Zed Books, 2012), 1.
20. Kim TallBear, *Native American DNA: Tribal Belonging and the False Promise of Genomic Science* (Minneapolis: University of Minnesota Press, 2013); James Clifford, *Returns: Becoming Indigenous in the Twenty-First Century* (Cambridge, MA: Harvard University Press, 2013).
21. Rebecca Skloot, "Your Cells. Their Research. Your Permission," *New York Times*, 30 December 2015.

Bibliography

Abu El-Haj, Nadia. *The Genealogical Science: The Search for Jewish Origins and the Politics of Epistemology*. Chicago: University of Chicago Press, 2012.

Adams, Vincanne, Michelle Murphy, and Adele Clarke. "Anticipation: Technoscience, Life, Affect, Temporality." *Subjectivity* 28 (2009): 246–65.

Adler, Antony. "The Ship as Laboratory: Making Space for Field Science at Sea." *Journal of the History of Biology* 47, no. 2 (2014): 333–62.

Allen, M. S., and E. Matisoo-Smith. "Pacific 'Babes': Issues in the Origins and Dispersal of Pacific Pigs and the Potential of Mitochondrial DNA Analysis." *International Journal of Osteoarchaeology* 11, nos. 1–2 (2001): 4–13.

Allison, A. C. "The Distribution of the Sickle-Cell Trait in East Africa and Elsewhere, and Its Apparent Relationship to the Incidence of Subtertian Malaria." *Transactions of the Royal Society for Tropical Medicine and Hygiene* 48 (1954): 312–18.

———. "Notes on Sickle-Cell Polymorphism." *Annals of Human Genetics* 19 (1954): 39–57.

———. "Protection Afforded by Sickle-Cell Trait against Subtertian Malarial Infection." *British Medical Journal* 1 (1954): 290–94.

Allison, A. C., E. W. Ikin, A. E. Mourant, and A. B. Raper. "Blood Groups in Some East African Tribes." *Journal of the Royal Anthropological Institute of Great Britain and Ireland* 82, no. 1 (1952): 55–61.

Almeling, Rene. *Sex Cells: The Medical Market for Sperm and Eggs*. Berkeley: University of California Press, 2011.

American Anthropological Association. *El Dorado Task Force Papers*. Vol. 2. 18 May 2002. http://anthroniche.com/darkness_documents/0599.pdf, accessed 22 June 2016.

Amrith, Sunil S. *Decolonizing International Health: India and Southeast Asia, 1930–65*. New York: Palgrave Macmillan, 2006.

Amsterdamska, Olga. “Standardizing Epidemics: Infection, Inheritance, and Environment in Prewar Experimental Epidemiology.” In *Heredity and Infection: The History of Disease Transmission*, edited by Jean-Paul Gaudillière and Ilana Lowy, 135–79. London: Routledge, 2001.

Anderson, Oscar Edward. *Refrigeration in America: A History of a New Technology and Its Impact*. Princeton, NJ: published for the University of Cincinnati by Princeton University Press, 1953.

Anderson, Warwick. “The Case of the Archive.” *Critical Inquiry* 39, no. 3 (2013): 532–47.

———. *The Collectors of Lost Souls: Turning Kuru Scientists into Whitemen*. Baltimore: Johns Hopkins University Press, 2008.

———. “The Frozen Archive, or Defrosting Derrida.” *Journal of Cultural Economy* 8 (2015), 379–87.

———. “Hybridity, Race, and Science: The Voyage of the Zaca, 1934–1935.” *Isis* 103, no. 2 (2012): 229–53.

———. “Natural Histories of Infectious Disease: Ecological Vision in Twentieth-Century Biomedical Science.” *Osiris* 19 (2004): 39–64.

———. “Objectivity and Its Discontents.” *Social Studies of Science* 43 (2013): 557–76.

———. “The Possession of Kuru: Medical Science and Biocolonial Exchange.” *Comparative Studies in Society and History* 42 (2000): 713–44.

Anderson, Warwick, and Ian R. MacKay. “Fashioning the Immunological Self: The Biological Individuality of F. Macfarlane Burnet.” *Journal of the History of Biology* 47, no. 1 (2014): 147–75.

Andrews, Lori B., and Dorothy Nelkin. *Body Bazaar: The Market for Human Tissue in the Biotechnology Age*. New York: Crown, 2001.

Appel, Toby A. *Shaping Biology: The National Science Foundation and American Biological Research, 1945–1975*. Baltimore: Johns Hopkins University Press, 2000.

Aronova, Elena, Karen S. Baker, and Naomi Oreskes. “Big Science and Big Data in Biology: From the International Geophysical Year to the International Biological Program to the Long Term Ecological Research (LTER) Network, 1957–Present.” *Historical Studies in the Natural Sciences* 40, no. 2 (2010): 183–24.

Arbour, Laura, and Doris Cook. “DNA on Loan: Issues to Consider When Carrying Out Genetic Research with Aboriginal Families and Communities.” *Community Genetics* 9 (2006): 153–60.

Arendt, Hannah. *The Human Condition*. 2nd ed. Chicago: University of Chicago Press, 1998.

Baker, P. T., and J. S. Weiner, eds. *The Biology of Human Adaptability*. Oxford: Clarendon, 1966.

Bamford, Sandra. *Biology Unmoored: Melanesian Reflections on Life and Biotechnology*. Berkeley: University of California Press, 2007.

Bangham, Derek R. *History of Biological Standardization: Characterization and Measurement of Complex Molecules Important in Clinical and Research Medicine;*

Contributions from the UK, 1900–1995. Bristol: published with the assistance of the Society for Endocrinology, 2000.

Bangham, Jenny. "Blood Groups and the Rise of Human Genetics in Mid-Twentieth Century Britain." PhD diss., Cambridge University, 2013.

———. "Writing, Printing , Speaking: Rhesus Bloodgroup Genetics and Nomenclatures in the Mid-twentieth Century." *British Journal for the History of Science* 47, no. 2 (2014): 335–61.

Bangham, Jenny, and Soraya de Chadarevian. "Human Heredity after 1945: Moving Populations Centre Stage." *Studies in History and Philosophy of Science, Part C* 47 (September 2014): 45–49.

Bashford, Alison. "Global Biopolitics and the History of World Health." *History of the Human Sciences* 19, no. 1 (2006): 67–88.

———. "Population, Geopolitics and International Organizations in the Mid-Twentieth Century." *Journal of World History* 19 (2008): 327–47.

Bauer, Susanne. "Mining Data, Gathering Variables and Recombining Information: The Flexible Architecture of Epidemiological Studies." *Studies in History and Philosophy of Biological and Biomedical Sciences* 39 (2008): 415–28.

Baust, John G., and John M. Baust, eds. *Advances in Biopreservation*. Boca Raton, FL: CRC, 2007.

Beatty, John. "Scientific Collaboration, Internationalism, and Diplomacy: The Case of the Atomic Bomb Casualty Commission." *Journal of the History of Biology* 26, no. 2 (1993): 205–31.

Beck, Ulrich. *Risk Society: Towards a New Modernity*. London: Sage, 1992.

Becker, Ernest. *The Denial of Death*. New York: Simon and Schuster, 1973.

Belanger, Dian Olson. *Deep Freeze: The United States, the International Geophysical Year, and the Origins of Antarctica's Age of Science*, Boulder: University Press of Colorado, 2007.

Belasco, Warren James, and Roger Horowitz. *Food Chains: From Farmyard to Shopping Cart*. Philadelphia: University of Pennsylvania Press, 2009.

Benedict XVI. *Faith and the Future*. Chicago: Franciscan Herald Press, 1971.

Benson, Keith Rodney, Jane Maienschein, and Ronald Rainger. *The Expansion of American Biology*. New Brunswick, NJ: Rutgers University Press, 1991.

Bernard, Claude. *An Introduction to the Study of Experimental Medicine*. 1927. Reprint, New York: Dover, 1957.

Berndt, Catherine H. "The Concept of Primitive." *Sociologus* 10, no. 1 (1960): 50–69.

Berson, Joshua. "Linguistic Liberalism: Ethnography, Property, Northern Australia, and the Making of the Endangered Language, 1919–1992." PhD diss., University of Pennsylvania, 2009.

Bhattacharya, Sanjoy. "WHO-Led or WHO-Managed? Re-Assessing the Smallpox Eradication Program in India, 1960–1980." In *Medicine at the Border: Disease, Globalization and Security, 1850 to the Present*, edited by Alison Bashford, 60–75. New York: Palgrave Macmillan, 2006.

Biolsi, Thomas. "Imagined Geographies: Sovereignty, Indigenous Space, and American Indian Struggle." *American Ethnologist* 32, no. 2 (2005): 239–59.

"Biomedical Challenges Presented by the American Indian." Paper presented at the Pan American Health Organization Advisory Committee on Medical Research, Washington, DC, 1968.

Birn, Anne-Emanuelle. *Marriage of Convenience: Rockefeller International Health and Revolutionary Mexico*. Rochester, NY: University of Rochester Press, 2006.

Black, Francis. "Serological Epidemiology in Measles." *Yale Journal of Biology and Medicine* 32, no. 1 (September 1959): 44–50.

Blair, W. F. *Big Biology: The US/IBP*. Stroudsburg, PA: Dowden, Hutchinson, and Ross, 1977.

Bliss, Catherine Anne. *Race Decoded: The Genomic Fight for Social Justice*. Palo Alto, CA: Stanford University Press, 2012.

Blumberg, Baruch S. *Hepatitis B: The Hunt for a Killer Virus*. Princeton, NJ: Princeton University Press, 2002.

———. *Proceedings of the Conference on Genetic Polymorphisms and Geographic Variations in Disease*. New York: Grune and Stratton, 1961.

Bolnick, Deborah, Duana Fullwiley, Troy Duster, Richard S. Cooper, Joan H. Fujimura, Jonathan Kahn, Jay S. Kaufman, et al. "The Science and Business of Genetic Ancestry Testing." *Science* 318, no. 5849 (October 2007): 399–400.

Borofsky, Robert. "Cook, Lono, Obeyesekere, and Sahlins." *Current Anthropology* 38, no. 2 (1997): 255–82.

———. *Yanomami: The Fierce Controversy and What We Might Learn from It*. California Series in Public Anthropology 12. Berkeley: University of California Press, 2005.

Borowy, Iris. *Coming to Terms with World Health: The League of Nations Health Organization, 1921–1946*. New York: Peter Lang, 2009.

———. "International Health Work: The Beginnings." *Michael* 8 (2011): 210–21.

Bowker, Geoffrey C. *Memory Practices in the Sciences*. Inside Technology. Cambridge, MA: MIT Press, 2005.

Bowler, Peter J. "From 'Savage' to 'Primitive': Victorian Evolutionism and the Interpretation of Marginalized Peoples." *Antiquity* 66 (1992): 721–29.

Boyd, David. "A Tale of 'First Contact': The Hagahai of Papua New Guinea." *Research in Melanesia* 20 (1996): 103–40.

Blumberg, Baruch S. *Hepatitis B: The Hunt for a Killer Virus*. Princeton, NJ: Princeton University Press, 2002.

Brandt, Allen. *No Magic Bullet: A Social History of Venereal Disease in the United States since 1880*. Expanded ed. New York: Oxford University Press, 1987.

Brass, Tom. *Labor Regime Change in the Twenty-First Century: Unfreedom, Capitalism and Primitive Accumulation*. Leiden: Brill, 2011.

Brenneis, Donald. "On the El Dorado Taskforce Papers." *Anthropology News* 43, no. 6 (2002): 8.

Bresalier, Michael. "Uses of a Pandemic: Forging the Identities of Influenza and Virus Research in Interwar Britain." *Social History of Medicine* 25, no. 2 (2011): 400–424.

Briggs, Charles L., and Clara Mantini-Briggs. *Stories in the Time of Cholera: Racial Profiling during a Medical Nightmare*. Berkeley: University of California Press, 2004.

Bristow, Adrian F., Trevor Barrowcliffe, and Derek R. Bangham. "Standardization of Biological Medicines: The First Hundred Years, 1900–2000." *Notes and Records of the Royal Society of London* 60, no. 3 (2006): 271–89.

Brockington, Fraser. *World Health*. Baltimore: Penguin Books, 1958.

Brown, David. "Asking Old Human Tissue to Answer New Scientific Questions." *Washington Post*, 16 April 2012.

Brown, Harrison. *The Challenge of Man's Future: An Inquiry Concerning the Condition of Man during the Years That Lie Ahead*. 1954. Reprint, New York: Viking, 1963.

Brown, Kendall W. *History of Mining in Latin America: From the Colonial Era to the Present*. Albuquerque: University of New Mexico Press, 2012.

Browne, Janet. *The Secular Ark: Studies in the History of Biogeography*. New Haven, CT: Yale University Press, 1983.

Brues, Alice M. "Selection and Polymorphism in the A-B-O Blood Groups." *American Journal of Physical Anthropology* 12 (1954): 559–97.

Brush, Stephen G. *Choosing Selection: The Revival of Natural Selection in American Evolutionary Biology, 1930–1970*. Philadelphia: American Philosophical Society, 2009.

———. *The Temperature of History: Phases of Science and Culture in the Nineteenth Century*. New York: B. Franklin, 1978.

Bunning, Jonny. "The Freezer Program: Value after Life." In *Cryopolitics: Frozen Life in a Melting World*, edited by Joanna Radin and Emma Kowal. Cambridge, MA: MIT University Press, forthcoming.

Bunton, Robin, and Alan Petersen. *Genetic Governance: Health, Risk and Ethics in the Biotech Era*. New York: Routledge, 2005.

Burke, John C. "The Wild Man's Pedigree: Scientific Method and Racial Anthropology." In *The Wild Man Within: An Image in Western Thought from the Renaissance to Romanticism*, edited by Edward Dudley and Maximillian E. Novak: 277–78. Pittsburgh: University of Pittsburgh Press, 1972.

Burn, J. H., D. J. Finney, and L. G. Goodwin. *Biological Standardization*. 2nd ed. London: Oxford University Press, 1950.

Burnet, F. Macfarlane. "Men or Molecules: A Tilt at Molecular Biology." *Lancet* 1 (1966): 37–39.

Burnet, F. Macfarlane, and David O. White. *Natural History of Infectious Disease*. Cambridge: Cambridge University Press, 1972.

Bynum, Caroline Walker. *Wonderful Blood: Theology and Practice in Late Medieval Northern Germany and Beyond*. Philadelphia: University of Pennsylvania Press, 2007.

Byrd, Jodi A. *The Transit of Empire: Indigenous Critiques of Colonialism*. Minneapolis: University of Minnesota Press, 2011.

Camporesi, Pierre. *Juice of Life: The Symbolic and Magic Significance of Blood*. New York: Continuum, 1995.

Campos, Luis. *Radium and the Secret of Life*. Chicago: University of Chicago Press, 2015.

Canguilhem, Georges. "The Living Being and Its Environment (Milieu)." In *La connaissance de la vie*, 129–54. 1952. Reprint, Paris: J. Vrin, 1980.

———. *The Normal and the Pathological*. New York: Zone Books, 1989.

Cann, H. M., C. De Toma, L. Cazes, M. F. Legrand, V. Morel, L. Piouffre, J. Bodmer, W. F. Bodmer, et al. "A Human Genome Diversity Cell Line Panel." *Science* 296, no. 5566 (2002): 261–62.

Cann, Rebecca L., and J. Koji Lum. "Dispersal Ghosts in Oceania." *American Journal of Human Biology* 16 (2004): 440–51.

Capshew, James H., and Karen A. Rader, "Big Science: Price to the Present," *Osiris* 7 (1992): 2–25.

Carrel, Alexis. "Latent Life of Arteries." *Journal of Experimental Medicine*, July 23, 1910.

Carrel, Alexis, and Charles Claude Guthrie. "Results of the Biterminal Transplantation of Veins." *American Journal of the Medical Sciences* 132, no. 3 (1906): 415–22.

Carse, Ashley. "Nature as Infrastructure: Making and Managing the Panama Canal Watershed." *Social Studies of Science* 42, no. 4 (2012), 539–63.

Casper, Monica, and Adele Clarke. "Making the Pap Smear into the 'Right Tool' for the Job: Cervical Cancer Screening in the USA, Circa 1940–95." *Social Studies of Science* 28, no. 2 (1998): 255–90.

Cavalli-Sforza, L. L. "Genes, Peoples and Languages." *Scientific American* 265, no. 5 (November 1991): 104–10.

———. "Population Structure and Human Evolution." *Proceedings of the Royal Society of London, Series B, Biological Sciences* 164 (1966): 362–79.

Cavalli-Sforza, L. Luca, Allan Wilson, Charles R. Cantor, Robert Cook-Deegan, and Mary-Claire King. "Call for a Worldwide Survey of Human Genetic Diversity: A Vanishing Opportunity for the Human Genome Project." *Genomics* 11, no. 2 (Summer 1991): 490–91.

Cave, Stephen. *Immortality: The Quest to Live Forever and How It Drives Civilization*. New York: Crown, 2012.

Chadarevian, Soraya de, and Harmke Kamminga. *Molecularizing Biology and Medicine: New Practices and Alliances, 1910s–1970s*. Studies in the History of Science, Technology and Medicine 6. Amsterdam: Harwood Academic, 1998.

Chambers, R., and H. P. Hale. "The Formation of Ice in Protoplasm." *Proceedings of the Royal Society* 110 (1932): 336–52.

Chan, C. W., D. Lynch, R. Spathis, F. W. Hombhanje, A. Kaneko, R. M. Garruto, and J. K. Lum. "Flashback to the 1960s: Utility of Archived Sera to Explore the Origin and Evolution of *Plasmodium falciparum* Chloroquine Resistance in the Pacific." *Acta Tropica* 99 (2006): 15–22.

Chang, Hasok. *Inventing Temperature: Measurement and Scientific Progress*. Oxford: Oxford University Press, 2004.

Charbonnier, Georges. *Conversations with Claude Lévi-Strauss*. Translated by John Weightman and Doreen Weightman. London: Jonathan Cape, 1969.

Chen, Mel. *Animacies: Biopolitics, Racial Mattering, and Queer Affect*. Durham, NC: Duke University Press, 2012.

Clarke, Adele. *Disciplining Reproduction: Modernity, American Life Sciences, and "the Problems of Sex"* Berkeley: University of California Press, 1998.

———. "Research Materials and Reproductive Science in the United States, 1910–1940." In *Physiology in the American Context, 1850–1940*, edited by Gerald Geison, 323–50. Bethesda, MD: American Physiological Society, 1987.

Clarke, Adele, Janet K. Shim, Laura Mamo, Jennifer Ruth Fosket, and Jennifer Fishman. *Biomedicalization: Technoscience, Health, and Illness in the U.S.* Durham, NC: Duke University Press, 2010.

Clarke, Bruce. *Energy Forms: Allegory and Science in the Era of Classical Thermodynamics*. Ann Arbor: University of Michigan Press, 2001.

Claw, Katrina, and Nánibaa Garrison. "Bringing Indigenous Researchers to the Forefront of Genomics." *SACNAS* (Winter 2015). http://sacnas.org/about/stories/sacnas-news/winter-2015-indigenous-genomics, accessed 10 June 2016.

Clemente, Jose C., et al. "The Microbiome of Uncontacted Amerindians." *Science Advances* 1, no. 3 (2015). http://advances.sciencemag.org/content/1/3/e1500183.short, accessed 1 January 2016.

Clifford, James. *Returns: Becoming Indigenous in the Twenty-First Century*. Cambridge, MA: Harvard University Press, 2013.

Cockburn, W. Chas. "The International Contribution to the Standardization of Biological Substances. I. Biological Standards and the League of Nations 1921–1946." *Biologicals* 19 (1991): 161–69.

Cold Spring Harbor Symposia on Quantitative Biology. Vol. 15, *Origin and Evolution of Man*. Edited by Katherine Brehme Warren. Cold Spring Harbor, NY: Biological Laboratory, 1950.

Coleman, David C. *Big Ecology: The Emergence of Ecosystem Science*. Berkeley: University of California Press, 2010.

Collins, K. J., and J. S. Weiner. *Human Adaptability: A History and Compendium of Research in the International Biological Programme*. London: Taylor and Francis, 1977.

Collopy, Peter Sachs. "Race Relationships: Collegiality and Demarcation in Physical Anthropology." *Journal of the History of the Behavioral Sciences* 51, no. 3 (2015): 237–60.

Comaroff, Jean, and John Comaroff. *Of Revelation and Revolution*. Vol. 1, *Christianity, Colonialism, and Consciousness in South Africa*. Chicago: University of Chicago Press, 1991

Comaroff, John, and Jean Comaroff. *Of Revelation and Revolution*. Vol. 2, *The Dialectics of Modernity on a South African Frontier*. Chicago: University of Chicago Press, 1997.

Comfort, Nathaniel. *The Science of Human Perfection: How Genes Became the Heart of American Medicine*. New Haven, CT: Yale University Press, 2012.

Commoner, Barry. *Science and Survival*. New York: Viking, 1963.

Conklin, Beth. "Shamans versus Pirates in the Amazonian Treasure Chest." *American Anthropologist* 104, no. 4 (2002): 1050–61.

Cook, Harold. "Time's Bodies." In *Merchants and Marvels: Commerce, Science and Art in Early Modern Europe*, edited by Pamela Smith and Paula Findlen, 223–47. New York: Routledge, 2002.

Cooper, Melinda. *Life as Surplus: Biotechnology and Capitalism in the Neoliberal Era*. Seattle: University of Washington Press, 2008.

Cooper, Melinda, and Catherine Waldby. *Clinical Labor: Tissue Donors and Research Subjects in the Global Bioeconomy*. Durham, NC: Duke University Press, 2014.

Coriell, Lewis L., Arthur E. Greene, and Ruth K. Silver. "Historical Development of Cell and Tissue Culture Freezing." *Cryobiology* 1, no. 1 (1964): 72–79.

Corrigan, Oonagh, and Richard Tutton. *Genetic Databases: Socio-Ethical Issues in the Collection and Use of DNA*. London: Routledge, 2004.

Cousins, Ewert H. *Hope and the Future of Man*. Philadelphia: Fortress, 1972.

Couzin-Frankel, Jennifer. "The Legacy Plan." *Science* 329 (July 9, 2010): 135–37.

Cowan, Ruth Schwartz. "How the Refrigerator Got Its Hum." In *The Social Shaping of Technology: How the Refrigerator Got Its Hum*, edited by Donald A. MacKenzie and Judy Wajcman, 202–18. Philadelphia: Open University Press, 1985.

Cowles, Henry M., et al. "Introduction." *Isis* 106, no. 3 (2015): 621–22.

Crane, Johanna T. "Unequal 'Partners': AIDS, Academia and the Rise of Global Health." *Behemoth* 3, no. 3 (2010): 78–97.

Creager, Angela N. H. "Adaptation or Selection? Old Issues and New Stakes in the Postwar Debates over Bacterial Drug Resistance." *Studies in History and Philosophy of Biological and Biomedical Sciences* 38, no. 1 (2007): 159–90.

———. *Life Atomic: A History of Radioisotopes in Science and Medicine*. Chicago: University of Chicago Press, 2013.

———. *The Life of a Virus: Tobacco Mosaic Virus as an Experimental Model, 1930–1965*. Chicago: University of Chicago Press, 2002.

———. "Producing Molecular Therapeutics from Human Blood: Edwin Cohn's Wartime Enterprise." In *Molecularizing Biology and Medicine: New Practices and Alliances, 1910s–1970s*, ed. Soraya de Chadarevian and Harmke Kamminga. Amsterdam: OPA, 1998: 99–128.

———. "Radioisotopes as Political Instruments, 1946–1953." *Dynamis* 29 (2009): 219–39.

Creager, Angela N. H., and Hannah Landecker. "Technical Matters: Method, Knowledge and Infrastructure in Twentieth-Century Life Science." *Nature Methods* 6, no. 10 (2009): 701–5.

Cronon, William. *Nature's Metropolis: Chicago and the Great West*. New York: W. W. Norton, 1991.

Cueto, Marcos. *Cold War, Deadly Fevers: Malaria Eradication in Mexico, 1955–1975*. Baltimore: Johns Hopkins University Press, 2007.

———. *Missionaries of Science: The Rockefeller Foundation and Latin America*. Bloomington: Indiana University Press, 1994.

Cunningham, Joan, and Terry Dunbar. "Consent for Long-Term Storage of Blood Samples by Indigenous Australian Research Participants: The Druid Study Experience." *Epidemiologic Perspectives and Innovations* 4, no. 7 (2007).

Daston, Lorraine, and Peter Louis Galison. *Objectivity*. New York: Zone, 2007.

De Chadarevian, Soraya. "Following Molecules: Hemoglobin between the Clinic and the Laboratory." In *Molecularizing Biology and Medicine: New Practices and Alliances, 1910s–1970s*, edited by Soraya de Chadarevian and H. Kamminga, 171–201. Amsterdam: OPA, 1998.

———. "Human Population Studies and the World Health Organization." *Dynamis* 35, no. 2 (2015): 359–88.

De la Cadena, Marisol, and Orin Starn. *Indigenous Experience Today*. New York: Berg, 2007.

De Solla Price, Derek. *Little Science, Big Science*. New York: Columbia University Press, 1963.

De Souza, Yvonne G., and John S. Greenspan. "Biobanking Past, Present and Future: Responsibilities and Benefits." *AIDS* 27, no. 3 (2013): 303–12.

Declich, S., and A. O. Carter. "Public Health Surveillance: Historical Origins, Methods and Evaluation." *Bulletin of the World Health Organization* 72, no. 2 (1994): 285–304.

"Deep-Freeze." *Time*, 28 April 1952, 68–71.

Derrida, Jacques. *Archive Fever: A Freudian Impression*. Chicago: University of Chicago Press, 1996.

Dessauer, Herbert C., and Mark S. Hafner, eds. *Collections of Frozen Tissues: Value, Management, Field and Laboratory Procedures, and Directory of Existing Collections*. Lawrence, KS: Association of Systematics Collections, 1984.

Diamond, Stanley. *In Search of the Primitive: A Critique of Civilization*. New Brunswick, NJ: Transaction Books, 1974.

———. "On the Origins of Modern Theoretical Anthropology." *American Anthropologist* 66 (1964): 127–29.

Dienel, Hans-Liudger. "Carl Linde and His Relationship with Georges Claude: The Cooperation between Two Independent Inventors in Cryogenics and Its Side Effects." In *History of Artificial Cold, Scientific, Technological and Cultural Issues*, edited by Kostas Gavroglu, 171–88. Dordrecht: Springer, 2014.

———. *Linde: History of a Technology Corporation, 1879–2004*. New York: Palgrave Macmillan, 2004.

Dobzhansky, Theodosius. "Evolution at Work." *Science* 127 (1958): 1091–98.

———. *Evolution, Genetics and Man*. New York: John Wiley and Sons, 1955.

Douglas, Mary. *Risk and Blame: Essays in Cultural Theory*. London: Routledge, 1992.

Doyle, Rodney. "Disciplined by the Future: The Promising Bodies of Cryonics." *Science as Culture* 6, no. 4 (1997): 582–616.

Dudley, Edward, and M. E. Novak, eds. *The Wild Man Within: An Image in Western Thought from the Renaissance to Romanticism*. Pittsburgh: University of Pittsburgh Press, 1972.

Dukepoo, F. "The Trouble with the Human Genome Diversity Project." *Molecular Medicine Today* 4 (1998): 242–43.

Duster, Troy. "Molecular Reinscription of Race: Unanticipated Issues in Biotechnology and Forensic Science." *Patterns of Prejudice* 40, nos. 4–5 (2006): 427–41.

Eckart, Wolfgang U. *Man, Medicine, and the State: The Human Body as an Object of Government Sponsored Medical Research in the 20th Century*. Stuttgart: Steiner, 2006.

Edgerton, David. *The Shock of the Old: Technology and Global History since 1900*. Oxford: Oxford University Press, 2007.

Edwards, Paul N. *The Closed World: Computers and the Politics of Discourse in Cold War America*. Cambridge, MA: MIT Press, 1996.

Edwards, R. H. "Primitive." *Current Anthropology* 2 (1962): 396.

Eiseman, Elisa, and Susanne Haga. *Handbook of Human Tissue Sources: A National Resource of Human Tissue Samples*. Santa Monica, CA: Rand, 1999.

Eiseman, Elisa, and Rand Corporation. *Case Studies of Existing Human Tissue Repositories: "Best Practices" for a Biospecimen Resource for the Genomic and Proteomic Era*. Santa Monica, CA: RAND, 2003.

Elser, William J., Ruth A. Thomas, and Gustav I. Steffen. "The Desiccation of Sera and Other Biological Products (Including Microorganisms) in the Frozen State with the Preservation of the Original Qualities of Products So Treated." *Journal of Immunology* 28, no. 6 (1935): 433–73.

"Embryo Is Frozen to Make Life." *Argus*, 14 February 1953.

Escobar, Arturo. *Encountering Development: The Making and Unmaking of the Third World*. Princeton, NJ: Princeton University Press, 1995.

Ettinger, Robert C. W. *The Prospect of Immortality*. Garden City, NY: Doubleday, 1964.

Evans, Alfred S. "Epidemiology and the Public Health Laboratory." *American Journal of Public Health* 57, no. 6 (1967): 1041–52.

———. "Serological Surveys: The Role of the WHO Reference Serum Banks." *World Health Organization Chronicle* 21 (1968): 185–90.

Evans, Alfred S., and Nancy E. Mueller. "The Past Is Prologue: Use of Serum Banks in Cancer Research." *Cancer Research* 57 (1992): 5557–60.

Eveleth, Rose. "Genetic Testing and Tribal Identity." *Atlantic*, 26 January 2015. http://www.theatlantic.com/technology/archive/2015/01/the-cultural-limitations-of-genetic-testing/384740/.

Fabian, Johannes. *Time and the Other: How Anthropology Makes Its Object*. New York: Columbia University Press, 1983.

Fairhead, J., M. Leach, and M. Small. "Where Techno-Science Meets Poverty: Medical Research and the Economy of Blood in the Gambia, West Africa." *Social Science and Medicine* 63, no. 4 (2006): 1109–20.

Farley, John. *Brock Chisholm, the World Health Organization, and the Cold War*. Vancouver: University of British Columbia Press, 2009.

———. *To Cast Out Disease: A History of the International Health Division of the Rockefeller Foundation (1913–1951)*. New York: Oxford University Press, 2004.

Farman, Abou. "Speculative Matter: Secular Bodies, Minds and Person." *Cultural Anthropology* 28, no. 4 (2013): 737–59.

Farman Farmaian, Abou Ali. "Secular Immortal." PhD diss., New York University, 2012.

Feldman, Simon D., and Alfred I. Tauber. "Sickle Cell Anemia: Reexamining the First 'Molecular Disease.'" *Bulletin of the History of Medicine* 71 (1997): 623–50.

Fforde, Cressida, Jane Hubert, and Paul Turnbull, eds. *The Dead and Their Possessions: Repatriation in Principle, Policy and Practice*. London: Routledge, 2002.

Fischer, Michael M. J. *Emergent Forms of Life and the Anthropological Voice*. Durham, NC: Duke University Press, 2003.

———. "In the Science Zone: The Yanomami and the Fight for Representation." *Anthropology Today* 17, no. 4 (2001): 9–14.

Fleck, A. C., and F. A. Ianni. "Epidemiology and Anthropology: Some Suggested Affinities in Theory and Method." *Human Organization* 16, no. 4 (1958): 38–40.

Flosdorf, Earl W., and Stuart Mudd. "Biologics Now Preserved by Drying from the Frozen State." *Refrigerating Engineering* 36, no. 6 (December 1938): 379–80.

Ford, E. B. "Polymorphism." *Biological Reviews* 20, no. 2 (1945): 73–88.

Fortun, Michael. "Mediated Speculations in the Genomics Futures Markets." *New Genetics and Society* 20, no. 2 (August 2001): 139–56.

———. *Promising Genomics: Iceland and DeCODE Genetics in a World of Speculation*. Berkeley: University of California Press, 2008.

Foucault, Michel. *The Archaeology of Knowledge*. London: Tavistock, 1972.

Fountain, Henry. "Alfred S. Evans, 78, Expert on Origins of Mononucleosis." *New York Times*, 25 January 1996.

Frankel, O. H., and E. Bennett, eds. *Genetic Resources in Plants—Their Exploration and Conservation*. IBP Handbook 11. Philadelphia: F. A. Davis, 1970.

Franklin, Sarah. "Ethical Biocapital: New Strategies of Cell Culture." In *Remaking Life and Death: Toward an Anthropology of the Biosciences*, edited by Sarah Franklin and Margaret M. Lock, 97–128. Santa Fe, NM: School of American Research Press, 2003.

Freidburg, Susanne. *Fresh: A Perishable History*. Cambridge, MA: Belknap Press of Harvard University Press, 2009.

Friedlaender, Jonathan Scott. "Commentary: Changing Standards of Informed Consent: Raising the Bar." In *Biological Anthropology and Ethics: From Repatriation to Genetic Identity*, edited by Trudy Turner, 263–74. Albany: State University of New York Press, 2005.

———, ed. *Genes, Language, and Culture History in the Southwest Pacific*. Human Evolution Series. Oxford: Oxford University Press, 2007.

———. "Genes, People, and Property: Furor Erupts over Genetic Research on Indigenous Groups." *Cultural Survival Quarterly* 20, no. 2 (1996): 22.

Friedlaender, Jonathan Scott, W. W. Howells, and John G. Rhoads. *The Solomon Islands Project: A Long-Term Study of Health, Human Biology, and Culture Change*. New York: Oxford University Press, 1987.

Friedlaender, Jonathan Scott, and Joanna Radin. *From Anthropometry to Genomics: Reflections of a Pacific Fieldworker*. Bloomington, IN: IUniverse, 2009.

Froment, Alain. "Anthropobiological Surveys in the Field: Reflections on the Bioethics of Human Medical and DNA Surveys." In *Centralizing Fieldwork: Critical Perspectives from Primatology, Biological, and Social Anthropology*, edited by Jeremy MacClancy and Agustin Fuentes, 186–99. New York: Berghahn, 2011.

Fujimura, Joan H., Deborah Bolnick, Ramya Rajagopalan, Jay S. Kaufman, Richard C. Lewontin, Troy Duster, Pilar Ossorio, and Jonathan Marks. "Clines without Classes: How to Make Sense of Human Variation." *Sociological Theory* 32, no. 3 (2014): 208–27.

Fuller, Barry J., Nick Lane, and Erica E. Benson, eds. *Life in the Frozen State*. Boca Raton, FL: CRC, 2004.

Fullwiley, Duana. *The Enculturated Gene: Sickle Cell Health Politics and Biological Difference in West Africa*. Princeton, NJ: Princeton University Press, 2011.

Gajdusek, D. Carleton. *Journal of a Medical and Population Genetic Survey Expedition of the Research Vessel Alpha Helix to the Banks and Torres Islands of the New Hebrides, Southern Islands of the British Solomon Islands Protectorate and Pingelap Atoll, Eastern Caroline Islands, September 8, 1972 to November 26, 1972*. Bethesda, MD: Study of Child Growth and Development and Disease Patterns in Primitive Cultures, National Institute of Neurological and Communicative Disorders and Stroke, 1985.

———. "Urgent Opportunistic Observations: The Study of Changing, Transient and Disappearing Phenomena of Medical Interest in Disrupted Primitive Human Communities." In *Health and Disease in Tribal Societies* Ciba Foundation Symposium 49, 69–102. Amsterdam: Elsevier, 1977.

Gajdusek, D. Carleton, and V. Zigas. "Kuru: Clinical, Pathological and Epidemiological Study of an Acute Progressive Degenerative Disease of the Central Nervous System among Natives of the Eastern Highlands of New Guinea." *American Journal of Medicine* 26, no. 3 (1959): 442–69.

Galison, Peter Louis, and Bruce William Hevly. *Big Science: The Growth of Large-Scale Research*. Stanford, CA: Stanford University Press, 1992.

Garfield, Seth. *Indigenous Struggle at the Heart of Brazil: State Policy, Frontier Expansion, and the Xavante Indians, 1937–1988*. Durham, NC: Duke University Press, 1988.

Garner, W. E. "Gustav Tammann, 1861–1938." *Journal of the Chemical Society* (1952), 1961–73.

Gaudillière, Jean-Paul. "From Propaganda to Scientific Marketing: Schering, Cortisone, and the Construction of Drug Markets." *History and Technology* 29, no. 2 (2013): 188–209.

Gavroglu, Kostas. *History of Artificial Cold, Scientific, Technological and Cultural Issues*. Dordrecht: Springer, 2014.

Gieryn, Thomas F. *Cultural Boundaries of Science: Credibility on the Line*. Chicago: University of Chicago Press, 1999.

———. "Three Truth Spots." *Journal of the History of the Behavioral Sciences* 38, no. 2 (2002): 113–32.

Gil-Riaño, Sebastián. "Historicizing Anti-Racism: UNESCO's Campaigns against Race Prejudice in the 1950s." PhD diss., University of Toronto, 2014.

Gooday, Graeme. "Placing or Replacing the Laboratory in the History of Science?" *Isis* 99, no. 4 (2008): 783–95.

Gordon, Avery. *Ghostly Matters: Haunting and the Sociological Imagination*. New University of Minnesota Press ed. Minneapolis: University of Minnesota Press, 2008.

Gould, Stephen Jay. *Rock of Ages: Science and Religion in the Fullness of Life*. New York: Ballantine, 1999.

Gradmann, Christoph. "Sensitive Matters: The World Heath Organisation and Antibiotic Resistance Testing, 1945–1975." *Social History of Medicine* 26, no. 3 (2013): 555–74.

Gradmann, Christoph, and Jonathan Simon, eds. *Evaluating and Standardizing Therapeutic Agents, 1890–1950*. New York: Palgrave Macmillan, 2010.

Greely, Henry T. "Genes, Patents, and Indigenous Peoples: Biomedical Research and Indigenous Peoples' Rights." *Cultural Survival Quarterly* 20, no. 2 (1996): 54.

Greenaway, Frank. *Science International: A History of the International Council of Scientific Unions*. Cambridge: Cambridge University Press, 1996.

Grosz, Elizabeth. *The Nick of Time: Politics, Evolution, and the Untimely*. Sydney: Allen and Unwin, 2004.

Grubb, Rune. "Interactions between Immunology and Genetics—Blood Group Systems as Important Early Models and Tools." In *Immunology, 1930–1980: Essays on the History of Immunology*, edited by Pauline H. Mazumdar, 131–42, Toronto: Wall and Thompson, 1989.

Gruber, Jacob. "Ethnographic Salvage and the Shaping of Anthropology." *American Anthropologist* 72, no. 6 (1970): 1289–99.

Guldi, Jo, and David Armitage. *The History Manifesto*. Cambridge: Cambridge University Press, 2014.

Gumbrecht, Hans Ulrich. *After 1945: Latency as Origin of the Present*. Stanford, CA: Stanford University Press, 2013.

Gunson, Niel. "British Missionaries and Their Contribution to Science in the Pacific Islands." In *Darwin's Laboratory: Evolutionary Theory and Natural History in the Pacific*, edited by Roy Macleod and Philip F. Rehbock, 283–316. Honolulu: University of Hawaii Press, 1994.

Guthe, T. "Freezing and Transport of Sera in Liquid Nitrogen at –150c to –196c." *Bulletin of the World Health Organization* 33, no. 6 (1965): 864–67.

Hacking, Ian. "Canguilhem amid the Cyborgs." *Economy and Society* 27 (1998): 202–16.

———. *Historical Ontology*. Cambridge, MA: Harvard University Press, 2002.

———. "Making People Up." In *The Science Studies Reader*, edited by Mario Biagioli,, 161–71. New York: Routledge, 1999.

Hagen, Joel. *An Entangled Bank: The Origins of Ecosystem Ecology*. New Brunswick, NJ: Rutgers University Press, 1992.

Haldane, J. B. S. "The Rate of Mutation of Human Genes." *Proceedings of the International Congress on Genetic Heredity* 35 (1949): 267–73.

Hamblin, Jacob Darwin. *Arming Mother Nature: The Birth of Catastrophic Environmentalism*. New York: Oxford University Press, 2013.

———. "Exorcising Ghosts in the Age of Automation: United Nations Experts and Atoms for Peace." *Technology and Culture* 47, no. 4 (2006): 734–56.

Hamilton, Shane. "Cold Capitalism: The Political Ecology of Frozen Concentrated Orange Juice." *Agricultural History* 77, no. 4 (Autumn 2003): 557–81.

Haraway, Donna Jeanne. *Modest_Witness@Second_Millennium. FemaleMan©_Meets_OncoMouse*TM*: Feminism and Technoscience*. New York: Routledge, 1997.

———. "Remodeling the Human Way of Life: Sherwood Washburn and the New Physical Anthropology." In *Bones, Bodies, Behavior: Essays on Biological Anthropology*, edited by George W. Stocking, 206–60. Madison: University of Wisconsin Press, 1988.

———. *Simians, Cyborgs, and Women: The Reinvention of Nature*. New York: Routledge, 1991.

———. "Situated Knowledges: The Science Question in Feminism and the Privilege of Partial Perspective." *Feminist Studies* 14 (1988): 575–99.

———. *When Species Meet*. Minneapolis: University of Minnesota Press, 2008.

Hård, Mikael. *Machines Are Frozen Spirit: The Scientification of Refrigeration and Brewing in the 19th Century; A Weberian Interpretation*. Boulder, CO: Westview, 1994.

Hardin, Garrett. *Nature and Man's Fate*. New York: New American Library, 1959.

Harding, Sandra. *Sciences from Below: Feminisms, Postcolonialities, and Modernities*. Durham, NC: Duke University Press, 2008.

Harmon, Amy. "Where'd You Go with My DNA?" *New York Times*, 24 April 2010.

Harris, D. Fraser. "Latent Life, or, Apparent Death." *Scientific Monthly* 14, no. 5 (1922): 429–40.

Harry, Debra. "Acts of Self-Determination and Self-Defense: Indigenous Peoples' Responses to Biocolonialism." In *Rights and Liberties in the Biotech Age: Why We Need a Genetic Bill of Rights*, edited by Sheldon Krimsky and Peter Shorett, 87–98. Lanham, MD: Rowan and Littlefield, 2005.

Hayden, Cori. "A Biodiversity Sampler for the Millennium." In *Reproducing Reproduction: Kinship, Power and Technological Innovation*, edited by Sarah Franklin and Helena Ragone, 173–206. Philadelphia: University of Pennsylvania Press, 1996.

———. "Suspended Animation: A Brine Shrimp Essay." In *Remaking Life and Death: Toward an Anthropology of the Biosciences*, edited by Sarah Franklin and Margaret Lock, 193–225. Santa Fe, NM: School for Advanced Research Press, 2003.

———. *When Nature Goes Public: The Making and Unmaking of Bioprospecting in Mexico*. In-Formation Series. Princeton, NJ: Princeton University Press, 2003.

Hayes, Vanessa. "Indigenous Genomics." *Science* 332, no. 6030 (2011): 639.

Health and Disease in Tribal Societies. Ciba Foundation Symposium 49. Amsterdam: Elsevier, 1977.

Healy, Kieran. *Last Best Gifts: Altruism and the Market for Human Blood and Organs*. Chicago: University of Chicago Press, 2007.

Healy, Melissa. "FDA Wins High-Profile Support in Consumer Genetics Kerfuffle." *Los Angeles Times*, 13 March 2014.

Hecht, Gabrielle. *Being Nuclear: Africans and the Global Uranium Trade*. Cambridge, MA: MIT Press, 2012.

———. "On the Fallacies of Cold War Nostalgia." In *Entangled Geographies: Empire and Technopolitics in the Global Cold War*, edited by Gabrielle Hecht, 75–99. Cambridge, MA: MIT Press, 2011.

Hecht, Susanna B. *Scramble for the Amazon and the "Lost Paradise" of Euclides da Cuhna*. Chicago: University of Chicago Press, 2013.

Heidegger, Martin. *The Question Concerning Technology, and Other Essays*. New York: Garland, 1977.

Helmreich, Stefan. *Sounding the Limits of Life: Essays in the Anthropology of Biology and Beyond*. Princeton, NJ: Princeton University Press, 2016.

———. "Species of Biocapital." *Science as Culture* 17, no. 4 (2008): 463–78.

———. "Trees and Seas of Information: Alien Kinship and the Biopolitics of Gene Transfer in Marine Biology and Biotechnology." *American Ethnologist* 30, no. 3 (2003): 340–58.

———. "What Was Life? Answers from Three Limit Biologies." *Critical Inquiry* 37, no. 4 (2011): 671–96.

Hemley, Robin. *Invented Eden: The Elusive, Disputed History of the Tasaday*. New York: Farrar, Straus and Giroux, 2003.

Hemming, John. *Die If You Must: Brazilian Indians in the Twentieth Century*. London: Pan Macmillan, 2003.

Hess, David. "Science in an Era of Globalization: Alternative Pathways." In *The Postcolonial Science and Technology Studies Reader*, edited by Sandra Harding, 419–39. Durham, NC: Duke University Press, 2011.

Hess, Volker, and J. Andrew Mendelsohn. "Case and Series: Medical Knowledge and Paper Technology, 1600–1900." *History of Science* 48, nos. 3–4 (2010): 287–314.

Hewlett, Richard G., and Jack M. Holl. *Atoms for Peace and War, 1953–1961: Eisenhower and the Atomic Energy Commission*. Berkeley: University of California Press, 1989.

Hilliard, David L. "Colonialism and Christianity: The Melanesian Mission in the Solomon Islands." *Journal of Pacific History* 9 (1974): 93–116.

Hiratsuka, V. Y., J. E. Brown, T. J. Hoeft, and D. A. Dillard. "Alaska Native People's Perceptions, Understandings, and Expectations for Research Involving Biological Specimens." *International Journal for Circumpolar Health* 71 (2012): 18642.

Hoagland, Hudson, and Gregory Pincus. "Revival of Mammalian Sperm after Immersion in Liquid Nitrogen." *Journal of General Physiology* 25, no. 3 (1942): 337–44.

Hoeyer, Klaus. *Exchanging Human Bodily Materials: Rethinking Bodies and Markets*. Dordrecht: Springer, 2013.

Hoffman, Adina, and Peter Cole. *Sacred Trash: The Lost and Found World of the Cairo Geniza*. New York: Shocken, 2011.

Holland, Walter W. "Karel Raska—the Development of Modern Epidemiology. The Role of the IEA." *Central European Journal of Public Health* 18, no. 1 (2010): 57–60.

Hollinger, David A. *Science, Jews, and Secular Culture: Studies in Mid-Twentieth-Century American Intellectual History*. Princeton, NJ: Princeton University Press, 1996.

Holmes, Douglas R., and George E. Marcus. "Collaboration Today and the Re-Imagination of the Classic Scene of Fieldwork Encounter." *Collaborative Anthropologies* 1 (2008): 81–101.

Holmes, Frederic L. *Claude Bernard and Animal Chemistry: The Emergence of a Scientist*. Cambridge, MA: Harvard University Press, 1985.

Hooper, Edward. *The River: A Journey to the Source of HIV and AIDS*. Boston: Little, Brown, 1999.

Howells, William W., and William L. Moss. "Anthropometry and Blood Types in Fiji and Solomon Islands." *Anthropological Papers of the American Museum of Natural History* 33, no. 4 (1933): 279–339.

Hsu, Francis L. K. "Rethinking the Concept 'Primitive.'" *Current Anthropology* 179, no. 3 (June 1964): 169–78.

Huber, Valeska. "The Unification of the Globe by Disease? The International Sanitary Conferences of 1851–1894." *Historical Journal* 49, no. 2 (2006): 453–76.

Hughes, Thomas P. *Networks of Power: Electrification in Western Society, 1880–1930*. Baltimore: Johns Hopkins University Press, 1983.

Huntelmann, Axel. "Seriality and Standardization in the Production of'606.'" *History of Science* 48, no. 161 (2010): 435–60.

Huxley, Aldous. *Brave New World*. New York: Doubleday, 1932.

Iriye, Akira. *Global Community: The Role of International Organizations in the Making of the Contemporary World*. Berkeley: University of California Press, 2002.

Itano, Harvey A., and James V. Neel. "A New Inherited Abnormality of Human Hemoglobin." *Proceedings of the National Academy of Sciences* 36 (1950): 613–17.

Jasanoff, Sheila. *States of Knowledge: The Co-production of Science and Social Order*. New York: Routledge, 2004.

Jasanoff, Sheila, and Sang-Hyun Kim. *Dreamscapes of Modernity: Sociotehnical Imaginaries and the Fabrication of Power*. Chicago: University of Chicago Press, 2015.

Jolly, Margaret. "Imagining Oceania: Indigenous and Foreign Representations of a Sea of Islands." *Contemporary Pacific* 19 (2007): 508–45.

Jones, David S. "How Personalized Medicine Became Genetic, and Racial: Werner Kalow and the Formations of Pharmacogenetics." *Journal of the History of Medicine and Allied Sciences* 68, no. 1 (2012): 1–48.

———. *Rationalizing Epidemics: Meanings and Uses of American Indian Mortality since 1600*. Cambridge, MA: Harvard University Press, 2004.

———. "Virgin Soils Revisited." *William and Mary Quarterly* 60, no. 4 (2003): 703–42.

Kakaliouras, Ann. "An Anthropology of Repatriation: Contemporary Physical Anthropology and Native American Ontologies of Practice." *Current Anthropology* 53, no. S5 (2012): S210–S221.

Kauanui, J. Kēhaulani. *Hawaiian Blood: Colonialism and the Politics of Sovereignty and Indigeneity*. Durham, NC: Duke University Press, 2008.

Kavaler, Lucy. *Cold against Disease*. New York: John Day, 1971.

Kearns, Rick. "Yanomami of Brazil Honor Return of Stolen Blood." *Indian Country Today*, 10 April 2015.

Keating, Peter. "Holistic Bacteriology: Ludwick Hirszfeld's Doctrine of Serogenesis between the Two World Wars." In *Greater Than the Parts: Holism in Biomedi-*

cine 1920–1950, edited by Christopher Lawrence and George Weisz, 283–302. New York: Oxford University Press, 1998.

Keck, Frederic. "Stockpiling as a Technique of Preparedness: Conserving the Past for an Unpredictable Future." In *Cryopolitics: Frozen Life in a Melting World*, edited by Joanna Radin and Emma Kowal. (Cambridge, MA, MIT Press, forthcoming).

Keesing, Roger M. *'Elota's Story: The Life and Times of a Solomon Islands Big Man*. 1978. Reprint, New York: Holt, Rinehart and Winston, 1983.

Keesing, Roger M., and Peter Corris. *Lightning Meets the West Wind: The Malaita Massacre*. New York: Oxford University Press, 1980.

Keilin, David. "The Leeuwenhoek Lecture: The Problem of Anabiosis or Latent Life: History and Current Concept." *Proceedings of the Royal Society of London, Series B, Biological Sciences* 150, no. 939 (1959): 149–91.

Keilin, David, and Y. L. Wang. "Stability of Haemoglobin and of Certain Endo-erythrocytic Enzymes *in Vitro*." *Biochemical Journal* 41, no. 1 (1947): 491–500.

Kendrick, Brigadier General Douglas B. *Medical Department, U.S. Army, Blood Program in World War II*. Washington, DC: Office of the Surgeon General, 1989.

Kevles, Daniel J. "Big Science and Big Politics in the United States: Reflections on the Death of the SSC and the Life of the Human Genome Project." *Historical Studies in the Physical and Biological Sciences* 2 (1997): 269–97.

———. *In the Name of Eugenics: Genetics and the Uses of Human Heredity*. Berkeley: University of California Press, 1986.

Kevles, Daniel J., and European Group on Ethics in Science and New Technologies to the European Commission. *A History of Patenting Life in the United States with Comparative Attention to Europe and Canada: A Report to the European Group on Ethics in Science and New Technologies*. Luxembourg: Office for Official Publications of the European Commission, 2002.

Kimmelman, Barbara. "The American Breeders' Association: Genetics and Eugenics in an Agricultural Context, 1903–13." *Social Studies of Science* 13, no. 2 (1983): 163–204.

Kimura, Motoo. *The Neutral Theory of Molecular Evolution*. New York: Cambridge University Press, 1983.

Kimura, Motoo, and James F. Crow. "The Number of Alleles That Can Be Maintained in a Finite Population." *Genetics* 49 (1964): 725–38.

King, Nicholas B. "The Scale Politics of Emerging Disease." *Osiris* 19 (2004): 62–76.

Kingsland, Sharon. *The Evolution of American Ecology, 1890–2000*. Baltimore: Johns Hopkins University Press, 2005.

Kirk, R. G. W. "Wanted—Standard Guinea Pigs: Standardization and the Experimental Animal Market in Britain ca. 1919–1947." *Studies in History and Philosophy of Science, Part C* 39, no 3. (2008): 280–91.

Kirksey, S. Eben, and Stefan Helmreich. "The Emergence of Multispecies Ethnography." *Cultural Anthropology* 25, no. 4 (2010): 545–76.

Kirsch, S. "Lost Tribes: Indigenous People and the Social Imaginary." *Anthropological Quarterly* 70, no. 2 (1997): 58–67.

Kleinman, Daniel Lee. *Impure Cultures: University Biology and the World of Commerce*. Science and Technology in Society. Madison: University of Wisconsin Press, 2003.

Knauft, B. M. *From Primitive to Postcolonial in Melanesia and Anthropology*. Ann Arbor: University of Michigan Press, 1999.

Knorr-Cetina, K. *Epistemic Cultures: How the Sciences Make Knowledge*. Cambridge, MA: Harvard University Press, 1999.

Kohler, Robert E. *Lords of the Fly: Drosophila Genetics and the Experimental Life*. Chicago: University of Chicago Press, 1994.

Kolata, Gina. *Flu: The Story of the Great Influenza Pandemic of 1918 and the Search for the Virus That Caused It*. New York: Farrar, Straus and Giroux, 1999.

Koselleck, Reinhart. *Futures Past: On the Semantics of Historical Time*. Translated by Keith Tribe. New York: Columbia University Press, 2004.

Kowal, Emma. "Orphan DNA: Indigenous Samples, Ethical Biovalue and Postcolonial Science." *Social Studies of Science* 43, no. 5 (2013): 577–97.

Kowal, Emma, Ashley Greenwood, and Rebekah E. McWhirter. "All in the Blood: A Review of Aboriginal Australians' Cultural Beliefs about Blood and Implications of Biospecimen Research." *Journal of Empirical Research on Human Research Ethics* 10, no. 4 (2015): 347–59.

Kowal, Emma, and Joanna Radin. "Indigenous Biospecimens and the Cryopolitics of Frozen Life." *Journal of Sociology* 51, no. 1 (2015): 63–80.

Kowal, Emma, Joanna Radin, and Jenny Reardon. "Indigenous Body Parts, Mutating Temporalities, and the Half-Lives of Postcolonial Technoscience." *Social Studies of Science* 43, no. 4 (2013): 465–83.

Krementsov, Nikolai. *Revolutionary Experiments: The Quest for Immortality in Bolshevik Science and Fiction*. New York: Oxford University Press, 2014.

Krech, Shepard. *The Ecological Indian: Myth and History*. New York: W. W. Norton, 2000.

Krige, John. *American Hegemony and the Postwar Reconstruction of Science in Europe*. Cambridge, MA: MIT Press, 2006.

Kuklick, Henrika. "Personal Equations: Reflections on the History of Fieldwork, with Special Reference to Sociocultural Anthropology." *Isis* 102 (2011): 1–33.

Kuper, Adam. *The Reinvention of Primitive Society: Transformations of a Myth*. London: Routledge, 2005.

———. "The Return of the Native." *Current Anthropology* 44, no. 3 (2003).

Kurlansky, Mark. *Birdseye: The Adventures of a Curious Man*. New York: Anchor, 2012.

Kwa, Chunglin. "Representations of Nature Mediating between Ecology and Science Policy: The Case of the International Biological Programme." *Social Studies of Science* 17, no. 3 (1987): 413–42.

Lakoff, Andrew. "The Generic Biothreat, or, How We Became Unprepared." *Cultural Anthropology* 23, no. 3 (2008): 399–428.

Lakoff, Andrew, and S. J. Collier. "Infrastructure and Event: The Political Technology of Preparedness" in *Political Matter: Technoscience, Democracy, and Public*

Life, edited by B. Braun, S. J. Whatmore, and I. Stengers, 243–66. Minneapolis: University of Minnesota Press, 2010.

Landecker, Hannah. "Antibiotic Resistance and the Biology of History." *Body and Society*, published online before print (2015).

———. "Building a New Type of Body in Which to Grow a Cell: Tissue Culture at the Rockefeller Institute, 1910–1914." In *Creating a Tradition of Biomedical Research: Contributions to the History of the Rockefeller University*, edited by Darwin Stapelton, 151–74. New York: Rockefeller University Press, 2004.

———. *Culturing Life: How Cells Became Technologies*. Cambridge, MA: Harvard University Press, 2007.

———. "Living Differently in Biological Time: Plasticity, Temporality, and Cellular Biotechnologies." *Culture Machine* 7 (2005). http://www.culturemachine.net/index.php/cm/article/view/26/33.

———. "On Beginning and Ending with Apoptosis: Cell Death and Biomedicine." In *Remaking Life and Death: Toward an Anthropology of the Biosciences*, edited by Sarah Franklin and Margaret M. Lock, 23–59. Santa Fe, NM: School of American Research Press, 2003.

Landecker, Hannah, and Angela N. H. Creager. "Technical Matters: Method, Knowledge and Infrastructure in Twentieth-Century Life Science." *Nature Methods* 6, no. 10 (2009): 701–5.

Langill, Ellen D. *Sub-Zero at Fifty: A History of the Sub-Zero Company, Incorporated, 1945–1995*. Madison, WI: Sub-Zero Freezer, 1995.

Langmuir, Alexander D. "The Surveillance of Communicable Diseases of National Importance." *New England Journal of Medicine* 268 (1963): 182–92.

Lasker, Gabriel Ward. "Human Biological Adaptability." *Science* 166, no. 3912 (1964): 1480–86.

Lasker, Gabriel Ward, and Michael A. Little. *Happenings and Hearsay: Experiences of a Biological Anthropologist*. Detroit: Savoyard Books, 1999.

Latham, Michael E. *The Right Kind of Revolution: Modernization, Development and U.S. Foreign Policy from the Cold War to the Present*. Ithaca, NY: Cornell University Press, 2011.

Latour, Bruno. *Face à Gaïa: huit conférences sur le nouveau régime climatique*. Paris: La Découverte, 2015.

———."Give Me a Laboratory and I Will Raise the World." In *Science Observed: Perspectives on the Social Study of Science*, edited by Michael Mulkay and K. Knorr-Cetina, 141–70. London: Sage, 1983.

———. *Pandora's Hope: Essays on the Reality of Science Studies*. Cambridge, MA: Harvard University Press, 1999.

———. *Science in Action: How to Follow Scientists and Engineers through Society*. Milton Keynes, Philadelphia: Open University Press, 1987.

———. *We Have Never Been Modern*. Cambridge, MA: Harvard University Press, 1993.

———. "Why Has Critique Run out of Steam? From Matters of Fact to Matters of Concern." *Critical Inquiry* 30, no. 2 (2004): 225–48.

Laughlin, William, ed. *Papers of the Physical Anthropology of the American Indian.* New York: Viking Fund, 1949.

Law, Jules David. *The Social Life of Fluids: Blood, Milk, and Water in the Victorian Novel.* Ithaca, NY: Cornell University Press, 2010.

Lederer, Susan E. *Flesh and Blood: Organ Transplantation and Blood Transfusion in Twentieth-Century America.* New York: Oxford University Press, 2008.

———. *Subjected to Science: Human Experimentation in America before the Second World War.* Baltimore: Johns Hopkins University Press, 1995.

Lemov, Rebecca. *Database of Dreams: The Lost Quest to Catalog Humanity.* New Haven, CT: Yale University Press, 2015.

Lerner, Daniel. *The Passing of Traditional Society: Modernizing the Middle East.* Glencoe, IL: Free Press, 1958.

Lerner, I. Michael. *Genetic Homeostasis.* London: Oliver and Boyd, 1954.

Lévi-Strauss, Claude. "Race et culture." *Revue internationale des sciences sociales* 23, no. 4 (1971): 647–66.

———. *The Savage Mind.* Chicago: University of Chicago Press, 1968.

———. *Tristes Tropiques.* New York: Penguin, 1992.

Liloqula, Ruth. "Value of Life: Saving Genes versus Saving Indigenous Peoples." *Cultural Survival Quarterly* 20, no. 2 (1996): 42–45.

"Linde Cryobiology News: Report No. 2 from Linde Company, Division of Union Carbide." *Science* 139, no. 3549 (1963): 3.

Lindee, M. Susan. "The Repatriation of Atomic Bomb Victim Body Parts to Japan: Natural Objects and Diplomacy." *Osiris* 13 (1998): 376–409.

———. "Scaling Up: Human Genetics as a Cold War Network." *Studies in the History and Philosophy of Biological and Biomedical Sciences* 47 (2014): 185–90.

———. *Suffering Made Real: American Science and the Survivors at Hiroshima.* Chicago: University of Chicago Press, 1994.

———. "Voices of the Dead: James Neel's Amerindian Studies." In *Lost Paradises and the Ethics of Research and Publication*, edited by Francisco M. Salzano and A. Magdalena Hurtado, 27–48. New York: Oxford University Press, 2004.

Lindell, Bo, and R. Lowry Dobson. *Ionizing Radiation and Health.* Public Health Papers 6, Geneva: World Health Organization, 1961.

Lindquist, Bosse, director. *The Genius and the Boys*." TV movie. SVT Sales, 2009.

Link, Adrianna. "Salvaging a Record for Humankind: Urgent Anthropology at the Smithsonian Institution, 1964–1984." PhD diss., Johns Hopkins University, 2016.

Liotta, Lance, Mauro Ferrari, and Emanuel Petricoin. "Critical Proteomics: Written in Blood." *Nature* 425, no. 6961 (2003): 905.

Little, Michael. "Human Population Biology in the Second Half of the Twentieth Century." *Current Anthropology* 53, no. S5 (2012): S126–S138.

Little, Michael, and Kenneth R. Kennedy, eds. *Histories of American Physical Anthropology in the Twentieth Century.* New York: Lexington Books, 2010.

Livingstone, F. B. *Abnormal Hemoglobins in Human Populations.* Chicago: Aldine, 1967.

———. "Anthropological Implications of Sickle Cell Gene Distribution in West Africa." *American Anthropologist* 60 (1958): 533–62.

Livingstone, F. B., Henry Gershowitz, James V. Neel, W. W. Zuelzer, and Marvin D. Solomon. "The Distribution of Several Blood Group Genes in Liberia, the Ivory Coast and Upper Volta." *American Journal of Physical Anthropology* 18 (1960): 161–78.

Lock, Margaret M. "The Alienation of Body Tissue and the Biopolitics of Immortalized Cell Lines." *Body and Society* 7, nos. 2–3 (2001): 63–91.

———. "Interrogating the Human Genome Diversity Project." *Social Science and Medicine* 39 (1994): 603–6.

Lock, Margaret M., and Judith Farquhar. *Beyond the Body Proper: Reading the Anthropology of Material Life*. Body, Commodity, Text. Durham, NC: Duke University Press, 2007.

Love, Spencie. *One Blood: The Death and Resurrection of Charles R. Drew*. Chapel Hill: University of North Carolina Press, 1996.

Lovelock, James. *Gaia: A New Look at Life on Earth*. New York: Oxford University Press, 1979.

———. *Homage to Gaia: The Life of an Independent Scientist*. Oxford: Oxford University Press, 2000.

———. "The Mechanism of Protective Action of Glycerol against Haemolysis by Freezing and Thawing." *Biochemica et Biophysica Acta II* 11 (1953): 28–36.

———. "A Physical Basis for Life Detection Experiments." *Nature* 207, no. 7 (1965): 568–70.

———. "The Physical Instability of Human Red Blood Cells." *Biochemical Journal* 60 (1955): 692–96.

Lovelock, James, and M. W. H. Bishop. "Prevention of Freezing Damage to Living Cells by Dimethyl Sulfoxide." *Nature* 183 (1959): 1394–95.

Lozina-Lozinskii, L. K. *Studies in Cryobiology: Adaptation and Resistance of Organisms and Cells to Low Temperature*. Translated by P. Harry. New York: Wiley, 1974.

Lum, K. "Contributions of Population Origins and Gene Flow to the Diversity of Neutral and Malaria Selected Autosomal Genetic Loci of Pacific Island Populations." In *Genes, Language, and Culture History in the Southwest Pacific,* edited Jonathan Scott Friedlaender, 219–30. New York: Oxford University Press, 2007.

Luyet, B. J. "The Effects of Ultra-Violet, X-, and Cathode Rays on the Spores of Mucoraceae." *Radiology* 18 (1932): 1019–22.

———. "Human Encounters with Cold, from Early Primitive Reactions to Modern Experimental Modes of Approach" *Cryobiology* 51 (1964): 4–10.

———. "Some Basic Considerations on the Preservation of Biological Materials at Low Temperature." In *Long-Term Preservation of Red Blood Cells,* edited by Mary T. Sproul, 3–17. Washington, DC: National Academies, 1965.

———. "Working Hypotheses on the Nature of Life." *Biodynamica* 1, no. 1(1934): 1–7.

Luyet, B. J., and M. P. Gehenio. *Life and Death at Low Temperatures*. Normandy, MO: Biodynamica, 1940.

Luyet, B. J., and M. C. Gibbs. "On the Mechanism of Congelation and of Death in the Rapid Freezing of Epidermal Plant Cells." *Biodynamica* 1 (1937): 1–18.

Luyet, B. J., and E. Hodapp. "Revival of Frog Spermatozoa Vitrified in Liquid Air." *Proceedings of the Society for Experimental Biology* 39 (1938): 433–44.

Lynch, Michael. "Archives in Formation: Privileged Spaces, Popular Archives and Paper Trails." *History of the Human Sciences* 12, no. 2 (1999): 65–87.

MacArthur, R, and E. O. Wilson. *The Theory of Island Biogeography*. Princeton, NJ: Princeton University Press, 1967.

Malinin, Theodore I. *Surgery and Life: The Extraordinary Career of Alexis Carrel*. New York: Harcourt Brace Jovanovich, 1979.

Marcus, George E., ed. *Para-Sites: A Casebook against Cynical Reason*. Chicago: University of Chicago Press, 2000.

Marks, Jonathan. *Human Biodiversity: Genes, Race, and History*. New York: Alaine de Gruyter, 1995.

———. "The Human Genome Diversity Project: Good for If Not Good as Anthropology?" *Anthropology Newsletter* 36 (April 1995): 72.

———. "The Legacy of Serological Studies in American Physical Anthropology." In *History and Philosophy of the Life Sciences*, 18 (1996): 345–62.

———. "The Origins of Anthropological Genetics." *Current Anthropology* 53, no. S5 (2012): S161–S172.

Marramao, Giacomo. *Kairos: Towards an Ontology of "Due Time."* Aurora, CO: Davies Group, 2007.

Marx, Karl. *Capital: A Critique of Political Economy*. Vol. 1. New York: Penguin, 1976.

———. *Capital: A Critique of Political Economy*. Vol. 2, *The Process of Circulation of Capital*. Edited by Friedrich Engels, translated by Ernest Untermann. Chicago: Charles H. Kerr, 1909.

Masco, Joseph. "Atomic Health, or How the Bomb Altered American Notions of Death." In *Against Health: How Health Became the New Morality*, edited by Jonathan M Metzl and Anna Kirkland, 133–56. New York: New York University Press, 2010.

———. "Mutant Ecologies: Radioactive Life in Post–Cold War New Mexico." *Cultural Anthropology* 19, no. 4 (2004): 517–50.

———. *Nuclear Borderlands: The Manhattan Project in Post–Cold War New Mexico*. Princeton, NJ: Princeton University Press, 2006.

———. "On the Fallacies of Cold War Nostalgia." In *Entangled Geographies: Empire and Technopolitics in the Global Cold War*, edited by Gabrielle Hecht, 75–94. Cambridge, MA: MIT Press, 2011.

Matisoo-Smith, E. "Animal Translocations, Genetic Variation, and the Human Settlement of the Pacific." In *Genes, Language, and Culture History in the Southwest Pacific*, edited by Jonathan Scott Friedlaender, 157–70. Oxford: Oxford University Press, 2007.

Maybury-Lewis, David. *Akwe-Shavante Society*. Oxford: Clarendon, 1967.

———. *The Savage and the Innocent*. New York: Beacon, 1958.

Mazumdar, Pauline H. "'In the Silence of the Laboratory': The League of Nations Standardizes Syphilis Tests." *Social History of Medicine* 16, no. 3 (2003): 437–58.

McDonald, H. "East Kimberley Concepts of Health and Illness: A Contribution to Intercultural Health Programs in Northern Australia." *Australian Aboriginal Studies* 2 (2006): 86–97.

McLaren, Angus. *Reproduction by Design: Sex, Robots, Trees, and Test-Tube Babies in Interwar Britain*. Chicago: University of Chicago Press, 2012.

McLuhan, Marshall. *Understanding Media: The Extensions of Man*. New York: McGraw-Hill, 1964.

Mead, Aroha Te Pareake. "Genealogy, Sacredness, and the Commodities Market." *Cultural Survival Quarterly* 20, no. 2 (1996): 46–51.

Medawar, Peter B. Foreword to *Human Biology: An Introduction to Human Evolution, Variation, and Growth*, by G. A. Harrison et al., v–vi. New York: Oxford University Press, 1964.

———. *The Future of Man*. BBC Reith Lectures. New York: Basic Books, 1959.

———. "Problems of Adaptation." *New Biology* 11 (1951): 10–26.

Mendelsohn, Andrew J. "From Eradication to Equilibrium: How Epidemics Became Complex after World War I." In *Greater Than the Parts: Holism in Biomedicine, 1920–1950*, edited by Christopher Lawrence and George Weisz, 303–34. New York: Oxford University Press, 1998.

Merriwether, D. Andrew. "Freezer Anthropology: New Uses for Old Blood." *Philosophical Transactions: Biological Sciences* 354, no. 1379 (1999): 121–29.

Meryman, Harold T. "Basile J. Luyet: In Memoriam." *Cryobiology* 12, no. 4 (1975): 285–92.

———. "Mechanics of Freezing in Living Cells and Tissues." *Science* 124, no. 3321 (1956): 515–21.

Midgley, Mary. *Science as Salvation: A Modern Myth and Its Meaning*. New York: Routledge, 1992.

Miller, Mark Edwin. *Forgotten Tribes: Unrecognized Indians and the Federal Acknowledgment Process*. Lincoln: University of Nebraska Press, 2004.

Minkove, Judith F. "Safe Keeping." *Hopkins Medicine*, 1 October 2013, 20–23.

Mitchell, Robert. *Experimental Life: Vitalism in Romantic Science and Literature*. Baltimore: Johns Hopkins University Press, 2013.

Mitchell, Robert, and Catherine Waldby. "National Biobanks: Clinical Labor, Risk Production and the Creation of Biovalue." *Science, Technology and Human Values* 35, no. 3 (2010): 330–55.

Mitman, Gregg. "In Search of Health: Landscape and Disease in American Environmental History." *Environmental History* 20, no. 2 (2005): 184–210.

Mol, Annemarie. *The Body Multiple: Ontology in Medical Practice*. Durham, NC: Duke University Press, 2002.

Montagu, Ashley. "The Fallacy of the Primitive." *Journal of the American Medical Association* 179, no. 12 (1962): 962–63.

Montalenti, Giuseppe. "Comment on Haldane, JBS. Disease and Evolution." *Rice Science* 19, suppl (1949): 333–34.

———. *Infectious Diseases as Selective Agents*. Edinburgh: Oliver and Boyd, 1965.

Montgomery, Charles. "The Octopus: Can the Myths of the Lau Survive Their Preservation?" *Walrus*, May 2006, 52–59.

Morton, Newton. "Problems and Methods in the Genetics of Primitive Groups." *American Journal of Physical Anthropology* 28 (1968): 191–202.

Motulsky, Arno G., et al. "Population Genetic Studies in the Congo." *American Journal of Human Genetics* 18, no. 6 (1966): 514–37.

Mourant, A. E. *The Distribution of the Human Blood Groups and Other Polymorphisms*. Oxford: Blackwell.

———. "The Use of Blood Groups in Anthropology." *Journal of the Royal Anthropological Institute of Great Britain and Ireland* 77, no. 2 (1947): 139–44.

Mukerji, Chandra. *A Fragile Power: Scientists and the State*. Princeton, NJ: Princeton University Press, 1989.

Muller, H. J. "Our Load of Mutations." *American Journal of Human Genetics* 2 (1950): 111–76.

Müller-Wille, Staffan. "Claude Lévi-Strauss on Race, History and Genetics." *BioSocieties* 5, no. 3 (2010): 330–47.

Müller-Wille, Staffan, and Isabelle Charmantier. "Lists as Research Technologies." *Isis* 103, no. 4 (2012): 743–52.

Murphy, Michelle. "Chemical Infrastructures of the St. Clair River." In *Toxicants, Health and Regulation since 1945*, edited by Soraya Boudia and Nathalie Jas, 103–15. London: Pickering and Chatto, 2013.

Nahmias, A. J., et al. "Evidence for Human Infection with HTLV III/LAV-like virus in Central Africa." *Lancet* 327, no. 8492 (1986): 1279–80.

Nance, John. *The Gentle Tasaday: A Stone Age People in the Philippine Rain Forest*. New York: Harcourt Brace Jovanovich, 1975.

Nash, Linda. *Inescapable Ecologies: A History of Environment, Disease, and Knowledge*. Berkeley: University of California Press, 2006.

Neel, James V. "The American Indian in the International Biological Program." Paper presented at the Biomedical Challenges Presented by the American Indian, Pan American Sanitary Bureau, Regional Office of the WHO, 1968.

———. "The Inheritance of Sickle-Cell Anemia." *Science* 110, no. 2846 (1949): 64–66.

———. "Lessons from a 'Primitive' People: Do Recent Data Concerning South American Indians Have Relevance to Problems of Highly Civilized Communities?" *Science* 170, no. 3960 (1970): 815–22.

———. "On Some Pitfalls in Developing an Adequate Genetic Hypothesis." *American Journal of Human Genetics* 7, no. 1 (1955): 1–14.

———. *Physician to the Gene Pool: Genetic Lessons and Other Stories*. New York: J. Wiley, 1994.

———. "'Private' Genetic Variants and the Frequency of Mutation among South American Indians." *Proceedings of the National Academy of Sciences* 70 (1973): 3311–15.

———. "The Study of Human Mutation Rates." *American Naturalist* 86, no. 828 (1952): 129–44.

———. "The Study of Natural Selection in Primitive and Civilized Human Populations." *Human Biology* 3 (1958): 43–72.

Neel, James V., and Francisco M. Salzano. "A Prospectus for Genetic Studies on the American Indians." In *The Biology of Human Adaptability*, edited by Paul Baker and J. S. Weiner, 245–74. Oxford: Clarendon, 1966.

Neel, James V., Francisco M. Salzano, P. C. Junqueira, F. Keiter, and David Maybury-Lewis. "Studies on the Xavante Indians of the Brazilian Matto Grosso." *American Journal of Physical Anthropology* 16, no. 1 (1964): 52–140

Neel, James V. and William Schull. *The Effect of Exposure to the Atomic Bombs on Pregnancy Termination in Hiroshima and Nagasaki*. Washington, DC: National Research Council, 1956.

Nelson, Alondra. *The Social Life of DNA: Race, Reparations, and Reconciliation after the Genome*. New York: Beacon, 2015.

Niebuhr, Reinhold. *The Nature and Destiny of Man*. New York: Charles Scribner's Sons, 1964.

Niezen, Ronald. *The Origins of Indigenism: Human Rights and the Politics of Identity*. Berkeley: University of California Press, 2003.

November, Joseph. *Biomedical Computing: Digitizing Life in the United States*. Baltimore: Johns Hopkins University Press, 2012.

O'Donnell, John. *Coriell: The Coriell Institute for Medical Research and a Half Century of Science*. Canton, MA: Science History, 2002.

Oldenziel, Ruth. "Islands: The United States as a Networked Empire." In *Entangled Geographies*, edited by Gabrielle Hecht, 13–42. Cambridge, MA: MIT Press, 2011.

Oliver, Douglas. *Black Islanders: A Personal Perspective of Bougainville, 1937–1991*. Honolulu: University of Hawaii Press, 1991.

"Ominous Refrigeration Need." *Refrigeration Engineering* 56 (December 1948): 495.

Ortner, Sherry B. "Resistance and the Problem of Ethnographic Refusal." *Comparative Studies in Society and History* 37, no. 1 (1995): 173–93.

Packard, Randall. *Making of a Tropical Disease: A Short History of Malaria*. Baltimore: Johns Hopkins University Press, 2007.

Palmer, Stephen. *Launching Global Health: The Caribbean Odyssey of the Rockefeller Foundation*. Ann Arbor: University of Michigan Press, 2010.

Parkes, A. S. *Off-Beat Biologist*. Cambridge: Galton Foundation, 1985.

———. "Some Biological Effects of Low Temperatures." *Advanced Science* 58 (September 1958): 1–8.

Parry, Bronwyn. "Technologies of Immortality: The Brain on Ice." *Studies in History and Philosophy of Biological and Biomedical Sciences* 35, no. 2 (2004): 391–413.

———. *Trading the Genome: Investigating the Commodification of Bio-Information*. New York: Columbia University Press, 2004.

Paul, John R. *Clinical Epidemiology*. Chicago: University of Chicago Press, 1958.

———. "Clinical Epidemiology." *Journal of Clinical Investigation* 17 (1938): 539–41.

———. *History of Poliomyelitis*. New Haven, CT: Yale University Press, 1971.

———. "The Story to Be Learned from Blood Samples: Its Value to the Epidemiologist." *Journal of the American Medical Association* 175, no. 7 (1961): 147–51.

Paul, John R., J. T. Riordan, and Lisabeth Kraft. "Serological Epidemiology: Antibody Patterns in North Alaskan Eskimos." *Journal of Immunology* 66 (1951): 695–713.

Pauling, L., A. Itano, S. J. Singer, and I. C. Wells. "Sickle-Cell Anemia: A Molecular Disease." *Science* 110 (1949): 543–47.

Pauly, Philip J. *Controlling Life: Jacques Loeb and the Engineering Ideal in Biology*. New York: Oxford University Press, 1987.

Payne, Anthony M. "Serum Surveys." *Milbank Memorial Fund Quarterly* 43, no. 2 (1965): 345–50.

Pendergrast, Mark. *Inside the Outbreaks: The Elite Medical Detectives of the Epidemic Surveillance Service*. New York: Mariner, 2010.

Pépin, Jacques. *The Origin of AIDS*. New York: Cambridge University Press, 2011.

Peters, John Fred. *Life Among the Yanomami: The Story of Change Among the Xilixana on the Mucajai River in Brazil*. Petersborough, ON: Broadview, 1998.

Petrick, Gabriella. "The Arbiters of Taste: Producers, Consumers and the Industrialization of Taste in America, 1900–1960." PhD diss., University of Delaware, 2007.

———. "'Like Ribbons of Green and Gold': Industrializing Lettuce and the Quest for Quality in the Salinas Valley, 1920–1965." *Agricultural History* 80, no. 3 (2006): 269–95.

Petryna, Adriana. "Ethical Variability: Drug Development and the Globalization of Clinical Trials." *American Ethnologist* 32, no. 2 (2005): 183–97.

———. "What Is a Horizon? Navigating Thresholds in Climate Change Uncertainty." In *Modes of Uncertainty: Anthropological Cases*, edited by Limor Samimian-Darash and Paul Rabinow, 147–64. Chicago: University of Chicago Press, 2015.

———. *When Experiments Travel: Clinical Trials and the Global Search for Human Subjects*. Princeton, NJ: Princeton University Press, 2009.

Piper, Liza. "Chesterfield Inlet, 1949, and the Ecology of Epidemic Polio." *Environmental History* 101, no. 1 (2015): 1–28.

Plotz, David. *The Genius Factory: The Curious History of the Nobel Prize Sperm Bank*. New York: Random House, 2005.

Polge, C. "Sir Alan Sterling Parkes, 10 September 1900–17 July 1990: Elected FRS 1933." *Biographical Memoirs of Fellows of the Royal Society* 52 (2006): 263–83.

Polge, C., and James Lovelock. "The Preservation of Bull Semen at –79c." *Veterinary Record* 64 (1952): 396.

Polge, C., A. Smith, and A. Parkes. "Revival of Spermatozoa after Vitrification and Dehydration at Low Temperatures." *Nature* 164 (1949): 666.

Pols, Jeanette, and I. Moser. "Cold Technologies versus Warm Care? On Affective and Social Relations with and through Care Technologies." *European Journal of Disability Research* 3, no. 2 (2009): 159–78.

Porter, Dorothy. "Calculating Health and Social Change: An Essay on Jerry Morris and Late-Modernist Epidemiology." *International Journal of Epidemiology* 36 (2007): 1180–84.

Pottage, A. "The Inscription of Life in Law: Genes, Patents, and Bio-Politics." *Modern Law Review* 61 (1998): 740–65.

Povinelli, Beth. *Labor's Lot: The Power, History, and Culture of Aboriginal Action.* Chicago: University of Chicago Press, 1993.

Pratt, Mary Louise. *Imperial Eyes: Travel Writing and Transculturation.* New York: Routledge, 1992.

Provine, William. "The Neutral Theory of Molecular Evolution in Historical Perspective." In *Population Biology of Genes and Molecules*, edited by Naoyuki Takahata and James F. Crow., 17–31. Tokyo: Baifukan, 1990.

Quirke, Viviane, and Jean-Paul Gaudillière. "The Era of Biomedicine: Science, Medicine, and Public Health in Britain and France after the Second World War." *Medical History* 52, no. 4 (2008): 441–52.

Qureshi, Sadiah. *Peoples on Parade: Exhibitions, Empire, and Anthropology in Nineteenth Century Britain.* Chicago: University of Chicago Press, 2012.

Rabinbach, Anson. *The Human Motor: Energy, Fatigue, and the Origins of Modernity.* New York: Basic Books, 1990.

Rabinow, Paul. *Making PCR: A Story of Biotechnology.* Chicago: University of Chicago Press, 1996.

Rader, Karen A. *Making Mice: Standardizing Animals for American Biomedical Research, 1900–1955.* Princeton, NJ: Princeton University Press, 2004.

Radin, Joanna. "Latent Life: Concepts and Practices of Human Tissue Preservation in the International Biological Program." *Social Studies of Science* 43, no. 4 (2013): 483–508.

———. "Planned Hindsight: Vital Valuations of Frozen Tissue at the Zoo and the Natural History Museum." *Journal of Cultural Economy* 8, no. 3 (2015): 361–78.

———. "Unfolding Epidemiological Stories: How the WHO Made Frozen Blood into a Flexible Resource for the Future." *Studies in History and Philosophy of Biological and Biomedical Sciences* 47 (2014): 62–73.

Radin, Joanna, and Emma Kowal. "Indigenous Blood and Ethical Regimes in the United States and Australia since the 1960s." *American Ethnologist* 42, no. 4 (2015): 749–65.

———. "The Politics of Low Temperature." in *Cryopolitics: Frozen Life in a Melting World*, edited by Joanna Radin and Emma Kowal. Cambridge, MA: MIT University Press, forthcoming).

Raffles, Hugh. *In Amazonia: A Natural History.* Princeton, NJ: Princeton University Press, 2002.

Rankin, William J. "Infrastructure and the International Governance of Economic Development, 1950–1965." In *Internationalization of Infrastructures*, edited by Jean-François Auger, Jan Jaap Bouma, and Rolf Kunneke, 61–75. Delft: Delft University of Technology.

Raska, Karel. "Epidemiologic Surveillance in the Control of Infectious Diseases." *Reviews of Infectious Diseases* 5, no. 5 (1983):1112–17.

———. "Epidemiological Surveillance with Particular Reference to the Use of Immunological Surveys." *Proceedings of the Royal Society of Medicine* 64 (1971): 12–16.

Rasmussen, Nicolas. *Picture Control: The Electron Microscope and the Transformation of Biology in America, 1940–1960.* Stanford, CA: Stanford University Press, 1997.

Reardon, Jenny. "The 'Persons' and 'Genomics' of Personal Genomics." *Personalized Medicine* 8, no. 1 (2011): 95–107.

———. *Race to the Finish: Identity and Governance in an Age of Genomics*. In-Formation Series. Princeton, NJ: Princeton University Press, 2005.

Reardon, Jenny, and Kim TallBear. "'Your DNA Is Our History': Genomics, Anthropology, and the Construction of Whiteness as Property." *Current Anthropology* 53, no. S5 (2012): S233–S245.

Reddy, Deepa. "Good Gifts for the Common Good: Blood and Bioethics in the Market of Genetics Research." *Cultural Anthropology* 22, no. 3 (2007): 429–72.

Rees, Jonathan. *Refrigeration Nation: A History of Ice, Appliances, and Enterprise*. Baltimore: Johns Hopkins University Press, 2013.

"Refrigeration Makes Artificial Insemination Possible." *Refrigerating Engineering* 58 (May 1950).

Reingold, Nathan, and Marc Rothenberg. *Scientific Colonialism: A Cross-Cultural Comparison*. Washington, DC: Smithsonian Institution Press, 1987.

"Research on Human Population Genetics: Report of a WHO Scientific Group." *Current Anthropology* 11, no. 2 (1970): 225–33.

"Response to Allegations against James V. Neel in *Darkness in El Dorado*, by Patrick Tierney." *American Journal of Human Genetics* 70, no. 1 (2002): 1–10.

Rheinberger, Hans-Jörg. *An Epistemology of the Concrete: Twentieth-Century Histories of Life*. Durham, NC: Duke University Press, 2010.

———. *Toward a History of Epistemic Things: Synthesizing Proteins in the Test Tube*. Stanford, CA: Stanford University Press, 1997.

Rhodes, Rosamund, Nada Gligorov, and Abraham Schwab, eds. *The Human Microbiome: Ethical, Legal, and Social Concerns*. Oxford: Oxford University Press, 2013.

Richards, V., et al. "Initial Clinical Experiences with Liquid Nitrogen Preserved Blood, Employing PVP as a Protective Additive." *American Journal of Surgery* 108, no. 2 (1964): 313–22.

Roberts, D. F., and G. A. Harrison, eds. *Natural Selection in Human Populations*. New York: Pergamon, 1959.

Roberts, Dorothy. *Fatal Invention: How Science, Politics, and Big Business Re-Create Race in the Twenty-First Century*. New York: New Press, 2011.

Roberts, Leslie. "A Genetic Survey of Vanishing Peoples." *Science* 252 (1991): 1614–17.

Rogers, Naomi. "Polio Can Be Cured: Science and Health Propaganda in the United States from Polio Polly to Jonas Salk." In *Silent Victories: The History and Practice of Public Health in Twentieth Century America*, edited by John Ward and Christopher Warren, 81–101. New York: Oxford University Press, 2007.

Romain, Tiffany. "Extreme Life Extension: Investing in Cryonics for the Long, Long Term." *Medical Anthropology* 29, no. 2 (2010): 194–215.

Roosth, Sophia. "Life, Not Itself: Inanimacy and the Limits of Biology." *Grey Room* 57 (2014): 56–81.

Rose, Nikolas S. *The Politics of Life Itself: Biomedicine, Power, and Subjectivity in the Twenty-First Century*. Princeton, NJ: Princeton University Press, 2007.

Rostand, Jean. "Glycerine et resistance du sperme aux basses temperatures." *Comptes Rendus de l'Académie des Sciences* 222 (1946): 1524–25.

Rous, Peyton, and J. R. Turner. "The Preservation of Living Red Blood Cells in Vitro. I. Methods of Preservation." *Journal of Experimental Medicine* 23, no. 2 (1916): 219–37.

Rural Advancement Fund International (RAFI). "The Patenting of Human Genetic Material" *RAFI Communique*, January–February 1994, 1–12. http://www.etcgroup.org/sites/www.etcgroup.org/files/publication/492/02/raficom36humangenetic.pdf, accessed 10 June, 2016.

———. "Patents, Indigenous Peoples, and Human Genetic Diversity." *RAFI Communique*, May 1993, 1–6, http://www.etcgroup.org/sites/www.etcgroup.org/files/publication/pdf_file/raficom31patents.pdf, accessed 10 June 2016.

Safier, Neil. "Global Knowledge on the Move: Itineraries, Amerindian Narratives, and Deep Histories of Science." *Isis* 101, no. 1 (2010): 133–45.

Sahota, Puneet Chawla. "Body Fragmentation: Native American Community Members' Views on Specimen Disposition in Biomedical/Genetics Research." *AJOB Empirical Bioethics* 5, no. 3 (2014): 19–30.

Said, Edward. *Orientalism*. New York: Vintage, 1979.

Salzano, Francisco M., ed. *The Role of Natural Selection in Human Evolution*. New York: Elsevier, 1975.

Salzano, Francisco M., and A. Magdalena Hurtado, eds. *Lost Paradises and the Ethics of Research and Publication*. New York: Oxford University Press, 2004.

Santos, Ricardo Ventura. "Indigenous Peoples, Postcolonial Contexts, and Genomic Research in the Late Twentieth Century: A View from Amazonia (1960–2000)." *Critique of Anthropology* 22, no. 1 (2002): 81–104.

———. "Pharmacogenomics, Human Genetic Diversity and the Incorporation and Rejection of Color/Race in Brazil." *BioSocieties* 10 (2014): 48–69.

Santos, Ricardo Ventura, M. Susan Lindee, and Vanderlei Sabastiao de Souza. "Varieties of the Primitive: Human Biological Diversity Studies in Cold War Brazil (1962–1970)." *American Anthropologist* 116, no. 4 (2014): 723–35.

Saparelli, Beatrice Pellegrini, Thomas Antoniette, and Jacques Dubochet. *Basile Luyet: Un vie pour la science, 1897–1974*. Sion, Switzerland: Editions de Musées Cantonaux du Valais, Sion, 1997.

Sauvy, Alfred. "Trois mondes, une planete." *L'Observateur* 118 (1952): 5.

Shachtman, Tom. *Absolute Zero and the Conquest of Cold*. Boston: Houghton Mifflin, 1999.

Schaffer, Simon, Lissa Roberts, Kapil Raj, and James Delbourgo, eds. *The Brokered World: Go-Betweens and Global Intelligence, 1770–1820*. Sagamore Beach, MA: Science History, 2009.

Schmidt, Paul J. "Basile J. Luyet and the Beginnings of Transfusion Cryobiology." *Transfusion Medicine Reviews* 20, no. 3 (2006): 242–46.

Schmidt-Nielsen, Knut. *Per Scholander, 1905–1980: A Biographical Memoir*. Washington, DC: National Academy of Sciences, 1987.

Schneider, David. *American Kinship: A Cultural Account*. Englewood Cliffs, NJ: Prentice Hall, 1968.

Schneider, William. "Blood Transfusion between the Wars." *Journal of the History of Medicine and Allied Sciences* 58, no. 2 (2003): 187–224.

———. "The History of Research on Blood Group Genetics: Initial Discovery and Diffusion." *History and Philosophy of the Life Sciences* 18 (1996): 277–303.

Scholander, P. F. *Enjoying a Life in Science: The Autobiography of P. F. Scholander*. Fairbanks: University of Alaska Press, 1990.

Schwarz, M. T. "Emplacement and Contamination: Mediation of Navajo Identity through Excorporated Blood." *Body and Society* 15, no. 2 (2009): 145–68.

Scott, James C. *Seeing Like a State: How Certain Schemes to Improve the Human Condition Have Failed*. New Haven, CT: Yale University Press, 1998.

Scurlock, R. G. *History and Origins of Cryogenics*. Oxford: Clarendon Press; Oxford University Press, 1992.

———. "A Matter of Degrees: A Brief History of Cryogenics." *Cryogenics* 30 (June 1990): 483–500.

Seidel, Robert. "A Home for Big Science: The Atomic Energy Commission's Laboratory System." *Historical Studies in Physical and Biological Sciences* 16, no. 1 (1985): 135–75.

Selcer, Perrin. "Beyond the Cephalic Index." *Current Anthropology* 53, no. S5 (2012): S173–S184.

———. "The View from Everywhere: Disciplining Diversity in Post–World War II International Social Science." *Journal of the History of the Behavioral Sciences* 45, no. 4 (2009): 309–29.

"Seroepidemiology." *Journal of Parasitology*, Section 2, Part 3: Supplement: Second International Congress of Parasitology, Technical Reviews 56, no. 4 (1970): 552–57.

Serres, Michel. *The Parasite*. Translated by Lawrence R. Schehr. Minneapolis: University of Minnesota Press, 2007.

Shannon, John. "The Role of Liquid Nitrogen and Refrigeration at the American Type Culture Collection." In *Roundtable Conference on the Cryogenic Preservation of Cell Cultures*, edited by A. P. Rinfret and B. LaSalle, 1–8. Washington, DC: National Academy of Sciences, 1975.

Shapiro, T. Rees. "His Blood-Freezing Method Continues to Save Lives." *Washington Post*, 21 January 2010.

Sharp, Lesley A. *Bodies, Commodities, and Biotechnologies: Death, Mourning, and Scientific Desire in the Realm of Human Organ Transfer*. New York: Columbia University Press, 2007.

Sheppard, Philip. "Natural Selection and Some Polymorphic Characters in Man." In *Natural Selection in Human Populations*, edited by D. F. Roberts and G. A. Harrison, 35–48. New York: Pergamon, 1959.

Sherkow, Jacob S., and Henry T. Greely. "The History of Patenting Genetic Material." *Annual Review of Genetics* 49 (November 2015): 161–82.

Siddiqi, Javed. *World Health and World Politics: The World Health Organization and the UN System*. Columbia: University of South Carolina Press, 1995.

Silverman, Rachel. "The Blood Group 'Fad' in Post-War Racial Anthropology." In *Kroeber Anthropological Society Papers*, edited by Jonathan Marks, 11–27. Berkeley: University of California Press, 2000.

Simpson, Audra. "On Ethnographic Refusal: Indigeneity, 'Voice' and Colonial Citizenship." *Junctures: The Journal for Thematic Dialogue* 9 (December 2007): 67–80.

Sivasundaram, Sujit. "Science." In *Pacific Histories: Ocean, Land, People*, edited by David Armitage and Alison Bashford, 237–62. London: Palgrave Macmillan, 2014.

Skloot, Rebecca. *The Immortal Life of Henrietta Lacks*. New York: Random House, 2010.

———. "Your Cells. Their Research. Your Permission." *New York Times*, 30 December 2015.

Slater, Leo B. *War and Disease: Biomedical Research on Malaria in the Twentieth Century*. New Brunswick, NJ: Rutgers University Press, 2009.

Smillie, W. G. *Public Health: Its Promise for the Future*. New York: Macmillan, 1955.

Smith, A. "Prevention of Hemolysis during Freezing and Thawing of Red Blood Cells." *Lancet* 2 (1950): 910–11.

Smith, Audrey Ursula. *Biological Effects of Freezing and Supercooling*. Baltimore: Williams and Wilkins, 1961.

Smith, Crosbie. *The Science of Energy: A Cultural History of Energy Physics in Victorian Britain*. Chicago: University of Chicago Press, 1998.

Smith, Crosbie, and M. Norton Wise. *Energy and Empire: A Biographical Study of Lord Kelvin*. Cambridge: Cambridge University Press, 1989.

Smith, Jenny Leigh. "Empire of Ice Cream: How Life Became Sweeter in the Postwar Soviet Union." In *Food Chains: From Farmyard to Shopping Cart*, edited by Warren Belasco and Roger Horowitz, 142–57. Philadelphia: University of Pennsylvania Press, 2009.

Smith, Linda Tuhiwai. *Decolonizing Methodologies: Research and Indigenous Peoples*. 2nd ed. London: Zed Books, 2012.

Smocovitis, Vassiliki Betty. "Humanizing Evolution: Anthropology, the Evolutionary Synthesis and the Prehistory of Biological Anthropology, 1927–1962." *Current Anthropology*, 53, no. S5 (2012): S108–S125.

Snowden, Frank. *The Conquest of Malaria: Italy, 1900–1962*. New Haven, CT: Yale University Press, 2006.

Sommer, Marianne. "Biology as a Technology of Social Justice in Interwar Britain: Arguments from Evolutionary History, Heredity, and Human Diversity," *Science, Technology, and Human Values* 39, no. 4 (2014): 561–86.

———. *History Within: The Science, Culture, and Politics of Bones, Organisms, and Molecules*. Chicago: University of Chicago Press, 2016.

Sorrenson, Richard. "Ship as a Scientific Instrument in the Eighteenth Century." *Osiris* 11 (1996): 221–36.

Soto Laveaga, Gabriela. *Jungle Laboratories: Mexican Peasants, National Projects, and the Making of the Pill*. Durham, NC: Duke University Press, 2009

Spencer, Frank, ed. *A History of American Physical Anthropology, 1930–1980*. New York: Academic, 1982.

Sproul, Mary T., ed. *Long-Term Preservation of Red Blood Cells*. Washington, DC: National Academy of Sciences–National Research Council, 1964.

Squier, Susan Merrill. *Liminal Lives: Imagining the Human at the Frontiers of Biomedicine*. Durham, NC: Duke University Press, 2004.

Stallones, Reuel A. *Environment, Ecology, and Epidemiology*. Fourth PAHO/WHO Lecture on the Biomedical Sciences. Washington, DC: Pan American Health Organization, 1971.

Staples, Amy L. S. "Constructing International Identity: The World Bank, Food and Agriculture Organization, and World Health Organization, 1945–1965." PhD diss., Ohio State University, 1998.

Star, Susan Leigh. "The Ethnography of Infrastructure." *American Behavioral Scientist* 43 (1999): 377–91.

Star, Susan Leigh, and Karen Ruhleder. "Steps Towards an Ecology of Infrastructure: Design and Access for Large Information Spaces." *Information Systems Research* 7, no. 1 (1996): 111–34.

Stark, Laura. *Behind Closed Doors: IRBs and the Making of Ethical Research*. Chicago: University of Chicago Press, 2011.

Stark, Laura, and Nancy Campbell. "Stowaways in the History of Science: The Case of Simian Virus 40 and Research on Federal Prisoners at the U.S. National Institutes of Health, 1960." *Studies in History and Philosophy of Biological and Biomedical Sciences* 48 (2014): 218–30.

Starr, Douglas P. *Blood: An Epic History of Medicine and Commerce*. New York: Alfred A. Knopf, 1998.

Stebbins, G. Ledyard. "International Biological Program." *Science* 137, no. 3532 (1962): 768–70.

———. "International Horizons in the Life Sciences." *AIBS Bulletin* 12, no. 6 (1962): 13–19.

Steedman, Carolyn. *Dust: The Archive and Cultural History*. New Brunswick, NJ: Rutgers University Press, 2002.

Steele, James H. "Karel Raska, 1909–1987, a Tribute." *Journal of Infectious Diseases* 158, no. 5 (1988): 915–16.

Steinberg, Karen K., Eric J. Sampson, Geraldine McQuillan, and Muin J. Khoury. "Use of Stored Tissue Samples for Genetic Research in Epidemiologic Studies." In *Stored Tissue Samples: Ethical, Legal and Public Implications*, edited by Robert F. Weir, 82–88. Iowa City: University of Iowa Press, 1998.

Stepan, Nancy. *Eradication: Ridding the World of Diseases Forever?* Ithaca, NY: Cornell University Press, 2011.

Stevens, Hallam. *Life Out of Sequence: A Data-Driven History of Bioinformation*. Chicago: University of Chicago Press, 2013.

Stocking, George W. *Race, Culture, and Evolution: Essays in the History of Anthropology*. New York: Free Press, 1968.

———. *Victorian Anthropology*. New York: Free Press, 1987.

Stoler, Ann Laura. *Along the Archival Grain: Epistemic Anxieties and Colonial Common Sense*. Princeton, NJ: Princeton University Press, 2009.

——, ed. *Imperial Debris: On Ruins and Ruination*. Durham, NC: Duke University Press, 2013.

"Stopping the Biological Clock." *AIBS Bulletin* 12, no. 5 (1962): 112.

Strasser, Bruno. "The Experimenter's Museum: Genbank, Natural History, and the Moral Economies of Biomedicine." *Isis* 102, no. 1 (2011): 60–96.

——. "Laboratories, Museums, and the Comparative Perspective: Alan A. Boyden's Quest for Objectivity in Serological Taxonomy, 1924–1962." *Historical Studies in the Natural Sciences* 40, no. 2 (2010): 149–82.

——. "Linus Pauling's 'Molecular Diseases': Between History and Memory." *American Journal of Medical Genetics* 115 (2002): 83–93.

Strathern, Marilyn. *Kinship, Law and the Unexpected: Relatives Are Always a Surprise*. New York: Cambridge University Press, 2005.

——. *Property, Substance, and Effect: Anthropological Essays on Persons and Things*. London: Athlone, 1999.

Strong, D. Michael. "The US Navy Tissue Bank: 50 Years on the Cutting Edge." *Cell and Tissue Banking* 1, no. 1 (2004): 9–16.

Strong, Pauline Turner. "Fathoming the Primitive." *Ethnohistory* 33 (1986): 175–94.

Sturdy, Steve. "Reflections: Molecularization, Standardization and the History of Science." In *Molecularizing Biology and Medicine: New Practices and Alliances, 1910s–1970s*, edited by Soraya de Chadarevian and Harmke Kamminga, 254–71. Amsterdam: OPA, 1998.

Sturm, Circe Dawn. *Blood Politics: Race, Culture, and Identity in the Cherokee Nation of Oklahoma*. Berkeley: University of California Press, 2002.

Sulloway, Frank J. *Freud, Biologist of the Mind: Beyond the Psychoanalytic Legend*. Cambridge, MA: Harvard University Press, 1992.

Sunder Rajan, Kaushik. *Biocapital: The Constitution of Postgenomic Life*. Durham, NC: Duke University Press, 2006.

Surgenor, Douglas M. "Blood." *Scientific American* 190, no. 2 (1954): 54–64.

Swanson, Kara W. *Banking on the Body: The Market in Blood, Milk, and Sperm in Modern America*. Cambridge, MA: Harvard University Press, 2014.

——. "Human Milk as Technology and Technologies of Human Milk: Medical Imaginings in the Early Twentieth-Century United States." *Women's Studies Quarterly* 37, nos. 1–2 (2009): 20–37.

TallBear, Kim. "Beyond the Life / Not Life Binary: A Feminist-Indigenous Reading of Cryopreservation, Interspecies Thinking and the New Materialisms." In *Cryopolitics: Frozen Life in a Melting World*, edited by Joanna Radin and Emma Kowal. Cambridge, MA: MIT University Press, forthcoming.

——. *Native American DNA: Tribal Belonging and the False Promise of Genomic Science*. Minneapolis: University of Minnesota Press, 2013.

Tanabe, K., et al. "*Plasmodium falciparum* Accompanied the Human Expansion out of Africa." *Current Biology* 20, no. 14 (2010): 1283–89.

Tapper, Melbourne. *Sickle Cell Anemia and the Politics of Race*. Philadelphia: University of Pennsylvania Press, 1999.

Tauali'i, M., E. L. Davis, K. L. Braun, J. A. Tsark, N. Brown, M. Hudson, and W. Burke. "Native Hawaiian Views on Biobanking." *Journal of Cancer Education* 29 (2014): 570–76.

Taubenberger, Jeffery K., Johan V. Hultin, and David M. Morens. "Discovery and Characterization of the 1918 Pandemic Influenza in Historical Context." *Antiviral Therapy* 12, no. 4, part B (2007): 581–91.

Taubes, G. "Scientists Attacked for 'Patenting' Pacific Tribe." *Science* 270, no. 17 (November 1995): 1112.

Taussig, Michael. *The Devil and Commodity Fetishism in South America*. Chapel Hill: 1980. Reprint, University of North Carolina Press, 2010.

———. *Shamanism, Colonialism, and the Wild Man: A Study in Terror and Healing*. Chicago: University of Chicago Press, 1986.

Terris, M., and S. Blatt. "Differences in Serum Cholesterol in Young White and Negro Adults." *American Journal of Public Health* 54, no. 12 (1964): 1996–2008.

Thevenot, Roger. *A History of Refrigeration throughout the World*. Translated by J. C. Fidler. Paris: International Institute of Refrigeration, 1979.

Thomas, Nicholas. *Entangled Objects: Exchange, Material Culture, and Colonialism in the Pacific*. Cambridge, MA: Harvard University Press, 1991.

Thomas, William L. *Man's Role in Changing the Face of the Earth*. Chicago: University of Chicago Press, 1956.

Thompson, Laura, Paul Baker, Betty Bell, John W. Bennett, George F. Carter, John J. Honigmann, Carleton S. Coon, et al. "Steps toward a Unified Anthropology [and Comments and Reply]." *Current Anthropology* 8, nos. 1–2 (1967): 67–91.

Tierney, Patrick. *Darkness in El Dorado: How Scientists and Journalists Devastated the Amazon*. New York: W. W. Norton, 2000.

Tilley, Helen. *Africa as a Living Laboratory: Empire, Development, and the Problem of Scientific Knowledge, 1870–1950*. Chicago: University of Chicago Press, 2011.

Timmermans, Stefan, and Marc Berg. "Standardization in Action: Local Universality through Medical Protocols." *Social Studies of Science* 27, no. 2 (1997): 273–305.

Tippett, A. R. *Solomon Islands Christianity: A Study in Growth and Obstruction*. London: Lutterworth, 1967.

Tirard, Stephane. *Histoire de la vie latente: Des animaux ressuscitants du XVIIIe Siècle aux embryons congelés du XXe siècle*. Paris: Vuibert, 2010.

Tomba, Massimiliano. *Marx's Temporalities*. Translated by Peter D. Thomas and Sara R. Farris. Chicago: Haymarket Books, 2013.

Tresch, John. *The Romantic Machine: Utopian Science and Technology after Napoleon*. Chicago: University of Chicago Press, 2012.

———. "Technological World-Pictures: Cosmic Things and Cosmograms." *Isis* 98, no. 1 (2007): 84–99.

Tsosie, Rebecca. "Cultural Challenges to Biotechnology: Native American Genetic Resources and the Concept of Cultural Harm." *Journal of Law, Medicine, and Ethics* 35, no. 3 (2007): 396–411.

Tuofo, Zhu, et al. "An African HIV-1 Sequence Form 1959 and Implications for the Origin of the Epidemic." *Nature* 391, no. 6667 (1998): 594–97.

Turnbaugh, Peter J., et al. "The Human Microbiome Project: Exploring the Microbial Part of Ourselves in a Changing World." *Nature* 449, no. 7164 (2007): 804.

Turner, Arthur R. *Frozen Blood: A Review of the Literature, 1949–1968*. New York: Gordon and Breach, 1970.

Turner, Victor. "Betwixt and Between: The Liminal Period in Rites of Passage." In *Betwixt and Between: Patterns of Masculine and Feminine Initiation*, edited by Louise Carus Madhi, Steven Foster and Meredith Little, 3–22. Peru, IL: Open Court, 1987.

———. "Variations on a Theme of Liminality." In *Secular Ritual*, edited by Sally Falk Moore and Barbara Myerhoff, 36–52. Amsterdam: Van Gorcum, Assen, 1977.

Twilley, Nicola. "The Coldscape." *Cabinet* 47 (2012): 78–84.

Ulijaszek, Stanley, and Rebecca Huss-Ashmore. *Human Adaptability: Past, Present, and Future*. Oxford: Oxford University Press, 1997.

Van Vleck, Jenifer. *Empire of the Air: Aviation and the American Ascendancy*. Cambridge, MA: Harvard University Press, 2013.

Vandepitte, J. M., W. W. Zuelzer, James V. Neel, and J. Colaert. "Evidence Concerning the Inadequacy of Mutation as an Explanation of the Frequency of the Sickle Cell Gene in the Belgian Congo." *Blood* 10, no. 4 (1955): 341–50.

Vaught, Jim. *Testimony before the Subcommittee on Investigations and Oversight Committee on Science and Technology United States House of Representatives: Biorepository Policies and Practices of the National Cancer Institute*. Edited by Office of Biorepositories and Biospecimen Research. Bethesda, MD: National Cancer Institute, National Institutes of Health, 2008.

Waddington, Conrad. *The Scientific Attitude*. London: Penguin, 1948.

Wailoo, Keith. *Drawing Blood: Technology and Disease Identity in Twentieth-Century America*. Baltimore: Johns Hopkins University Press, 1997.

———. *Dying in the City of the Blues: Sickle Cell Anemia and the Politics of Race and Health*. Chapel Hill: University of North Carolina Press, 2001.

Wailoo, Keith, and Stephen Gregory Pemberton. *The Troubled Dream of Genetic Medicine: Ethnicity and Innovation in Tay-Sachs, Cystic Fibrosis, and Sickle Cell Disease*. Baltimore: Johns Hopkins University Press, 2006.

Wald, Priscilla. "What's in a Cell? John Moore's Spleen and the Language of Bioslavery." *New Literary History* 36 (2005): 205–25.

Waldby, Catherine. "Stem Cells, Tissue Cultures and the Production of Biovalue." *Health: An Interdisciplinary Journal for the Social Study of Health, Illness and Medicine* 6, no. 3 (2002): 305–23.

Waldby, Catherine, and Melinda Cooper. "The Biopolitics of Reproduction." *Australian Feminist Studies* 23, no. 55 (2008): 57–73.

Waldby, Catherine, and Robert Mitchell. *Tissue Economies: Blood, Organs, and Cell Lines in Late Capitalism*. Durham, NC: Duke University Press, 2006.

Ward, Ryk, B. L. Frazier, K. Dew-Jager, and S. Paabo. "Extensive Mitochondrial Diversity within a Single Amerindian Tribe." *Proceedings of the National Academy of Science, USA* 88 (1991): 8720–24.

Warren, Jonathan W. *Racial Revolutions: Antiracism and Indian Resurgence in Brazil.* Durham, NC: Duke University Press, 2001.

Washburn, Sherwood. "The New Physical Anthropology." *Transactions of the New York Academy of Sciences* 13, no. 7 (1951): 298–304.

Weber, Max. *From Max Weber: Essays in Sociology*. Edited by H. H. HH and C. W. Mills. New York: Oxford University Press, 1948.

Weinberg, Alvin. *Reflections on Big Science*. Cambridge, MA: MIT Press, 1967.

Weiner, J. S., and John Adam Lourie. *Human Biology: A Guide to Field Methods*. Oxford: Published for the International Biological Programme by Blackwell Scientific, 1969.Weir, Robert F., and Robert S. Olick with Jeffrey C. Murray. *The Stored Tissue Issue: Biomedical Research, Ethics, and Law in the Era of Genomic Medicine*. New York: Oxford University Press, 2004.

Weir, Robert F., and Robert S. Olick, with Jeffrey C. Murray. *The Stored Tissue Issue: Biomedical Research, Ethics, and Law in the Era of Genomic Medicine.* New York: Oxford University Press, 2004.

Weiss, Kenneth M., and Ranajit Chakraborty. "Genes, Populations, and Disease, 1930–1980: A Problem Oriented Review." In *A History of American Physical Anthropology, 1930–1980*, edited by Frank Spencer, 371–404. New York: Academic, 1982.

Westad, Odd Arne. *The Global Cold War: Third World Interventions and the Making of Our Times*. Cambridge: Cambridge University Press, 2005.

Widmer, Sandra. "Making Blood 'Melanesian': Fieldwork and Isolating Techniques in Genetic Epidemiology (1963–1976)." *Studies in History and Philosophy of Biological and Biomedical Sciences* 47 (2014): 118–29.

Williams, Edwin William. *Frozen Foods: A Biography of an Industry*. 1968. Reprint, Boston: Cahners 1970.

Wills, W., G. Saimot, C. Brochard, B.S. Blumberg, W. T. London, R. Dechene, and I. Millman. "Hepatitis B Surface Antigen (Australia Antigen) in Mosquitoes Collected in Senegal, West Africa." *American Journal of Tropical Hygiene* 25 (1976): 186–90.

Wilson, Eric. *The Spiritual History of Ice: Romanticism, Science, and the Imagination*. New York: Palgrave Macmillan, 2003.

Wise, M. Norton, and Crosbie Smith. "Work and Waste: Political Economy and Natural Philosophy in Nineteenth Century Britain." *History of Science* 27 (1989): 263–301.

Wiwchar, David. "Nuu-Chah-Nulth Blood Returns to West Coast." *Ha-Shilth-Sa* 31, no. 25 (2004): 1–4.

Wolf, Eric. *Europe and the People without History*. Berkeley: University of California Press, 1982.

Wolstenholme, Gordon, ed. *Man and His Future*. Boston: Little, Brown, 1963.

World Health Organization. *Immunological and Haematological Surveys*. Technical Report Series, no. 181. Geneva: World Health Organization, 1959.

———. *Multipurpose Serological Surveys and Serum Reference Banks*. WHO Technical Report Series, no. 454. Geneva: World Health Organization, 1970.

———. *Research in Population Genetics of Primitive Groups*. World Health Organization Technical Report Series, no. 279. Geneva: World Health Organization, 1964.

———. *Research on Human Population Genetics*. World Health Organization Technical Report Series, no. 387. Geneva: World Health Organization, 1968.

Woolsey, L. H. "The Leticia Dispute between Colombia and Peru." *American Journal of International Law* 29, no. 1 (1935): 94–99.

Worthington, E. B. *The Evolution of IBP*. Vol. 1. Cambridge: Cambridge University Press, 1975.

Yanagihara, Richard, R. M. Garruto, M. A. Miller, M. Leon-Monzon, P. O. Liberski, and D. Carleton Gajdusek. "Isolation of HTLV-1 from Members of a Remote Tribe in Papua New Guinea." *New England Journal of Medicine* 323 (1990): 993–94.

Yanagihara, Richard, Carol Jenkins, S. S. Alexander, C. A. Mora, and R. M. Garruto. "Human T Lymphotrophic Virus Type I Infection in Papua New Guinea: High Prevalence among the Hagahai Confirmed by Western Analysis." *Journal of Infectious Disease* 162 (1990): 649–54.

Yaqoob, Waseem. "The Archimedean Point: Science and Technology in the Thought of Hannah Arendt." *Journal of European Studies* 44, no. 3 (2014): 199–224.

Zakariya, Nasser. "Is History Still a Fraud?" *Historical Studies in the Natural Sciences* 43, no. 5 (2013): 631–41.

Zammito, John. "Koselleck's Philosophy of Historical Time(s) and the Practice of History." *History and Theory* 43, no. 1 (2004): 124–35.

Index